COMPUTER APPLICATIONS IN ELECTRICAL ENGINEERING SERIES

Franklin F. Kuo, editor

BOWERS & SEDORE
SCEPTRE: A Computer Program for Circuit and Systems Analysis

CADZOW & MARTENS
Discrete-Time and Computer Control Systems

DAVIS
Computer Data Displays

FRIEDMAN & MENON
Fault Detection in Digital Circuits

JENSEN & LIEBERMAN
IBM Circuit Analysis Program: Techniques and Applications

KUO & MAGNUSON
Computer Oriented Circuit Design

LIN
An Introduction to Error-Correcting Codes

SIFFERLEN AND VARTANIAN
Digital Electronics with Engineering Applications

STOUTEMYER
PL/1 Programming for Engineering and Science

...SCEPTRE

A Computer Program For Circuit and Systems Analysis

PRENTICE-HALL INTERNATIONAL, INC., *London*
PRENTICE-HALL OF AUSTRALIA, PTY. LTD., *Sydney*
PRENTICE-HALL OF CANADA, LTD., *Toronto*
PRENTICE-HALL OF INDIA PRIVATE LIMITED, *New Delhi*
PRENTICE-HALL OF JAPAN, INC., *Tokyo*

...SCEPTRE

A Computer Program For Circuit and Systems Analysis

JAMES C. BOWERS

College of Engineering
University of South Florida
Tampa, Florida

STEPHEN R. SEDORE

IBM Corporation
Owego, New York

Prentice-Hall, Inc., Englewood Cliffs, New Jersey

Current printing (last digit):

10 9 8 7 6 5 4 3 2 1

13-791590-X

Library of Congress Catalog Card Number: 71-150225

Printed in the United States of America

This book is dedicated to the

ASTRONAUTS of APOLLO 8

who, on that Christmas Eve in 1968, thrilled us all when they made man's first broadcast from the vicinity of the moon and reminded the anxious world that,

"In the beginning God . . ."

The Apollo 8 commemorative stamp issued by the United States Post Office Department in 1969.

PREFACE

This book is a complete single source for all information concerning the SCEPTRE program. It is expected to be useful as a college text for engineering students, as well as a handy reference book for the practicing engineer. It also could be very valuable to the technician, in that it gives him the tools necessary to solve sophisticated engineering problems.

SCEPTRE is an automatic circuit analysis program capable of determining initial conditions, transient and steady-state responses of large networks. It is therefore a tool that can be used by an engineer to assist in the design and analysis of various circuits. SCEPTRE solves circuit equations from a simple description of circuit topology and component values. A free format, easy-to-learn, engineer oriented language is used. SCEPTRE provides for complete flexibility in modeling, to handle almost any conceivable system description. In addition, it has a Defined Parameter mode which allows it to solve state equations which might be written to describe any type of system. For that matter, the Defined Parameter mode can solve general equations whether or not they relate to physical systems. There are virtually no limitations on the flexibility which the program offers. An automatic computer analysis program can be defined as one that produces engineering solutions without requiring the user to write any type of equation or do any real programming. The SCEPTRE program completely meets this definition. Among the several large general purpose programs presently available, SCEPTRE represents a second generation, modern approach. While it will some day be superceded, there are at least two broad features which are time-less. The SCEPTRE capability to accommodate any type of device model configuration on a temporary or permanently stored basis is unique. Also the SCEPTRE approach of providing a completely user oriented, free – format is significant. These two features will no doubt be included in all future programs. Therefore, the individual who learns the SCEPTRE approach will be in an ideal position to "pick-up" the third generation programs as they become available in the years to come.

From an academic standpoint, there are perhaps two approaches to the Computer-Aided-Analysis field. One is to study the field in general, discussing briefly all of the various programs. The second, as suggested by this book, is to study one representative modern program in detail. The authors believe the latter approach is much more useful to the engineering student for several reasons. First, if the student gains a comprehensive knowledge of a particular program, he is then able to capably use the program. This is valuable to him as a student, especially when working on a project or a thesis; but it also gives him increased confidence on his first job, having the assurance that he can readily solve complicated engineering problems using SCEPTRE. Secondly, once the student realizes the ease with which a truly modern, user oriented program can handle his problems, he is then a confirmed "believer". He easily "picks-up" the later programs on his own, as they become available.

Chapter One of the book contains the basic theoretical formulations of the SCEPTRE program. It includes the mathematical "blueprint" that the programmers convert into a series of executable steps that are performed by the computer. While a knowledge of this material is not necessary to become a proficient user of SCEPTRE, it is the foundation of the program. Chapter Two is essentially the present Air Force Weapons Laboratory User's Manual, but organized differently. All of the rules and guidelines for the program are set-out in clear, tabular form, so the user may have a ready reference source. Chapter Three presents, for the first time in print, the complete step-by-step procedures for obtaining model parameter data for six types of solid-state devices. This Chapter includes over 100 Figures clearly showing the methods and techniques necessary to obtain accurate data for the Diode, Transistor, Zener Diode, JFET, MOSFET, and Unijunction devices. The specific models are presented (such as the Ebers-Moll for the transistor) and then are followed by the procedures mentioned above, including the actual test equipment used and where each dial is set. All of the Models are used in practical examples to further illustrate the complete "picture". Chapter Four develops a dozen practical examples which are purposely selected to illustrate all of the key points in the SCEPTRE program. The authors are quite aware of the questions which commonly occur, and this Chapter is presented to remove all of these potential uncertainties. Chapter Five presents the SCEPTRE approach to Control System Analysis. This is a relatively new area for SCEPTRE. As Systems and Control engineers become aware of the ease with which SCEPTRE directly handles control problems, this area could greatly expand. Since SCEPTRE has unmatched modeling and non-linear flexibility, the type of control problem which can be accommodated is almost limitless. Various examples are fully illustrated, including linear, non-linear and sample data problems. Chapter Six presents another often overlooked area; that is, using SCEPTRE to solve problems in fields such as heat transfer, mechanics, fluids, etc. Several detailed examples are given in this Chapter which illustrate the procedures.

The authors believe that the material presented in this book should be introduced to the engineering student as soon in their study as possible. The solution of engineering problems using SCEPTRE should be as common place as using the slide rule. More practical and realistic problems can be assigned if the student is not burdened with lengthy hand calculations. Also more class room time can be spent teaching the basic theory rather than "turn – crank" procedures. The book can be used as a separate course in the engineering program or as a supplement to the first circuits or electronics course. With the present day trend of inter-discipline study, the book could be very useful as a core subject to be taught to all engineers. One of the big problems in training today's engineer is giving him significant background in the other disciplines. For example, it is very difficult in a limited number of hours to give the Mechanical Engineering major enough background in electronics to be meaningful to him. However, by learning the SCEPTRE approach the Mechanical Engineer could readily solve most any electronics problem with little trouble. This applies to all of the majors. It would open up to each discipline the ability to work practical problems in all areas, thus aiding the understanding. Many practicing Electrical Engineers now use SCEPTRE to aid them in their initial Thermal analysis calculations, before going to the specialist to complete the final design.

A one semester hour course could be used at the lower division level, where Chapters Two and Four would be covered in detail and Chapters Three, Five, and Six covered more briefly. At a senior or graduate level, Chapters One, Three, and Five would become significant.

In conclusion, it has been the authors' experience with students and practicing engineers that the biggest problem is the individual's fear of the unknown. Most of them assume that to use a Computer Analysis Program requires a broad background in programming and computer technology. While this is true for many of the programs, it is definitely not the case for SCEPTRE. By spending just a few hours reading Chapters Two and Four of this book, the individual can immediately begin to set-up and solve practical engineering problems, with no previous background beyond a Sophomore circuits or electronics course.

Appendix G contains useful information about where SCEPTRE tapes may be obtained, what computer versions are available and where to obtain technical advice. Appendix I includes tabulated model data for a number of solid-state devices. In fact, the reader is advised to study carefully all of the topics covered in the appendices, since much supplementary information has been included.

September 1970

James C. Bowers
Stephen R. Sedore

ACKNOWLEDGMENTS

In a book of this type there are many individuals deserving acknowledgment. One group for developing the SCEPTRE program and the second group for their efforts in supporting the writing of this text. The program was developed by I.B.M. and was funded by the Air Force Weapons Laboratory at Kirtland Air Force Base, New Mexico. The primary architects of the SCEPTRE program were Dr. Ralph M. Warten, John R. Sents and the second named author. Later Program developments were assisted by Gerald Minturn, Allen Wright, Jerry Patterson and John Melton. Throughout the project, Captain Gary Prichard and later Captain John Anderson bore the responsibility of approving every key feature of the program.

In support of the book, the authors wish to thank Mr. John M. Murray of Honeywell, Inc., St. Petersburg, Florida. His contributions to Chapter Three were extensive. In addition, the authors are grateful to Honeywell, Inc. for allowing Mr. Dave Goodrich and his colleagues to pursue portions of this work in such a timely fashion, and also to I.B.M. for the managerial encouragement that was given from the beginning. The authors greatly appreciate the many suggestions and corrections on Chapter Three furnished by Dynetics, Inc. of Bellevue, Washington. Majors C. M. Glass and J. P. Portasik contributed significant portions of Chapter Five. The reviews furnished by Mr. Walt Whipple and Dr. George Zobrist were certainly helpful. University of South Florida graduate students, George Gaffney and Ravindra Shah contributed in many ways. Finally the first named author wishes to express his personal appreciation to Mr. C. J. Cerulli of Honeywell for having the foresight to create the interest which made this book possible.

James C. Bowers
Stephen R. Sedore

CONTENTS

1

MATHEMATICAL FORMULATION

2

SCEPTRE FORMAT

3

SOLID-STATE MODELING

4

ELECTRONIC CIRCUIT EXAMPLES

5

ANALYSIS OF AUTOMATIC CONTROL SYSTEMS

6

ANALYSIS OF OTHER NON-ELECTRICAL SYSTEMS

APPENDICES

...SCEPTRE

A Computer Program For Circuit and Systems Analysis

1

MATHEMATICAL FORMULATION

1.1 INTRODUCTION

The purpose of this chapter is to completely define and derive the basic mathematical formulation of the SCEPTRE program. It is true that a user may become very efficient in the use of SCEPTRE without any knowledge of the material in this chapter; however, the mathematical formulation of any complex computer program is its very foundation. It is the mathematical blueprint that programmers convert into a series of executable steps that are performed by computers. Skillful programming may make the difference between efficient or inefficient operation of a given feature or program, but it never replaces what may have been left out of the original blueprint.

Three distinctly different types of circuit analyses exist; DC, AC, and transient. It is not suprising, therefore, that different formulation techniques, that are specifically designed for each of these types of circuit analysis, also exist. The object of a DC program is to determine the DC steady state of networks. Time never becomes an independent variable and the equations that result are always algebraic, both linear and nonlinear. The solution is almost always done via an iterative process that converges to the final result. An AC program, on the other hand, determines the AC steady state of networks at given frequencies. Here also, time does not enter as an independent variable. The usual technique is to construct a complex admittance or impedance matrix that is valid at a given frequency and to solve the resulting system of equations just once for that frequency. The results are suitable for the construction of the familiar Bode plots—amplitude and phase versus frequency. The general transient formulation differs most principally from the other two in that time does appear as the independent variable and that systems of first order differential equations are formed that are numerically integrated. SCEPTRE contains separate formulations for both DC and transient analysis. Since the latter is considerably more involved, its derivation is treated first and at greater length.

It should be pointed out that the SCEPTRE program does not contain an individual AC formulation. Solutions to AC problems can be carried out (see Sec. 4.9) using the transient algorithm, but never as efficiently as in one that is designed specifically for this type of problem. If, however, the user is interested in a transient solution component that may exist before the network assumes its steady state, then a program with only a pure AC formulation is not sufficient, and a program such as SCEPTRE is required. Also many times saturation or clipping exists in a circuit. The SCEPTRE transient solution indicates this important fact, while the pure AC type of formulation would not, since linear operation is assumed.

1.1.1 Definitions and Notation

The starting point for the derivation of the complete transient formulation is the definition of a tree [1]. A tree is a connected subgraph of the original network which includes every node of that network but contains no closed loops or meshes. It is obvious that no unique tree exists for any network. A specific tree may be defined as one in which elements are inserted in the following order of priority: voltage sources, capacitors, resistors, and inductors. All voltage sources are included in this tree and all current sources are excluded.* All elements that are included in the tree are called tree branches; all elements that are excluded from the tree are called links. It is apparent, for example, that capacitors are more likely to be tree branches than are inductors because of the priority ranking that is assigned.

It is convenient to divide all of the possible network elements into classes. The passive elements, resistance, capacitance, and inductance, are either tree branches or links. Thus, six classes are required for the passive elements. In addition, one class each is needed for voltage source tree branches and current source links. The eight resulting classes are listed below:

Class 1 - Capacitor links
Class 2 - Resistor links
Class 3 - Inductor links
Class 4 - Capacitor tree branches
Class 5 - Resistor tree branches
Class 6 - Inductor tree branches
Class 7 - Voltage source tree branches
Class 8 - Current source links

These class numbers appear repeatedly in the derivation to follow. For example, the single subscripted notation V_4 refers to a column matrix with elements which are the voltages across all Class 4 elements. In the same way I_3 identifies a column matrix with elements which are the currents through all Class 3 elements. Double subscripts are used to refer to the following matrices:

R_{22} — a diagonal matrix with element values which are the resistive values of the resistor links

R_{55} — a diagonal matrix with element values which are the resistive values of the resistor tree branches

$S_{11} = C_{11}^{\frac{1}{}}$ — a diagonal matrix with element values which are the elastance values of capacitor links

$S_{44} = C_{44}^{\frac{1}{}}$ — a diagonal matrix with element values which are the elastance values of capacitor tree branches

L_{33} — a symmetric matrix composed only of inductor links and mutual inductance between inductor links

L_{66} — a symmetric matrix composed only of inductor tree branches and mutual inductance between inductor tree branches

$L_{36} = L_{63}^{T}$ — a matrix composed only of mutual inductances between inductor links and inductor tree branches

1.2 STATE VARIABLE CONCEPT

The state variables of any system are defined as the minimum set of variables which, together with all inputs (and occasionally the time derivatives of certain inputs), are sufficient to determine all other system quantities at any particular instant of time. All system quantities at time t_n are dependent upon the values of the state variables and

*This definition of the tree prohibits any network containing voltage source loops or current source cut sets. For practical purposes, this constitutes no significant loss.

inputs that hold at time t_n and are independent of those at any other time [2]. It can be shown that the state variables of the general network are the set of capacitor tree branch voltages (V_4) and the set of inductor link currents (I_3).* The only stipulation necessary is that the tree appropriate to the general network be formed according to the rule of Sec. 1.1. This choice of state variables is equivalent to specifying the independent initial conditions of the network [3]. It may be further stated that this choice of state variables leads to a set of simultaneous first-order differential equations [4].

The concept is exhibited easily with the aid of the network shown in Fig. 1.1. Both the voltage source and the current source are known functions of time. The members of the tree are drawn with bold lines.

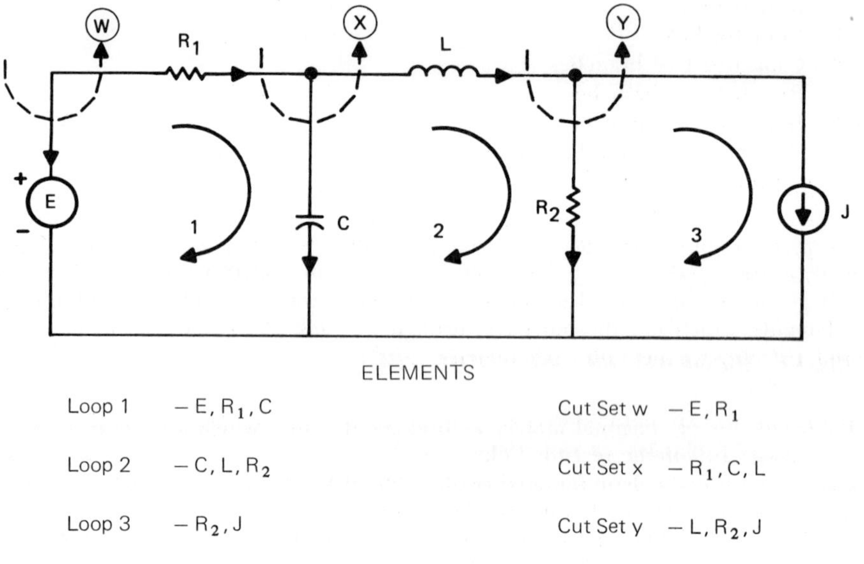

Fig. 1.1 Sample Network

The following set of equations apply:

$$\text{(a)}\quad I_{R_1}(t_o) = \frac{E(t_o) - V_C(t_o)}{R_1}$$

$$\text{(b)}\quad V_{R_1}(t_o) = R_1 I_{R_1}(t_o)$$

$$\text{(c)}\quad I_{R_2}(t_o) = I_L(t_o) - J(t_o)$$

$$\text{(d)}\quad V_{R_2}(t_o) = R_2 I_{R_2}(t_o)$$

*A proof of this statement is given in Sec. 1.6 where all network currents and voltages are derived from these variables.

$$\text{(e)}\quad I_C(t_o) = I_{R_1}(t_o) - I_L(t_o)$$

$$\text{(f)}\quad V_L(t_o) = V_C(t_o) - V_{R_2}(t_o)$$

If these equations are solved in the order of their appearance, the right sides are all known quantities because $V_C(t_o)$ and $I_L(t_o)$ are state variables and are supplied as initial conditions. At this point, the solutions at $t = t_o$ are in hand. The objective is now to determine these values at the first time increment beyond $t = t_o$. This is possible if a new set of "initial conditions" (or updated set of state variables) is obtained. This is done by determining the derivatives of these quantities at $t = t_o$. For this purpose, there are

$$\text{(g)}\quad \dot{V}_C(t_o) = \frac{I_C(t_o)}{C} \qquad \text{and}$$

$$\text{(h)}\quad \dot{I}_L(t_o) = \frac{V_L(t_o)}{L}$$

It is seen that the choice of state variables, V_C and I_L, led to the solution of all other currents and voltages for the network of Fig. 1.1 and further, led to the two first-order differential equations. It is always true that the number of first-order differential equations equals the number of state variables, or put another way, equals the number of independent capacitors and inductors in the network.

1.3 BLOCK DIAGRAM OF SOLUTION PROCESS

The solutions for all network currents and voltages of the general transient problem are continuous functions of time. The digital computer, however, does not present a continuous solution as does the analog computer. It instead must supply some number of solutions at distinct increments of time. Consider Fig. 1.2 in which V(t) represents any network voltage. The value $V(t_o)$ is either given as an initial condition or determined

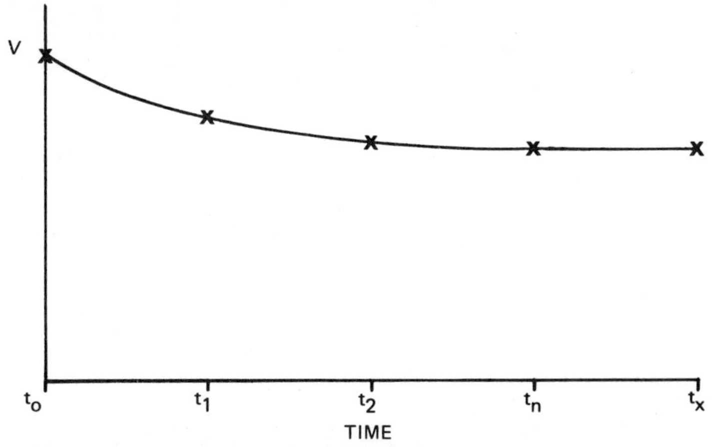

Fig. 1.2. Time increment solution.

from a knowledge of the network sources and initial conditions. It is required that a time history of V be determined up to the stipulated termination time, t_x. The solution is carried out repeatedly at the points $t = t_1, t_2, \ldots t_n, \ldots t_x$; where, $t_o \langle t_n \langle t_x$. It is not necessarily true that all the increments are equal. It is not unusual for hundreds or even thousands of time increments to be encompassed in the course of a transient problem. It is clear that all of the solution points may be connected by straight lines to give an approximation* to the continuous solution. This procedure is usually followed for the transient solution of any type of problem on a digital computer.

It is pointed out by example in Sec. 1.2 that the knowledge of the state variables V_4 and I_3 together with all sources and forcing functions are sufficient to solve for all network currents and voltages at any point in time. $V(t_o)$ is computed from the values of the state variables and inputs at $t = t_o$, but the general $V(t_n)$ requires updated values of the state variables and inputs at $t = t_n$. The values of the inputs present no problem for presumably they are all known quantities, but the state variables are repeatedly updated by the program. The process is best shown with the aid of the block diagram of Fig. 1.3. The process begins with the insertion of the independent initial conditions $V_4(t_o)$ and $I_3(t_o)$. All currents and voltages and state variable derivatives that are valid at $t = t_o$ are computed (see Eqs. (a) through (h) in the preceding example). The differential equations representing $\dot{V}_4(t_o)$ and $\dot{I}_3(t_o)$ are integrated by the integration routine to provide $V_4(t_1)$ and $I_3(t_1)$. Since $t_1 = t_o + \Delta t$, it is desirable that Δt be as large as possible for reasons of economy. A variable step integration routine [5] is most appropriate for this task. Once $V_4(t_1)$ and $I_3(t_1)$ are obtained, the process is repetitive and the program will proceed through to the desired termination time, t_x.

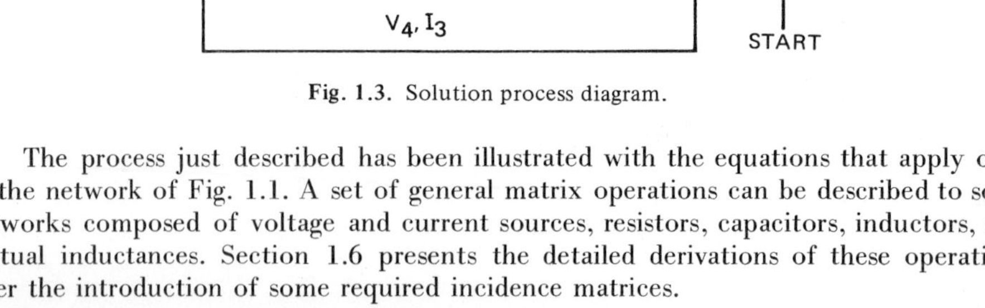

Fig. 1.3. Solution process diagram.

The process just described has been illustrated with the equations that apply only to the network of Fig. 1.1. A set of general matrix operations can be described to solve networks composed of voltage and current sources, resistors, capacitors, inductors, and mutual inductances. Section 1.6 presents the detailed derivations of these operations after the introduction of some required incidence matrices.

This process implies a certain degree of nonlinear capability. If a passive element (say a capacitor) is dependent upon the voltage across itself (which is a state variable), that capacitor is nonlinear. This presents no difficulty since the state variables are known quantities at the start of each solution step and; therefore, the capacitor value can be

*The term "approximation" is used in the sense that the intervals between the solution points cannot be exactly accurate regardless of how accurately the solution points themselves are computed.

updated without error. A different situation may exist if, for example, a network resistor is a function of the voltage across itself. The independent variable in this instance is not a state variable. In some cases (see Sec. 1.7) the solution is effected without error, but in other cases a "computational delay" occurs. That is, computation at the nth time step begins with independent variable values that were valid at the (n-1)th step. The amount of error introduced by these delays is difficult to assess. As a rule, the error is proportional to the degree of nonlinearity exhibited by the functional dependence. Computational delays never occur if the independent variables are either state variables or time. The validity of the use of other independent variables cannot be unambiguously stated; the status is topology dependent. The program always prints out a warning message if a computational delay occurs, provided that the recommended input formats are used.

1.4 INCIDENCE MATRICES

The topological makeup of any network may be described in a number of ways. Two methods are chosen here and then combined into a very useful form.

It is well known [1] that an incidence matrix, which represents the topological relationship between the elements and the fundamental cut sets of the network, may be constructed for any network. This matrix* is called the Q matrix** where:

$Q = [q_{ij}]$ is a matrix containing n-1 rows and b columns for the general connected network consisting of n nodes and b elements where:

$q_{ij} = +1$ if the direction of fundamental cut-set i coincides with the reference direction of element j in fundamental cut-set i.

$q_{ij} = -1$ if the direction of fundamental cut-set i is in opposition with the reference direction of element j in fundamental cut-set i.

$q_{ij} = 0$ if the fundamental cut-set i does not include element j.

The direction of each fundamental cut-set is customarily chosen to agree with that of its associated tree branch. It is convenient to partition Q according to links and tree branches as:

$$Q = [Q_L \;\vdots\; Q_T] \tag{1.1}$$

which, if the rows and columns are manipulated appropriately, become:

$$Q = [-F^T \;\vdots\; U]$$

---

*This matrix is often called the fundamental cut-set matrix. A cut-set is defined as that set of elements that would isolate the network into two groups of nodes if that set were "cut" out of the network.

**Capital letters with and without brackets will be used to denote matrices. The brackets will be omitted wherever they are not needed to avoid confusion.

where the (n-1) rows and columns of the unit matrix U correspond to the tree branches and the (b-n+1) columns of $-F^T$ correspond to the links associated with any tree. Figure 1.4 shows the incidence matrix Q and its partitioned form for the network of Fig. 1.1.

$$
Q = \begin{array}{c} \text{Cut} \\ \text{Sets} \end{array}
\begin{array}{c} \\ X \\ Y \\ W \end{array}
\begin{array}{c}
\begin{array}{ccc|ccc} R_1 & L & J & C & R_2 & E \end{array} \\
\left[\begin{array}{ccc|ccc}
-1 & 1 & 0 & 1 & 0 & 0 \\
0 & -1 & 1 & 0 & 1 & 0 \\
1 & 0 & 0 & 0 & 0 & 1
\end{array}\right]
\end{array}
$$

$$
Q = [-F^T \mid U]
$$

Fig. 1.4 Incidence Matrix Q

Another incidence matrix is defined which represents the topological relationship between the elements and the fundamental circuits or loops of the network. This matrix* is called the T matrix where:

$T = [t_{ij}]$ is a matrix containing (b-n+1) rows and b columns for the general connected network consisting of n nodes and b elements where:

$t_{ij} = +1$ if element j is part of fundamental circuit i, and if the reference directions of element j and fundamental circuit i coincide.

$t_{ij} = -1$ if element j is part of fundamental circuit i, and if the reference directions of element j and fundamental circuit i do not coincide.

$t_{ij} = 0$ if element j is not part of fundamental circuit i.

The direction of each fundamental circuit is customarily chosen to agree with that of its associated link. The number of rows in the T matrix corresponds to the fact that only (b-n+1) circuit equations are required for the solution of any network. If the columns of T are manipulated appropriately, T may be partitioned as:

$$
T = [U \mid F] \tag{1.2}
$$

where the (b-n+1) columns of the unit matrix U correspond to the links and the n-1 columns of F correspond to the tree branches of any tree [6]. Figure 1.5 shows the incidence matrix T and its partitioned form for the network of Fig. 1.1.

*This matrix is often called the fundamental circuit matrix.

		R_1	L	J	C	R_2	E
T = Loops	1	1	0	0	1	0	−1
	2	0	1	0	−1	1	0
	3	0	0	1	0	−1	0
T =		[U			F]		

Fig. 1.5 Incidence Matrix T

If the columns of Q and T are arranged in the same element order, it is well known* that:

$$Q\,T^T = 0 \tag{1.3}$$

Also, since

$$Q\,I_b = 0 \text{ is always true, a simple partition permits:} \tag{1.4}$$

$$[-F^T\ U]\begin{bmatrix} I_L \\ I_T \end{bmatrix} = 0 \tag{1.5}$$

leading to

$$I_T = F^T I_L \tag{1.6}$$

where I_L and I_T are the vectors of link currents and tree branch currents respectively. Equation (1.6) is significant because it isolates the tree branch currents of the general network in terms of the link currents.

In the same way, since:

$$T\,V_b = 0 \tag{1.7}$$

is always true, a simple partition permits:

$$[U\ F]\begin{bmatrix} V_L \\ V_T \end{bmatrix} = 0 \tag{1.8}$$

or

$$V_L = -F\,V_T \tag{1.9}$$

*See Appendix A.

where V_L and V_T are the vectors of link voltages and tree branch voltages respectively. Equation (1.9) is significant because it isolates the link voltages of the general network in terms of the tree branch voltages. The Q matrix and the T matrix have been eliminated.

1.5 SIGNIFICANCE OF THE F MATRIX

Equations (1.6) and (1.9) indicate that the F matrix can be used to transform the vector of tree branch voltages into link voltages and the vector of link currents into tree branch currents. If these vectors and the F matrix are partitioned approximately, according to the classification of Sec. 1.1, Eq. (1.9) becomes

$$\begin{bmatrix} V_1 \\ V_2 \\ V_3 \\ V_8 \end{bmatrix} = - \begin{bmatrix} F_{14} & 0 & 0 & F_{17} \\ F_{24} & F_{25} & 0 & F_{27} \\ F_{34} & F_{35} & F_{36} & F_{37} \\ F_{84} & F_{85} & F_{86} & F_{87} \end{bmatrix} \begin{bmatrix} V_4 \\ V_5 \\ V_6 \\ E_7 \end{bmatrix} \tag{1.10}$$

and Eq. (1.6) becomes

$$\begin{bmatrix} I_4 \\ I_5 \\ I_6 \\ I_7 \end{bmatrix} = \begin{bmatrix} F_{14}^T & F_{24}^T & F_{34}^T & F_{84}^T \\ 0 & F_{25}^T & F_{35}^T & F_{85}^T \\ 0 & 0 & F_{36}^T & F_{86}^T \\ F_{17}^T & F_{27}^T & F_{37}^T & F_{87}^T \end{bmatrix} \begin{bmatrix} I_1 \\ I_2 \\ I_3 \\ J_8 \end{bmatrix} \tag{1.11}$$

where *

$$F = \begin{bmatrix} F_{14} & 0 & 0 & F_{17} \\ F_{24} & F_{25} & 0 & F_{27} \\ F_{34} & F_{35} & F_{36} & F_{37} \\ F_{84} & F_{85} & F_{86} & F_{87} \end{bmatrix}$$

* See Appendix B for justification of the zero submatrices in the F matrix.

and

$$F^T = \begin{bmatrix} F_{14}^T & F_{24}^T & F_{34}^T & F_{84}^T \\ 0 & F_{25}^T & F_{35}^T & F_{85}^T \\ 0 & 0 & F_{36}^T & F_{86}^T \\ F_{17}^T & F_{27}^T & F_{37}^T & F_{87}^T \end{bmatrix}$$

Expansion of Eq. (1.10) yields:

$$V_1 = -F_{14}V_4 \qquad\qquad -F_{17}E_7 \tag{1.12}$$

$$V_2 = -F_{24}V_4 - F_{25}V_5 \qquad -F_{27}E_7 \tag{1.13}$$

$$V_3 = -F_{34}V_4 - F_{35}V_5 - F_{36}V_6 - F_{37}E_7 \tag{1.14}$$

$$V_8 = -F_{84}V_4 - F_{85}V_5 - F_{86}V_6 - F_{87}E_7 \tag{1.15}$$

Expansion of Eq. (1.11) yields:

$$I_4 = F_{14}^T I_1 + F_{24}^T I_2 + F_{34}^T I_3 + F_{84}^T J_8 \tag{1.16}$$

$$I_5 = \qquad F_{25}^T I_2 + F_{35}^T I_3 + F_{85}^T J_8 \tag{1.17}$$

$$I_6 = \qquad\qquad F_{36}^T I_3 + F_{86}^T J_8 \tag{1.18}$$

$$I_7 = F_{17}^T I_1 + F_{27}^T I_2 + F_{37}^T I_3 + F_{87}^T J_8 \tag{1.19}$$

The indicated partitions have led to the eight matrix equations listed above. These equations, along with others, are used in the next section to formally derive all network quantities in terms of known values.

It is profitable to examine a few aspects of the F matrix before the formal derivation begins. Since the r_{th} group of rows of the partitioned F matrix must be concerned with Class r elements (if r is 1, 2, 3, or 8) and the s_{th} group of columns must be concerned with Class s elements (if s is 4, 5, 6, or 7), the F matrix may be connotated as in Fig. 1.6. Equations (1.12) through (1.15) can effectively be "read" horizontally from the four groups of rows of F. Equations (1.16) through (1.19) can be "read" vertically from the four groups of columns of F. This process is better illustrated by consideration of the F matrix in Fig. 1.7 that arises specifically from the circuit of Fig. 1.1. Equations (1.13) and (1.14) correspond to:

$$V_{R_1} = -V_C + E \tag{1.13'}$$

$$V_L = V_C - V_{R_2} \tag{1.14'}$$

which is read horizontally from the F matrix.

	Class 4	Class 5	Class 6	Class 7
Class 1	F_{14}	0	0	F_{17}
Class 2	F_{24}	F_{25}	0	F_{27}
Class 3	F_{34}	F_{35}	F_{36}	F_{37}
Class 8	F_{84}	F_{85}	F_{86}	F_{87}

Fig. 1.6 Connotated F Matrix

		Class 4	Class 5	Class 7
		C	R_2	E
Class 2	R_1	1	0	−1
Class 3	L	−1	1	0
Class 8	J	0	− 1	0

Fig. 1.7 F Matrix for Sample Circuit

Equations (1.16) and (1.17) correspond to:

$$I_C = I_{R_1} - I_L \qquad (1.16')$$

and,

$$I_{R_2} = I_L - J \qquad (1.17')$$

which is read vertically from the F matrix. The last four equations match Eqs. (b), (f), (e), and (c) in Sec. 2.1. It is seen then that much network information is determined by simply "scanning" the F matrix. This fact is often exploited in the actual operation of the program to avoid or reduce the repeated matrix manipulation that would otherwise be required [7].

1.6 DERIVATION OF NETWORK QUANTITIES

The purpose of this section is to present the derivation of all unknown network quantities (all resistor currents and voltages, all capacitor currents, capacitor link voltages, all inductor voltages, inductor tree branch currents, and all state variable derivatives) in terms of known quantities (state variables (V_4, I_3), sources (E_7, J_8, E_y, J_x), and sub-matrices of the F matrix).

1.6.1 Resistive Quantities

All resistor currents and voltages in any network are included in the vectors I_2, I_5, V_2, and V_5. These four unknown quantities require four matrix equations for their solution. Two are immediately available from Eqs. (1.13) and (1.17). Two more arise from Ohm's law and are written as:

$$V_2 = R_{22} I_2 \qquad (1.20)$$

$$V_5 = R_{55} I_5 \qquad (1.21)$$

Note that the last two equations permit Eq. (1.13) to be rewritten as:

$$R_{22} I_2 = -F_{24} V_4 - F_{25} R_{55} I_5 - F_{27} E_7 \qquad (1.22)$$

Substitution of Eq. (1.17) into Eq. (1.22) yields:

$$R_{22} I_2 = -F_{24} V_4 - F_{25} R_{55} [F_{25}^T I_2 + F_{35}^T I_3 + F_{85}^T J_8] - F_{27} E_7$$

or

$$I_2 = M_R^{-1} \left\{ -F_{24} V_4 - F_{25} R_{55} [F_{35}^T I_3 + F_{85}^T J_8] - F_{27} E_7 \right\} \qquad (1.23)$$

where*

$$M_R = R_{22} + F_{25} R_{55} F_{25}^T \tag{1.24}$$

Equation (1.23) is important because it isolates the vector of resistor link currents explicitly in terms of quantities that are all known at the beginning of each solution interval. Now that I_2 is known, the remaining resistive quantities I_5, V_2, and V_5 are solved from Eqs. (1.17), (1.20), and (1.21) respectively.

An alternate approach to attain the same end is possible. Equations (1.20) and (1.21) become

$$I_2 = G_{22} V_2 \tag{1.25}$$

and $I_5 = G_{55} V_5$ $\quad$ (1.26)

where, $G_{22} = R_{22}^{-1}$ and $G_{55} = R_{55}^{-1}$ and are composed of conductances only.

Then Eqs. (1.25) and (1.26) into (1.17) result in:

$$G_{55} V_5 = F_{25}^T G_{22} V_2 + F_{35}^T I_3 + F_{85}^T J_8 \tag{1.27}$$

If Eq. (1.13) is inserted into (1.27),

$$G_{55} V_5 = F_{25}^T G_{22} [-F_{24} V_4 - F_{25} V_5 - F_{27} E_7] + F_{35}^T I_3 + F_{85}^T J_8$$

or

$$V_5 = M_G^{-1} \left\{ F_{25}^T G_{22} [-F_{24} V_4 - F_{27} E_7] + F_{35}^T I_3 + F_{85}^T J_8 \right\} \tag{1.28}$$

where

$$M_G = G_{55} + F_{25}^T G_{22} F_{25} \tag{1.29}$$

Equation (1.28) sets up the vector of resistor tree branch voltages in terms of all known quantities. Once V_5 is known, the remaining resistive quantities V_2, I_2, and I_5 are determined from Eqs. (1.13), (1.25) and (1.16) respectively.

The practical difference between the two approaches lies in the size of the matrix that must be inverted. The M_R matrix of Equ. (1.24) is of dimension n where n equals the number of resistor links in any network. The M_G matrix of Eq. (1.29) is of dimension m where m equals the number of resistor tree branches in any network. In practice it more often than not happens that $m \leq n$ when a significant number of voltage sources and capacitors are present. The reason for this is that the definition of the specific tree requires that these elements receive prior consideration when the network tree is formed. The tree is then often formed to a large extent by these elements thus allowing compara-

* See Appendix C for remarks concerning the general non-singularity of this matrix.

tively few of the resistors to be tree branches–therefore, most resistors are links. Consequently, the second method of obtaining resistive quantities often requires the inversion of a smaller matrix than does the first method. SCEPTRE checks for $m < n$ and then formulates the resistive equations accordingly.

1.6.2 Capacitive Quantities

All capacitor currents and voltages must be included in the vectors I_4, I_1, V_4, and V_1. The vector of state variables, V_4, is a known quantity at the beginning of each time step, but the vector of the derivatives of the state variables, $\dot{V}_4$, must be determined. The unknown capacitive quantities then are I_1, I_4, V_1, and $\dot{V}_4$ and they are computed in that sequence.

Two equations are available in (1.12) and (1.16). Others arise from the basic voltage – current relationship in capacitors as:

$$\dot{V}_4 = S_{44} I_4 \tag{1.30}$$

$$\dot{V}_1 = S_{11} I_1 \tag{1.31}$$

Another equation in $\dot{V}_1$ and $\dot{V}_4$ is needed and is obtained by differentiation of Eq. (1.12) to obtain:

$$\dot{V}_1 = -F_{14} \dot{V}_4 - F_{17} \dot{E}_7 \tag{1.32}$$

From Eqs. (1.31) and (1.32):

$$S_{11} I_1 = -F_{14} \dot{V}_4 - F_{17} \dot{E}_7 \tag{1.33}$$

From Eqs. (1.16) into (1.30):

$$\dot{V}_4 = S_{44} [F_{14}^T I_1 + F_{24}^T I_2 + F_{34}^T I_3 + F_{84}^T J_8] \tag{1.34}$$

Substitution of Eqs. (1.34) into (1.33) results in:

$$S_{11} I_1 = -F_{14} S_{44} [F_{14}^T I_1 + F_{24}^T I_2 + F_{34}^T I_3 + F_{84}^T J_8] - F_{17} \dot{E}_7$$

or

$$I_1 = M_C^{-1} \left\{ -F_{14} S_{44} [F_{24}^T I_2 + F_{34}^T I_3 + F_{84}^T J_8] - F_{17} \dot{E}_7 \right\} \tag{1.35}$$

where

$$M_C = S_{11} + F_{14} S_{44} F_{14}^T \tag{1.36}$$

The significance of Eq. (1.35) is that the vector of capacitor link currents is explicitly given in terms of known quantities. The rest of the capacitive quantities, I_4, V_1, and $\dot{V}_4$ are then determined from Eqs. (1.16), (1.12), and (1.30) respectively.

1.6.3 Inductive Quantities

All inductive currents and voltages must be included in the vectors I_3, I_6, V_3 , and V_6 . The vector of state variables, I_3, is a known quantity at the beginning of each time step, but the vector of the derivatives of the state variable, $\dot{I}_3$, must be determined. The unknown inductive quantities then are $\dot{I}_3$, V_3 , V_6 , and I_6, and they are computed in that sequence.

Two equations are available in (1.14) and (1.18); others arise from the basic voltage current relationship in inductors as:

$$V_3 = L_{33}\,\dot{I}_3 + L_{36}\,\dot{I}_6 \tag{1.37}$$

$$V_6 = L_{63}\,\dot{I}_3 + L_{66}\,\dot{I}_6 \tag{1.38}$$

Another equation in $\dot{I}_3$ and $\dot{I}_6$ is needed and is obtained by differentiation of Eq. (1.18) to obtain: *

$$\dot{I}_6 = F_{36}^T\,\dot{I}_3 + F_{86}^T\,\dot{J}_8 \tag{1.39}$$

From Eqs. (1.14) and (1.37):

$$L_{33}\,\dot{I}_3 + L_{36}\,\dot{I}_6 = -F_{34}V_4 - F_{35}\,V_5 - F_{36}\,V_6 - F_{37}E_7 \tag{1.40}$$

From Eqs. (1.38) and (1.39):

$$V_6 = L_{63}\,\dot{I}_3 + L_{66}\,[F_{36}^T\,\dot{I}_3 + F_{86}^T\dot{J}_8] \tag{1.41}$$

Substitution of Eqs. (1.41) into (1.40) yields,

$$L_{33}\,\dot{I}_3 + L_{36}\,\dot{I}_6 = -F_{34}\,V_4 - F_{35}\,V_5 - F_{36}\,[L_{63}\,\dot{I}_3 + L_{66}\,[F_{36}^T\dot{I}_3 + F_{86}^T\dot{J}_8]] - F_{37}E_7$$

If Eq. (1.39) is used again:

$$L_{33}\,\dot{I}_3 + L_{36}\,[F_{36}^T\dot{I}_3 + F_{86}^T\dot{J}_8] = -F_{34}\,V_4 - F_{35}\,V_5 - F_{36}\,L_{63}\,\dot{I}_3 - F_{36}\,L_{66}\,F_{36}^T\dot{I}_3 - F_{36}\,L_{66}\,F_{86}^T\,\dot{J}_8 - F_{37}\,E_7$$

or

$$\dot{I}_3 = M_L^{-1}\left\{-F_{34}\,V_4 - F_{35}\,V_5 - [F_{36}\,L_{66}\,F_{86}^T + L_{36}\,F_{86}^T]\,\dot{J}_8 - F_{37}\,E_7\right\} \tag{1.42}$$

where

$$M_L = L_{33} + L_{36}\,F_{36}^T + F_{36}\,L_{63} + F_{36}\,L_{66}\,F_{36}^T \tag{1.43}$$

* Note that the need for source derivatives has been introduced by Eqs. (1.32) and (1.39).

Equation (1.42) presents the vector of the derivatives of the inductor link currents. The remaining inductive quantities, V_3, V_6, and I_6 are then determined from Eqs. (1.37) and (1.39), (1.38) and (1.39), and (1.18) respectively.

1.6.4 Current Source Voltages and Voltage Source Currents

Now that all passive currents and voltages have been found, Eq. (1.15) yields the vector of current source voltages and Eq. (1.19) yields the vector of voltage source currents.

An exception to the method outlined above occurs when the problem under analysis contains one or more "linearly dependent sources." These sources are defined as those that are linearly dependent upon resistor currents or resistor voltages.

These sources require special treatment because their magnitudes are direct functions of quantities that are not state variables. Some provision must be made for sources of this type within the mathematical formulation if the "computational delay" problem indicated in Sec. 1.3 is to be avoided.

For this purpose, a class of dependent voltage sources is defined called Class Y elements such that:

$$E_y = k_1 V_2 + k_2 V_5 \tag{1.44}$$

where k_1 is a matrix of constants having a number of rows equal to the number of Class Y elements and a number of columns equal to the number of resistor links; k_2 has a number of rows equal to the number of Class Y elements and a number of columns equal to the number of tree branch resistors.

A class of dependent current sources called Class X elements is also defined such that:

$$J_x = k_3 I_2 + k_4 I_5 , \tag{1.45}$$

where the matrices k_3 and k_4 are described analogously to k_1 and k_2.

The addition of these element classes enlarges the general F matrix such that the construction of Fig. 1.6 now becomes that of Fig. 1.8. Equations (1.13) and (1.17) now become:

$$V_2 = -F_{24} V_4 - F_{25} V_5 - F_{27} E_7 - F_{2y} E_y \tag{1.13'}$$

$$I_5 = F_{25}^T I_2 + F_{35}^T I_3 + F_{85}^T J_8 + F_{x5}^T J_x \tag{1.17'}$$

If Eqs. (1.44) and (1.45) together with (1.20) and (1.26) are substituted into (1.13') and (1.17'),

$$R_{22} I_2 = -F_{24} V_4 - F_{25} V_5 - F_{27} E_7 - F_{2y} [k_1 V_2 + k_2 V_5] \tag{1.46}$$

$$G_{55} V_5 = F_{25}^T I_2 + F_{35}^T I_3 + F_{85}^T J_8 + F_{x5} [k_3 I_2 + k_4 I_5] \tag{1.47}$$

The last two equations may be consolidated as:

$$\begin{bmatrix} (R_{22} + F_{2y} k_1 R_{22}) & (F_{25} + F_{2y} k_2) \\ (-F_{25}^T - F_{x5}^T k_3) & (G_{55} - F_{x5}^T k_4 G_{55}) \end{bmatrix} \begin{bmatrix} I_2 \\ V_5 \end{bmatrix} = \begin{bmatrix} -F_{24} V_4 - F_{27} E_7 \\ F_{35}^T I_3 + F_{85}^T J_8 \end{bmatrix}$$

and the resistive quantities I_2 and V_5 are determined without computational delay via inversion of the matrix on the left hand side.

F_{14}	0	0	F_{17}	F_{1y}
F_{24}	F_{25}	0	F_{27}	F_{2y}
F_{34}	F_{35}	F_{36}	F_{37}	F_{3y}
F_{84}	F_{85}	F_{86}	F_{87}	F_{8y}
F_{x4}	F_{x5}	F_{x6}	F_{x7}	F_{xy}

Fig. 1.8 F Matrix with Linearly Dependent Sources

One other point of interest arises as a result of the enlarged F matrix of Fig. 1.8. Equations (1.12) and (1.18) in turn become enlarged leading to:

$$\dot{V}_1 = -F_{14} \dot{V}_4 - F_{17} \dot{E}_7 - F_{1y} \dot{E}_y \quad \text{and} \tag{1.32'}$$

$$\dot{I}_6 = F_{36}^T \dot{I}_3 + F_{86}^T \dot{J}_8 + F_{x6}^T \dot{J}_x \tag{1.39'}$$

The requirement for quantities $\dot{E}_y$ and $\dot{J}_x$ imply the need [through differentiation of Eqs. (1.44) and (1.45)] for quantities $\dot{V}_2$, $\dot{V}_5$, $\dot{I}_2$, and $\dot{I}_5$. The formulation contains no provisions for these quantities, and there is no way that the user could know them to supply them as input data. SCEPTRE automatically checks for the conditions* F_{1y}, $F_{x6} \neq 0$ and terminates any run if one occurs.

*These conditions are quite rare in practical circuit analysis. The situations can be recognized by the presence of any circuit loop composed solely of capacitors and at least one Class Y element, or any circuit-set composed solely of inductors and at least one Class X element.

1.7 CONSIDERATIONS IN TIME VARYING AND NONLINEAR NETWORKS

The formulation described in the preceding section may be applied to the class of networks in which one or more of the passive elements are variable functions of time. Whether compensation of one form or another is needed depends upon which passive elements are variable or, more basically, how the passive elements are defined in terms of network variables such as current, voltage, charge, etc. An effective way to proceed is to choose networks for which known closed form solutions can be obtained and then to compare these solutions with those that result from application of the automatic program which is based on this formulation.

Consider the network of Fig. 1.9 in which a ten volt step function is to be applied to an RC circuit in which the capacitance is constant and the resistor is a variable function of time. The network is initially relaxed. Equation (1.48) describes the network.

$$10 = R\frac{dq}{dt} + \frac{q}{C} \tag{1.48}$$

or

$$\frac{dq}{dt} + \frac{q}{(t+1)C} = \frac{10}{(t+1)}$$

The general solution is found by separation of variables as:

$$q = K\,(t+1)^{-1/C} + 10C$$

or

$$V_C(t) = \frac{K}{(t+1)} + 10$$

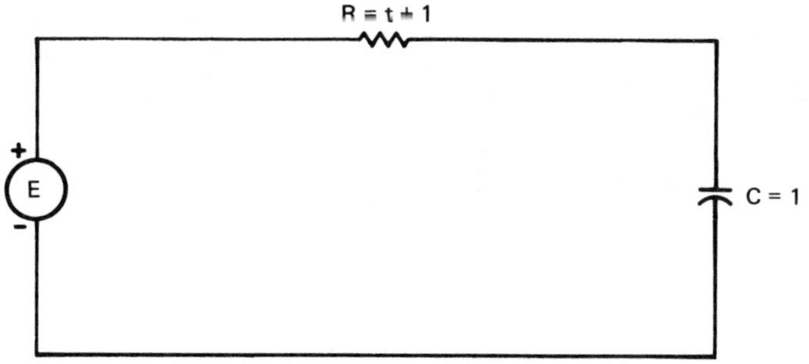

Fig. 1.9. Time varying resistor circuit.

The particular solution is evaluated from consideration of the initial conditions $V_C(o) = 0$ as:

$$V_C(t) = \frac{-10}{(t+1)} + 10$$

The SCEPTRE result is virtually indistinguishable from the closed form solution since the maximum error is about 0.02 volts. This error can be attributed to unavoidable roundoff in the computer and a certain amount of truncation error in the integration routine. It is clear, by illustration at least, that the formulation is valid for time varying resistors.

Attention is now directed to the problem of time varying capacitors. Consider the network of Fig. 1.10 in which a ten volt step function is applied to an RC circuit in which the resistance is constant and the capacitance is a variable function of time. The network is initially relaxed. Equation (1.48) is still valid but now becomes:

$$\frac{dq}{dt} + \frac{q}{R(t+5)} = \frac{10}{R}$$

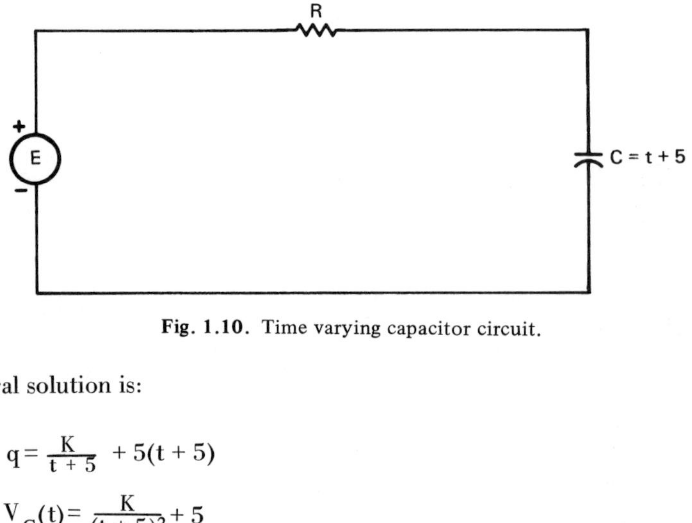

Fig. 1.10. Time varying capacitor circuit.

The general solution is:

$$q = \frac{K}{t+5} + 5(t+5)$$

or

$$V_C(t) = \frac{K}{(t+5)^2} + 5$$

Again using $V_C(0) = 0$, the particular solution is:

$$V_C(t) = \frac{-125}{(t+5)^2} + 5$$

Figure 1.11 presents plots of both the closed form solution and the SCEPTRE solution. A significant error is evident. The source of this error is determined by an examination of the manner in which state variable derivatives are computed. The relation $I = C\frac{dV}{dt}$ is implicit in Eq. (1.30). If, however, capacitance is defined as $C = \frac{Q}{V}$, then $Q = CV$, and:

$$I = \frac{dQ}{dt} = \frac{d}{dt}(CV) = C\frac{dV}{dt} + \frac{dC}{dt}V \tag{1.49}$$

A discrepancy between Eqs. (1.30) and (1.49) is apparent. The formulation has not included the $\frac{dC}{dt}$ V term, and this is the cause of the error pictured in Fig. 1.11. It is clear that this error is incurred whenever networks containing time varying capacitors are solved with this formulation.

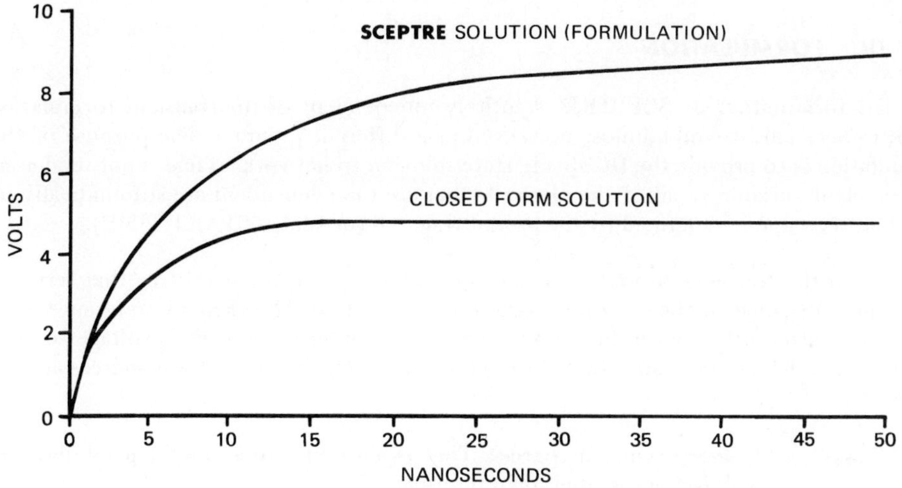

Fig. 1.11 Capacitor Voltage in Network of Fig. 1.10

Once the cause of the error is realized, a remedy is easily devised by the program user. If the time varying capacitor of Fig. 1.10 is shunted by a current generator J_C where $J_C = V_C \frac{dC}{dt}$, the total current into the node becomes $I_C = C\frac{dV}{dt} + \frac{dC}{dt}V$ which effects the necessary correction. A computer run with the additional current generator has been made for purposes of corroboration. The results were virtually identical with the closed form solution.

The interested reader may show that exactly the dual situation exists for time-varying inductors. That is, the SCEPTRE solution to any L(t)problem shows a significant disparity with the true solution. The situation is easily restored, however, with the insertion of a series voltage source equal to $\frac{dL}{dt} I_L$.

Another small, but very practical, class of non-linear networks which are accurately accommodated is now shown. Consider the situation in which one or more variable capacitors exist in a network such that C = C (V). If capacitance is defined as C = Q/V then Eq. (1.49) must be written as,

$$I = \frac{dQ}{dt} = [C(V) + V\frac{d}{dt}C(V)]\frac{dV}{dt} \tag{1.50}$$

Again there is a discrepancy with the formulation [Eq. (1.30)] which leads to error if C = C (V). If, however, these capacitances are instead defined as C = dQ/dV, then substitution into Eq. (1.30) yields simply I = dQ/dt which is certainly correct as it stands. The class of voltage dependent capacitors that are defined this way are the transition and diffusion capacitances associated with semiconductor junctions [8,9]. This definition allows the SCEPTRE formulation to accurately determine the transient response of transistor and diode networks without compensation!

1.8 DC FORMULATION

The DC formulation of SCEPTRE is entirely independent of the transient formulation given in Sec. 1.6; it could almost be viewed as a different program. The purpose of this formulation is to provide the DC steady state solution to networks. Time is not used as an independent variable so all of the elements that are time dependent are automatically set to whatever values are defined by the user at time = 0 (or time = START TIME).

As in the transient formulation, the starting point is the general tree that serves as the source of some of the equations used in the derivation. The element preference order is defined differently than in Sec. 1.1; however, the order of priority is voltage sources, inductors, resistors, and capacitors. Two additional categories of current sources are also defined:

Class 9 – "Primary" current sources. That is, current sources that depend upon the voltages across their own terminals.
Class 0 – "Secondary" current sources. That is, current sources that depend upon a primary current source.

The connotated form of the general F matrix as it would arise from the preceeding tree definition is shown in Fig. 1.12.* Horizontal "scans" of the Class 2 and 9 rows and a vertical scan of the Class 5 row lead to:

$$I_5 - F_{15}^T I_1 - F_{25}^T I_2 - F_{85}^T J_8 - F_{95}^T J_9 - F_{05}^T J_0 = 0 \tag{1.51}$$

$$V_2 + F_{25} V_5 + F_{26} V_6 + F_{27} E_7 = 0 \tag{1.52}$$

$$V_9 + F_{94} V_4 + F_{95} V_5 + F_{96} V_6 + F_{97} E_7 = 0 \tag{1.53}$$

Immediate simplifications may be made. The DC steady state of vectors I_1 and V_6 are obviously zero. If submatrix F_{94} is constrained to be zero and the following substitutions are made:

$$I_5 = G_{55} V_5$$

$$I_2 = G_{22} V_2$$

$$J_0 = \alpha J_9$$

* Note that the amended definition of element priority has led to a different pattern of zero valued submatrices from the transient F matrix of Fig. 1.6.

$J_9 = G_{99}V_9 + Q_9$

where α is a matrix containing the number of rows equal to the number of secondary current sources and the number of columns equal to the number of primary sources .

	Class 4	Class 5	Class 6	Class 7
Class 1	F_{14}	F_{15}	F_{16}	F_{17}
Class 2	0	F_{25}	F_{26}	F_{27}
Class 3	0	0	F_{36}	F_{37}
Class 8	F_{84}	F_{85}	F_{86}	F_{87}
Class 9	F_{94}	F_{95}	F_{96}	F_{97}
Class 0	F_{04}	F_{05}	F_{06}	F_{07}

Fig. 1.12 DC Formulation F Matrix

G_{55} and G_{22} are diagonal matrices containing only resistor conductances. G_{99} is a diagonal matrix containing only diode and transistor junction conductances.

Q is a vector of Q terms (see Appendix D), then Eqs. (1.47) to (1.49) become:

$$G_{55} V_5 - F_{25}^T G_{22} V_2 - F_{85}^T J_8 - [F_{95}^T + F_{05}\alpha] [G_{99} V_9 + Q_9] = 0 \qquad (1.51')$$

$$V_2 + F_{25} V_5 + F_{27} E_7 = 0 \qquad (1.52')$$

$$V_9 + F_{95} V_5 + F_{97} E_7 = 0 \qquad (1.53')$$

These three equations may be designated as $f_1(V_5, V_2, V_9)$, $f_2(V_5, V_2, V_9)$, $f_3(V_5, V_2, V_9)$. Application of a standard Newton-Raphson technique leads to the consolidated form:

$$\begin{bmatrix} f_1(V_5, V_2, V_9) \\ \\ f_2(V_5, V_2, V_9) \\ \\ f_3(V_5, V_2, V_9) \end{bmatrix} + \begin{bmatrix} \dfrac{\partial f_1}{\partial V_5} & \dfrac{\partial f_1}{\partial V_2} & \dfrac{\partial f_1}{\partial V_9} \\ \\ \dfrac{\partial f_2}{\partial V_5} & \dfrac{\partial f_2}{\partial V_2} & \dfrac{\partial f_2}{\partial V_9} \\ \\ \dfrac{\partial f_3}{\partial V_5} & \dfrac{\partial f_3}{\partial V_2} & \dfrac{\partial f_3}{\partial V_9} \end{bmatrix} \begin{bmatrix} \Delta V_5 \\ \\ \Delta V_2 \\ \\ \Delta V_9 \end{bmatrix} = 0$$

Since

$$V_5 = V_5(k+1) - V_5(k)$$

$$V_2 = V_2(k+1) - V_2(k)$$

$$V_9 = V_9(k+1) - V_9(k)$$

and evaluation of the Jacobian matrix from Eqs. (1.51') to (1.53') leads to:

$$\begin{bmatrix} V_5(k+1) \\ V_2(k+1) \\ V_9(k+1) \end{bmatrix} = \begin{bmatrix} V_5(k) \\ V_2(k) \\ V_9(k) \end{bmatrix} - \begin{bmatrix} G_{55} & -F_{25}^T G_{22} & -[F_{95}^T + F_{05}^T \alpha] G_{99} \\ F_{25} & I & 0 \\ F_{95} & 0 & I \end{bmatrix}^{-1} \begin{bmatrix} f_1(V_5, V_2, V_9) \\ f_2(V_5, V_2, V_9) \\ f_3(V_5, V_2, V_9) \end{bmatrix} \qquad (1.54)$$

Equation (1.54) describes an iterative computational sequence. Quantities V_5, V_2, and V_9 (the vectors of voltages across resistive tree branches, resistor links and primary current sources respectively) each have some assumed value (usually zero) to begin the computation at $k = 0$. All members of the right side of (1.54) that may be voltage dependent are updated. The left side is then computed and the first iteration ($k = 1$) is complete. The right side of Eq. 1.54 is re-evaluated on the basis of the results of the first iteration and the left side is again computed, thereby completing the second iteration ($k = 2$). The process continues until, after any iteration:

$$\left| V(k+1) - V(k) \right| \leq .0001 + .001 \left| V(k+1) \right|$$

If Eq. (1.55) is satisfied for the set of voltages in Eq. (1.54), the process is said to have converged and is terminated. If after 100 iterations, Eq. (1.55) has not been satisfied the process is terminated and an appropriate diagnostic message informs the user.

Once convergence has occurred, only capacitor voltages and inductor currents remain to be computed. From the F matrix in Fig. 1.12, the capacitor link voltages are expressed in terms of the tree branch voltages as

$$V_1 = -F_{14} V_4 - F_{15} V_5 - F_{17} E_7 \tag{1.56}$$

In addition, the principle of conservation of charge permits:

$$-F_{14}^T C_{11} V_1 + C_{44} V_4 = -F_{14}^T C_{11} V_1(o) + C_{44} V_4(o) \tag{1.57}$$

where V_1 (0) and V_4 (0) are initial voltages that may be specified by the user. Eqs. (1.56) and (1.57) lead to:

$$V_4 = M_c^{-1} \left\{ -F_{14}^T C_{11} F_{15} V_5 - F_{14}^T C_{11} F_{17} E_7 - F_{14}^T C_{11} V_1(o) + C_{44} V_4(o) \right\} \tag{1.58}$$

where

$$M_c = C_{44} + F_{14}^T C_{11} F_{14}$$

Once V_4 is determined, V_1 is found from Eq. (1.56). The first step in the solution of inductor currents utilizes:

$$I_2 = R_{22}^{-1} V_2$$

The inductor tree branch currents are then written from the F matrix in terms of link currents as:

$$I_6 = F_{26}^T I_2 + F_{36}^T I_3 + F_{86}^T J_8 + F_{96}^T J_9 + F_{06}^T J_0 \tag{1.59}$$

In addition, the flux relations around an inductor loop permit.

$$F_{36} L_{66} I_6 + L_{33} I_3 = 0 \tag{1.60}$$

Substitution of Eq. (1.59) into (1.60) leads to:

$$I_3 = M_L^{-1} \left\{ -F_{36} L_{66} [F_{26}^T I_2 + F_{86}^T J_8 + F_{96}^T J_9 + F_{06}^T J_0] \right\} \tag{1.61}$$

where

$$M_L = L_{33} + F_{36} L_{66} F_{36}^T$$

Once I_3 is determined, I_6 is found directly from Eq. (1.59). Only the trivial details of computing resistor tree branch currents, voltage source currents and current source voltages remain to complete the DC formulation.

1.9 UNIT SYSTEMS

Even the most trivial solution to any electrical network, regardless of the solution medium, is grossly inaccurate if an inconsistent set of units is used for the problem. When the program performs the operations implied in Sec. 1.6, it is assumed that the user has entered all input data in a mutually consistent system of units. The program is manipulating numbers; and if these numbers are improperly supplied due to an error in units, the results are quite disastrous. There are literally an infinite number of mutually consistent systems of units, but as a practical matter, the user seldom, if ever, deviates from the commonly used set shown in Chapter 2.

If, for any reason, it is desired to operate with a different set of units, the conversion is quite straightforward. The six most common quantities of interest in the transient solution of electrical networks are time, voltage, current, resistance, capacitance, and inductance. Three basic dimensional relations that combine to link all six quantities are:

$$\text{Voltage} = \frac{\text{Charge}}{\text{Capacitance}} = \frac{(\text{Current})\ (\text{Time})}{\text{Capacitance}} \tag{1.62}$$

$$\text{Time} = (\text{Resistance})\ (\text{Capacitance}) \tag{1.63}$$

$$\text{Voltage} = \frac{(\text{Inductance})\ (\text{Current})}{\text{Time}} \tag{1.64}$$

It is then possible to arbitrarily choose the desired units for the first three of the basic six quantities and then to systematically use Eqs. (1.62) through (1.64) to determine the units for the remaining three quantities. Assume that it is desired to operate with time in microseconds, current in amperes, and voltage in volts. Writing these quantities in terms of some multiple times absolute units, yields time = 10^{-6} seconds, current = 1 ampere, and voltage = 1 volt. Transposition of Eq. (1.62) indicates that the units of capacitance are:

$$C = \frac{It}{V} = \frac{(1)\ (10^{-6})}{1} = 10^{-6}$$ times absolute units or microfarads. Next, from (1.63)

$$R = \frac{t}{c} = \frac{10^{-6}}{10^{-6}} = 1$$ times absolute units or ohms. Finally, from (1.64)

$$L = \frac{Vt}{I} = (1)\ \frac{(10^{-6})}{1} = 10^{-6}$$ times absolute units or microhenries.

There is no particular advantage in efficiency to be gained by choosing one system of units over another. The only inducement is that of convenience or personal preference. If, for example, amperes of current are expected in a particular network, the user may well prefer numbers like 2.6 amperes rather than 2600 milliamperes.

1.10 STEP SIZE CONSIDERATIONS

In Section 1.3 it is illustrated that any transient solution in SCEPTRE is nothing else but a series of individual solutions that are carried out at automatically selected time intervals. It is at least intuitively clear that the amount of computer time required for any transient solution is proportional to the number of solution steps taken. All else being equal, a run that requires 2000 solution steps is more costly than a run that requires only 200 steps. It is also clear that the user would like to avoid the situations that force many solution steps to be taken. The purpose of this section is to show, in engineering language, the two factors that cause small step sizes to be taken.

The role of the integration routine is to compute $V_m(n+1)$ from the known quantities $V_m(n)$, $\dot{V}_m(n)$, and "h". Where V_m and $\dot{V}_m$ are the mth state variable and its derivative, the subscript n refers to the nth time step and h is the time step size to be attempted. The process may be interpreted pictorially from Fig. 1.2, where it is desired for example, to update the state variable from $V(t_o)$ to $V(t_1)$. It is now necessary to describe the method of numerical integration that is to be used. For that purpose, a basic two-pass trapezoidal method that is similar to one of the three methods that are optionally available in SCEPTRE is defined. Consider the computational sequence:

$$V_p = V(t_o) + \frac{h}{2}\dot{V}(t_o) \tag{1.65}$$

$$\dot{V}_p = f(V_p, t) \tag{1.66}$$

$$\frac{2}{3}h\left|\dot{V}(t_o) - \dot{V}_p\right| \leq .01 \tag{1.67}$$

$$V(t_1) = V(t_o) + \frac{h}{3}\left[2\dot{V}(t_o) + \dot{V}_p\right] \tag{1.68}$$

which can be followed with the aid of Fig. 1.3. Quantities h, the step size, and $V(t_o)$, the initial condition, are known before the first pass through the block labeled "Equation Solution" is begun. Quantity $\dot{V}(t_o)$ is picked up during the first pass through this block. Move next to the block labeled "Integration Routine", and the operation indicated in Eq. (1.65), is performed. Upon leaving the block, the path returns to the original block for the second pass and performs the operation of Eq. (1.66). Again proceed to the integration routine, and determine if the relation implied in Eq. (1.67) holds. * If so, the step is accepted and the state variable is updated as in Eq. (1.68). If however, the quantity on the left side of Eq. (1.67) is larger than the absolute error criteria of .01, the step is rejected, h is automatically halved, and the cycle is repeated.

In order to further clarify the preceding description, it is instructive to repeat the process with a differential equation that is derived from an actual network. In the interest of simplicity, a circuit is chosen that requires the integration of only one differential equation. Consider the network of Fig. 1.9, with the element values E = 1, R = 1, and C = 1. The interested reader may review the formulation of Sections 1.6.1 and 1.6.2 to

* This relation is based on an approximation to the second term of a Taylor Series expansion.

establish that the single differential equation that describes this network is:

$$\dot{V}_c = \frac{E - V_c}{RC}$$

Assume that $V_c(0) = 0$ and that an initial step size of $h = 0.2$ is to be attempted. Repeating the procedure given above, the first pass through "Equation Solution" evaluates:

$$\dot{V}_c(t_o) = \frac{1 - 0}{1} = 1$$

Eq. (1.65) is evaluated in the "Integration Routine" yielding:

$$V_{cp} = 0 + \frac{.2(1)}{2} = .1$$

The second pass through "Equation Solution" yields:

$$\dot{V}_{cp} = \frac{1 - .1}{1} = .9$$

At this point the process again shifts to "Integration Routine" for the error check given by Eq. (1.67). For this particular integration,

$$\frac{2}{3}(.2)\left|1 - .9\right| \cong .013 ,$$

which exceeds the stipulated value of .01. The error criterion has not been satisfied, and this step is rejected with the consequence that a smaller step size, and more solution steps are necessary.

It is well to pause to examine the computational sequence in order to determine the factors that force a small step size. The step size was rejected because the left side of Eq. (1.67) was too large. The left side of Eq. (1.67) was too large because of the successive derivative values that are computed, and these in turn owe their magnitudes to the parent differential equation. Therefore, large numerical values will occur if either the denominator is small or if the numerator is large.

The numerator can be large if the forcing function E is large. Note, however, that this effect can be mitigated if the capacitor voltage begins to approach the forcing function in magnitude. This frequently occurs in practice and is usually the reason why the program step size begins to increase significantly somewhere near the mid point of many transient runs. The denominator, however, depends only on the RC product, the network time constant. Therefore, a very small time constant can inhibit the step size throughout an entire run without any mitigating circumstance. This effect does not limit all integration methods equally. In many situations, another of the SCEPTRE integration routines (XPO) is able to take considerably larger step sizes, and because of this quality, has been selected as the default option.

The foregoing does not tell the whole story. If only small time constants are present in any given network, the solution step sizes taken by the program are indeed small, but the problem duration that is required would also be small since the network voltages and currents move to their steady states quite rapidly. The real problem arises when at least one large time constant is present in addition to one or more very small time constants. The large time constants require that a large problem duration be specified if the entire transient period is to be covered, and the small time constants force a comparatively small average step size. It is clear that the presence of both very small and large time constants can act to force a large number of solution steps in the course of a transient problem. This effect has long been recognized and is commonly referred to as the infamous "time constant spread" problem. A number of ways to potentially alleviate this problem are discussed in an example given in Sec. 4.11.

PROBLEMS

1.1 How many cut sets can be formed from the network of Fig. 1.1? How many loops?

1.2 Each fundamental cut set must include exactly one tree branch and at least one link element. How many fundamental cut sets are contained in the network of Fig. 1.1?

1.3 Each fundamental loop must include exactly one link and at least one tree branch. How many fundamental loops can be found in Fig. 1.1?

1.4 Would the elements E–R1–L–R2–J constitute a cut set? A fundamental cut set?

1.5 What is the visible relation between the F matrix and the fundamental loops and fundamental cut sets of the parent network?

1.6 What would be the most practical objection to the definition of voltage source links as an allowable class of element in the program formulation?

1.7 Construct the M_R matrix (Eq. 1.24) for the network of Fig. 1.1.

1.8 What is the general relation between the order of the matrix in Prob. 1.7 and the number of network resistor tree branches?

1.9 How would the specific tree of Fig. 1.1 be altered if two capacitors, Cx and C_y, were placed in parallel with R_2 in the parent network of Fig. 1.1?

1.10 Why is (Eq. 1.7) always true?

1.11 Why must a partition of the Q matrix of Sec. 1.4 always form a unit matrix if the member elements are properly ordered?

1.12 For some strange reason, a user wishes to use units of time in milliseconds, voltage in kilovolts and inductance in henries. What units should be used for capacitance, resistance, current and power?

1.13 The same user now wishes to use units of time in microseconds, current in amperes and voltage in volts. What units should be used for capacitance, resistance, inductance and power?

1.14 A connected network contains eight elements. What is the maximum and minimum number of nodes, tree branches and fundamental loops that could be contained in a network of this size in any possible configuration?

1.15 A network is composed of the following list of elements:

```
E1, 1-2 = 10
R1, 2-3 = 1
R2, 3-4 = 1
R3, 4-5 = 1
R4, 5-6 = 1
R5, 4-1 = 1
C1, 6-1 = 1
```

How many resistor tree branches and links are contained in this network?

1.16 A network contains two inductors, L1 and L2, which are both links. If a mutual inductance M exists between these inductors, then

$$M_L = \begin{bmatrix} L1 & M \\ M & L2 \end{bmatrix}$$

If L1 = 4, L2 = 1 and M = 2, why does this run have a short future?

1.17 What if M in Prob. 1.16 was given a value slightly greater than 2?

REFERENCES

1. Seshu, S. and N. Balabanian, *Linear Network Analysis*, Wiley and Sons, Inc., New York, 1959.

2. Yadeh, L. and C. Desoer, *Linear System Theory*, McGraw–Hill Book Company, New York, 1960.

3. Kuh, E. and R. Rohrer, "The State–Variable Approach to Network Analysis," *Proceedings of the IEEE,* pp. 672–686, (July, 1965).

4. Bashkow, T. R., "The A Matrix, New Network Description," *IRE Transaction on Circuit Theory*, Vol. CT–4, pp. 117–120, (September, 1957)

5. Warten, R. M., "Automatic Step Size Control for Runge–Kutta Integration," *IBM Journal of Research and Development*, Vol. 7, pp. 340–341, (October, 1963)

6. Seshu, S. and M. B. Reed, *Linear Graphs and Electrical Networks*, Addison–Wesley Publishing Company, Reading, Massachusetts, 1961

7. Sedore, S. R., "More Efficient Use of the F Matrix in Practical Circuit Analysis Programs, *IEEE Transactions on Computers*, Vol. C–17, May 1968.

8. Phillips, A., *Transistor Engineering*, McGraw–Hill Book Company, New York, 1962.

9. Nararati, R., *An Introduction to Semiconductor Electronics*, McGraw–Hill Book Company, New York, 1963.

10. Kreyszig, E., *Advanced Engineering Mathematics,* Wiley and Sons, Inc., New York 1962.

2

SCEPTRE FORMAT

2.1 INTRODUCTION

One of the distinguishing features of the SCEPTRE circuit analysis program is the ease with which the engineer can communicate with the computer. The circuit description language described in this chapter is used to convey all the necessary information to the SCEPTRE program. The user is not required to write network equations or to possess a knowledge of computer programming. In addition to the simplicity, the SCEPTRE program also provides a wide degree of flexibility and many significant and useful features.

2.1.1 Useful Features

The following is a partial list of useful features to give the user an indication of the versatility of the SCEPTRE program:

1. *INPUT CONVENIENCE:* Statements can be punched anywhere on the input program card (columns 1–72) with any desired spacing.
2. *STORED MODELS:* Any active element or interconnected group of elements that can be described as a combination of sources, passive elements and mutual inductance may be stored on tape by the user and called into use at any point in a network. Any element in the model can be described in mathematical form.
3. *LINEARLY DEPENDENT SOURCES:* Voltage and current sources that are linearly dependent on resistor voltages and currents respectively, can be accommodated without computational delay. This feature permits the extensive use of the family of small signal transistor equivalent circuits.
4. *PROGRAM LANGUAGE:* The program has been written entirely in FORTRAN IV to facilitate the task of adapting it to digital computers other than the IBM 7094. In fact, the SCEPTRE program is now available for a wide variety of machines.
5. *SUBPROGRAM CAPABILITY:* The user who is familiar with computer programming may write FORTRAN subroutines and insert them in otherwise conventional SCEPTRE runs. This option can be used to permit special purpose computation or to allow the user a greater degree of control over the program operation.
6. *FUNCTIONS:* Any variable quantity may be described either in equation form or tabular form, and any number of independent variables may be used.

7. *AUTOMATIC INITIAL CONDITIONS:* The user has the option of using a special portion of the program to determine the initial conditions of a network. He may then either use the transient section in the same run or accept the output of the initial condition section for inspection. Any run may use the initial condition mode only, the transient mode only, or may automatically combine the two modes.

8. *OUTPUT:* The output format allows selection of any or all source voltages and currents, element voltages and currents, defined parameters, step size, pass count, etc. The output is presented only in tabular form unless a plot is requested. Time or any other independent variable may be selected for plots.

9. *DEFINED PARAMETERS:* Special quantities may be defined that are available for output other than source and element voltages and currents.

10. *DIFFERENTIAL EQUATIONS:* The user may enter systems of first-order differential equations (state equations) that may or may not have anything to do with a particular electrical network. A slight variation of this technique permits the simulation of linear transfer functions.

11. *RERUN:* Multiple case rerun based on a single master run may be carried out automatically. The user supplies only the changes that differ from the master run for each repeated run.

12. *SAVE AND CONTINUE CAPABILITY:* Runs may be terminated and then subsequently continued after examination.

13. *AUTOMATIC TERMINATION:* Runs may be automatically terminated contingent on the behavior of specified network quantities.

14. *ECONOMICAL OPERATION:* A new solution approach has been employed to permit rapid solution of the network equations at each time step. In many cases this technique will lead to a comparatively small amount of computer time per run, especially when the rerun feature is used.

2.1.2 Sceptre Program Example

In order to demonstrate how easily a SCEPTRE program is written, the following complete program of a practical circuit is presented. The reader will then be able to write simple programs by following the steps and comments in this example. The fine details and the many other options available are discussed in the remainder of the chapter.

The circuit to be analyzed is shown in Fig. 2.1.

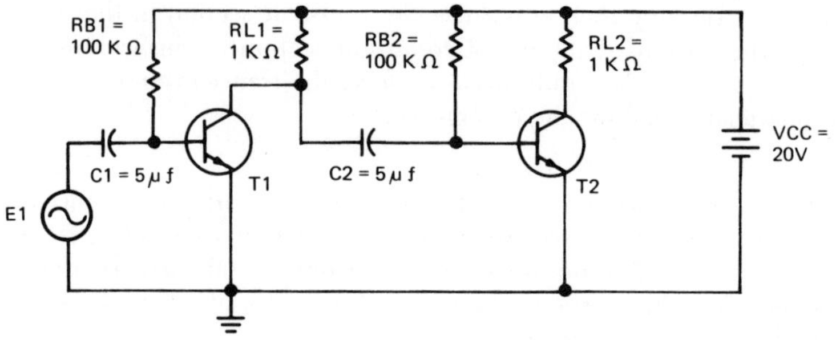

Fig. 2.1 Circuit schematic diagram.

The following steps are suggested to prepare the circuit for programming:

1. Draw an equivalent circuit consisting of only resistor, capacitors, inductors, and voltage and current sources. In many cases the original schematic may be sufficient.

2. Assign a name or number to all nodes in the circuit. The nodes may be labeled in any order.

3. Give a name to each circuit element. (*R*=resistor, *C*=capacitor, *L*=inductor, *E*=voltage source, *J*=current source, *M*=mutual inductance).

4. Assume arbitrary current flow directions in each passive circuit element. Show the current directions by arrows in the equivalent circuit.

5. Indicate the direction of positive current flow in each voltage or current source.

6. Choose and record circuit values in any consistent set of parameter units. The following set of units given in Table 2.1 is useful for high speed transistorized circuits.

Table 2.1. Units for high-speed transistorized circuits.

Parameter	Unit
Resistance	Kilohms
Capacitance	Picofarads
Inductance	Microhenries
Current	Milliamperes
Voltage	Volts
Time	Nanoseconds

The equivalent circuit developed by the preceding steps is shown in Fig. 2.2.

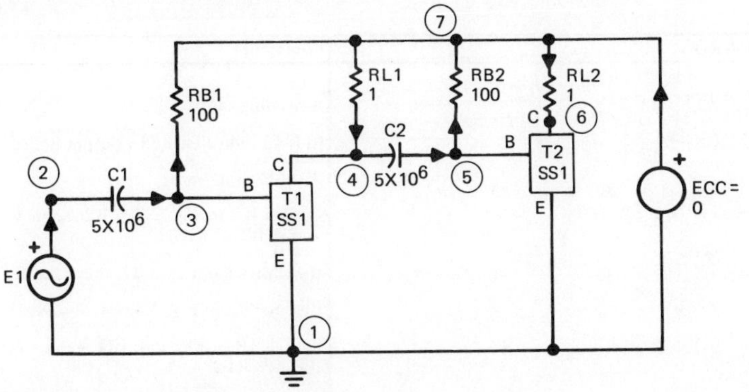

Fig. 2.2. Circuit in SCEPTRE form.

The transistors T1 and T2 have been replaced by blocks which contain the model circuit shown in Fig. 2.3. This model is a small signal h-Parameter equivalent circuit;* but it could be any other model, as complete as necessary to meet the analysis requirements. Any active element (Diode, FET, Integrated Circuit, etc.) may be modeled in this manner if it can be described as a combination of sources, passive or active elements, linear or non-linear devices, mutual inductances, equations and tables. Modeling is fully discussed in Chapter 3. Note that the model circuit is developed in SCEPTRE format in the same manner as the main circuit, showing node numbers, current arrows, etc.

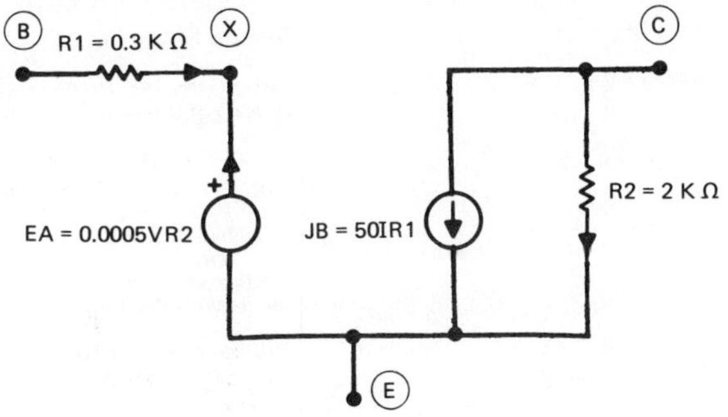

Fig. 2.3 Small signal h-parameter transistor model in SCEPTRE form

The SCEPTRE circuit description language is easy to learn and remember. The following are a few basic rules to keep in mind. Detailed rules are given throughout the chapter:

1. The headings to be used must come in a specific order (see Table 2.2).
2. Subheading and statements under each heading may come in any order.

---

*It should be remembered that this particular equivalent circuit is applicable only for a.c. small-signal analysis and is not valid for determining initial conditions or other d.c. quantities. It is only used here because of its familiarity. In Chapter 3, large signal models are discussed, which the reader will no doubt usually use.

PROGRAM CARDS	Comments
MODEL DESCRIPTION	A heading card
MODEL SS1 (B-E-C)	(B-E-C) Indicates model output nodes
ELEMENTS	A subheading card
EA, E-X = .0005 * VR2	Source EA in Fig. 2.3, VR2 stands for the voltage across R2
R1, B-X = .3	Resistor R1 in Fig. 2.3 (value = 300Ω)
R2, C-E = 2.0	Resistor R2 in Fig. 2.3 (value = 2K)
JB, C-E = 50. * IR1	Source JB in Fig. 2.3, IR1 stands for the current through R1
CIRCUIT DESCRIPTION	A heading card
ELEMENTS	A subheading card
E1, 1-2 = X1(.001*SIN(.000628*TIME))	The a.c. source E1 in Fig. 2.2
ECC, 1-7 = 0	The small signal a.c. equivalent for a battery is zero volts and ohms.
C1, 2-3 = 5E6	Capacitor C1 in Fig. 2.2 (C1 = 5E6 pf)
C2, 4-5 = 5E6	Capacitor C2 in Fig. 2.2
RB1, 3-7 = 100	Resistor RB1 in Fig. 2.2
RB2, 5-7 = 100	Resistor RB2 in Fig. 2.2
RL1, 7-4 = 1	Resistor RL1 in Fig. 2.2
RL2, 7-6 = 1	Resistor RL2 in Fig. 2.2
T1, 3-1-4 = MODEL SS1	(3-1-4) are the circuit nodes that connect to (B-E-C) of the model, in that order
T2, 5-1-6 = MODEL SS1	(5-1-6) are the circuit nodes that connect to (B-E-C) of the model, in that order
OUTPUTS	A subheading card
VC1, VC2	Voltage across C1 and C2 will be outputed in tabular form with time as the independent variable
VRL1, VRL2, PLOT	Voltage across RL1 and RL2 will be plotted with time as the independent variable (also in tabular form)
RUN CONTROLS	A subheading card
STOP TIME = 30,000	Problem duration has been set to 30,000 nanoseconds to accommodate three cycles of the a.c. output wave form
END	Always the last card

Fig. 2.4 SCEPTRE program card arrangement

3. Elements are listed by element name and then the node labels in the order of indicated current direction. This is followed by the element value.

4. Special characters such as comma, dash, parenthesis, and equal sign must appear according to the proper format (see Table 2.4).

5. Information can be punched anywhere on the data card (columns 1-72), with any desired spacing.

6. The voltage across any passive element is defined by placing a "V" in front of the element name, likewise for current using an "I".

7. A voltage source is EXX and a current source is JXX.

It is not expected that the reader will necessarily understand each step in the following example at this time; but rather the example serves to illustrate the approach, with the detailed rules to follow in subsequent paragraphs.

A valid SCEPTRE program for the circuit described in Figs. 2.1, 2.2, and 2.3 is shown in Fig. 2.4. Each line is a separate data card in this example. The comments listed in the column on the right are only for the reader's clarification.

This example is intended to illustrate the ease with which a SCEPTRE program is written for the determination of the response of a two stage linear RC coupled amplifier to a low amplitude 100 KHZ sinusoidal input. The output plots for the voltages across the two load resistors as a function of time are obtained. The remainder of the chapter will discuss all of the programming aspects in detail.

2.2 SCEPTRE CIRCUIT DESCRIPTION LANGUAGE

The initial conditions and transient solutions of large electrical networks are computed automatically when the circuit description language described herein is used to convey all the necessary information to the SCEPTRE program.

The SCEPTRE circuit description language is a structured free format language, the syntax of which is easy to learn and use. The language consists of descriptive statements constructed syntactically from user-derived component names, parameter names, node names, and value specifications. These are delimited by special characters such as comma, dash, parenthesis, and equal sign; thereby, allowing the program to interpret the statements properly. In general, several complete statements can be punched on a card, separated only by a comma. The rules for continuing a statement from one card to another generally require that the discontinuation be made immediately after delimiters, with the delimiter appearing as the last non-blank character on the card.

2.2.1 Headings and Subheadings

Networks are described in SCEPTRE language under the following headings and subheadings regardless of which mode of analysis is desired. Generally a given run only involves a portion of the many headings and subheadings that are available. The list of headings and subheadings in Table 2.2 is a complete list. Headings or subheadings which are not used are merely deleted.

Table 2.2. Complete heading and subheading list.

1. MODEL DESCRIPTION
   MODEL NAME (PERM OR TEMP) (NODE-NODE . . . . . NODE)
   (Comment or message cards, if any, up to 11 allowed)
   1.1 ELEMENTS
   1.2 DEFINED PARAMETERS
   1.3 OUTPUTS
   1.4 INITIAL CONDITIONS*
   1.5 FUNCTIONS

2. CIRCUIT DESCRIPTION
   (Comment or message cards, if any, up to 11 allowed)
   2.1 ELEMENTS
   2.2 DEFINED PARAMETERS
   2.3 OUTPUTS
   2.4 INITIAL CONDITIONS
   2.5 FUNCTIONS
   2.6 RUN CONTROLS

3. RERUN DESCRIPTION (N)
   (Comment or message cards, if any, up to 11 allowed)
   3.1 ELEMENTS
   3.2 DEFINED PARAMETERS
   3.3 INITIAL CONDITIONS
   3.4. FUNCTIONS
   3.5 RUN CONTROLS

4. CONTINUE
   4.1 RUN CONTROLS

5. RE-OUTPUT

6. END

*Rarely used under model heading, see Sec, 2.8.1

The MODEL DESCRIPTION heading is used when it is desired to store one or more models. The MODEL NAME card, comment cards (optional), and any or all of the five subheadings listed can be used for each model for either permanent or temporary storage under the MODEL DESCRIPTION heading. One or more models may be entered under one MODEL DESCRIPTION heading card.

The CIRCUIT DESCRIPTION heading is always used when any network is presented for analysis. Any or all of the six subheadings listed under the heading may be used.

The RERUN DESCRIPTION heading is used whenever the rerun feature is exercised. All changes to the master network must appear under this card. Any or all of the five subheadings listed under this heading may be used.

The CONTINUE heading is intended for use only when continued computation is desired after a problem has been originally run. The only subheading permitted under this heading is RUN CONTROLS. The only other heading that may appear together with CONTINUE in a run is END.

The RE-OUTPUT heading is used whenever the user desires output from a previously completed run without repeating that run. No subheadings are permitted under this heading. The only other heading that may appear with RE-OUTPUT is END.

The END heading is used to terminate every input data deck submitted to SCEPTRE. This heading must always be the last card in the problem deck.

The data supplied within each of the subheadings consists of descriptive statements, which are constructed as properly punctuated sequences of symbols.

The input data deck describing a network is formed simply by punching the heading card and the associated data sequence for each of the defined data groups. Remarks such as "title," "user name," and "date" may be supplied for output identification purposes by punching the desired remarks on cards following the heading card and preceding the ELEMENTS subheading card. The number of comment cards must not exceed 11, and the entire remark will appear as the title of the output listing and plots.

The headings and subheadings are now described in the order in which they appear in a program (see Table 2.2). It is important for the reader to bear in mind while reading all of the ground rules discussed in the following sections that much insight can be gained by referring to some of the complete examples given in Chapter 4.

It is necessary for organizational purposes to put all of the guide lines and rules in one compact chapter. This approach can have the effect of bewildering the new user; however, it is convenient for later reference. As the user becomes familiar with the procedures, he sees in retrospect that the rules are quite simple and direct. Probably the most <u>successful</u> way to quickly master the material is to read swiftly through Chapter 2. Then study carefully the examples of Chapter 4, while using Chapter 2 as a reference for the specific rules!

2.3 MODEL DESCRIPTION

All active devices must be represented in SCEPTRE by combinations of R, L, C, M, E, and J. Most users will find that repeated use is made of certain models of active devices or of standard combinations of passive elements, such as filter sections, etc. These circuits could be handled by inserting the components one by one in a CIRCUIT DESCRIPTION under ELEMENTS. A more convenient approach is to describe the model or network once, and store it for future use. The stored model feature is not "free;" its use requires extra tape manipulation, and hence, more computer time. However, the time saved in circuit preparation and the added flexibility will usually compensate for the extra time.

The stored model feature of SCEPTRE is especially flexible since the user stores whatever models meet his specific needs. A model may be stored permanently or temporarily. If it is stored permanently, the model may be called out for insertion in a main network as often as desired per run and for any number of runs. The user might desire to reserve this type of storage for proven models that will see reasonably frequent service. Temporary storage, on the other hand, stores a model for one run only plus whatever reruns are made during that run. This type of storage will probably prove most useful for experimental models or for models that are seldom used.

2.3.1 Permanent Storage

Any three-terminal transistor, for example, may be permanently stored on tape using the following format after the MODEL DESCRIPTION card:

```
MODEL NAME (PERM) (B-E-C)
```

The NAME may contain up to 18 alphanumeric characters and the B-E-C combination is used to correspond to the specific nodes connected to the base, emitter, and collector terminals of the transistor. Following the MODEL NAME card would be the subheadings ELEMENTS, DEFINED PARAMETERS, OUTPUTS, and FUNCTIONS. For the special case of the first model permanently stored on any individual tape, the first card must be MODEL DESCRIPTION (INITIAL).

With the word PRINT added to the MODEL DESCRIPTION card the program will generate a printed listing of all models stored on the permanent library tape. The word PRINT is omitted if no listing is desired (see Table 2.2).

The B, E, C nomenclature used to name the base, emitter, and collector nodes of the transistor model is not mandatory, but it is recommended as a systematic and orderly procedure for storing transistor models.

2.3.2 Temporary Storage

Each model intended for temporary storage must be described element by element under MODEL DESCRIPTION in the run in which it is to be used. The words INITIAL and

PRINT are never used on the MODEL DESCRIPTION card for temporary storage. The rest of the entry format is the same as for permanent storage except the work TEMP:

```
MODEL NAME (TEMP) (NODE-NODE-NODE)
```

or

```
MODEL NAME (NODE-NODE-NODE)
```

If neither TEMP or PERM is used, the program automatically assumes temporary storage.

2.3.3 Model Deletion

An entire model may be removed from the user's permanent model library tape. This can be done with an entry under MODEL DESCRIPTION of the following form:

```
MODEL 2N367A (DELETE)
```

2.3.4 Stored Model Illustration

The stored model feature is best illustrated by an example. Assume that the stored transistor model in Fig. 2.5b is to be inserted into the main circuit in Fig. 2.5a. The stored

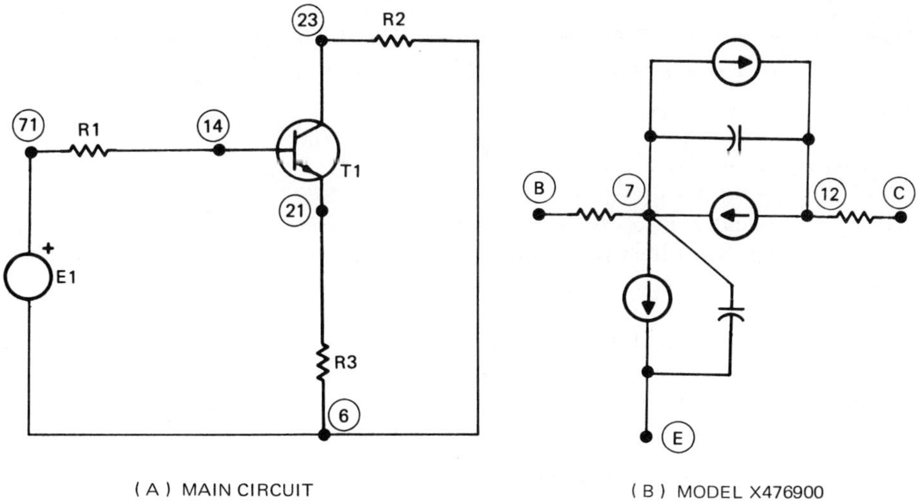

Fig. 2.5. Example of the stored model feature of SCEPTRE.

model contains terminal nodes B, E, and C and internal nodes 7 and 12. (The main circuit may also contain nodes 7 or 12 without danger of ambiguity). Assuming that the model was originally stored with the node sequence B, E, and C, the user would enter the following under ELEMENTS of the main program:

```
T1, 14-21-23 = MODEL X476900 (PERM)
```

This listing matches nodes 14, 21, and 23 of the main circuit with nodes B, E, and C of the stored model. The circuit designation of the stored model is T1 and the name of this particular stored model is X476900. The stored model is automatically transferred from the tape into the main circuit and nodes B, E, and C of the model become nodes 14, 21, and 23 of the main circuit. Internal nodes 7 and 12 become 7 T1 and 12 T1 in the main program because all internal nodes of stored models have appended to them the circuit designation of that model. Since no node name may contain more than six alphanumeric characters, the user should use as few characters as possible for the circuit designation names and internal nodes of any stored model.

The circuit designation format in the main program for temporary models differs slightly from the format used for permanent models. Where a permanently stored model is designated as:

```
T1, 14-21-23 = MODEL X476900 (PERM)
```

A temporarily stored model could be designated as:

```
T1, 14-21-23 = MODEL X476900
```

or

```
T1, 14-21-23 = MODEL X476900 (TEMP)
```

2.3.5 Changes to a Stored Model

The user who frequently makes use of the stored model feature of SCEPTRE often encounters the situation in which the topology of his stored model is satisfactory but the size of some of the model elements must be changed. Changes are effected easily for any individual run, but no permanent changes to the stored model are possible. (The user has the option of storing a second model which contains a different version of the original). All changes are indicated along with the statement that locates the stored model in the main circuit. This is always under the main circuit ELEMENTS subheading.

2.3.6 Changes in Elements or Defined Parameters in A Stored Model

Consider that a stored model has been called out at:

```
T1, 7-8-12 = MODEL 2N1734B
```

In this case, the model is inserted into the main circuit as it was originally stored. If a change is desired in one or more internal elements or defined parameters of the stored model, the proper entry is:

```
T1, 7-8-12 = MODEL 2N1734B (CHANGE CC = 50)
```

In this case, CC of the original model has been changed to 50 units of capacitance, regardless of its original form or size. All other elements in the stored model would remain as originally stored. Another practical situation is reflected by the entry:

```
T1, 7-8-12 = MODEL 2N1734B (CHANGE CC = TABLE 7(VCCT1),
JA = .2 * J2T1)
```

In this case, CC of the original stored model has been changed to a tabular function which must be represented, (TABLE 7 under the FUNCTIONS subheading of the main circuit). Also, JA has been changed to a different mathematical definition. The other elements of the stored model will remain in their original forms.

As a final example, assume that it is desired to change CC in a stored model to a different equation. Then:

```
T1, 7-8-12 = MODEL 2N1734B (CHANGE CC = EQUATION
5 (VCCT1, VCX))
```

In this case, CC of the stored model is changed to a closed form function which must be represented by EQUATION 5 under FUNCTIONS of the main circuit description. All elements or element voltages or currents (within a model) that are referenced on the right hand side of the changed expression or in the main circuit description must include the model designation as a suffix (VCC T1, J2T1). VCX is the voltage across a capacitor, CX, that is not in a model, and therefore, uses no suffix.

The reader will better understand the several paragraphs above after reading the sections on ELEMENTS and FUNCTIONS.

2.3.7 Changes in Functions in a Stored Model

This subsection describes the manner in which data listed under FUNCTIONS (i.e., tables and equations) in a stored model is changed. A table in a stored model is changed by the entry:

```
T1, 7-8-12 = MODEL 2N1741B (PERM, CHANGE TABLE 1
= TABLE 7)
```

which replaces TABLE 1 of the stored model with TABLE 7. TABLE 7 must be supplied under FUNCTIONS of the main circuit. No correlation is required between the number of point pairs contained in the original table and the point pairs in the new table.

An equation in a stored model is changed by the entry:

```
T1, 7-8-12 = MODEL 2N1741B (PERM, CHANGE Q1 = Q2)
```

which replaces the original EQUATION 1 of the stored model with EQUATION 2. EQUATION 2 must be supplied under FUNCTIONS of the main circuit. The only special stipulation is that the same independent variables that held for the original equation must also apply to the new equation. It is never possible to directly change an equation in a stored model to a table, or a table in a stored model to an equation, (CHANGE EQUATION 1 = TABLE 1) is not allowed. However, an element or a defined parameter in a stored model can be changed to a constant, table, equation, or expression, regardless of its original stored form.

2.3.8 N Terminal Model Storage

The storage of N terminal models differs from the storage of transistor models only in the number of terminal nodes (program limit 25 nodes) that are specified. Storage may be effected by an entry following the MODEL DESCRIPTION card as:

```
MODEL NAME (PERM OR TEMP) (NODE-NODE-....NODE)
```

As many internal nodes as desired may be included within the program limit of 301 nodes. The remainder of the information is entered under the appropriate subheadings. Regardless of the number of terminal nodes that are supplied, the user must take care that these nodes are referenced in the same sequence when called by the main program.

2.4 *SUBHEADING GENERAL INFORMATION*

Notice in Table 2.2 that the subheadings under the headings MODEL DESCRIPTION, CIRCUIT DESCRIPTION, and RE-RUN DESCRIPTION are essentially the same. After some general information, in this section, which is applicable to all subheadings, each of the subheadings (ELEMENTS, DEFINED PARAMETERS, OUTPUTS, INITIAL CONDITIONS, FUNCTIONS, and RUN CONTROLS) is discussed in Sec. 2.5 through Sec. 2.10, and the commentary is applicable to that particular subheading regardless of which of the specific headings is involved. The only variations occur under the RERUN DESCRIPTION heading. The variations are discussed separately in Sec. 2.11.

The SCEPTRE equivalent circuit may consist of resistors, capacitors, inductors, mutual inductance, voltage sources, and current sources, and/or stored models containing these elements. All of these elements may be linear or nonlinear; thus facilitating transistor circuit analysis. Furthermore, most equivalent circuits composed of these elements can be accommodated. This allows the use of either standard or complex experimental transistor equivalent circuits. The value, or behavior, of any equivalent circuit element may be defined by a numerical constant, tabular list, or mathematical expression.

After the equivalent circuit is drawn, each node is given an arbitrary alphanumeric designation consisting of six characters or less. A node is defined as the point of common potential (voltage) at the junction created by the connection of two or more network elements.

Next, each component or element in the circuit is given a unique name consisting of not more than five alphanumeric characters. The first character of each element name must be R, C, L, E, or J corresponding to the element type; i.e., resistor, capacitor, inductor, voltage source or current source, respectively. The letter M is used to designate mutual inductance in the same way. The only exception to this is the circuit designation for any stored model which has no specific rule for its first character, but should be limited to a total of three alphanumeric characters. It is normally very helpful to record on the equivalent circuit diagram the names and nodes chosen. Current flow directions, which are arbitrary, and source polarities are also indicated.

The circuit parameter values are specified in a consistent set of parameter units. Although any consistent set is acceptable, the system given in Table 2.1 is useful for high-speed transistorized circuits. If another system is desired, the most effective choice is units of voltage, current and time that correspond to the magnitude of those that are expected in the problems; then determine units of R, L, and C from the fundamental current-voltage relationships. This choice is always possible in a practical circuit, since the analyst should have some approximate idea of the range of variables. The reader is referred to the discussion of units in Sec. 1.9.

2.4.1 Subheading Symbols

Sequences of symbols and punctuation are used to convey information in each of the subheadings. Some of these symbols are not completely clear until the reader follows their usage in the examples of Chapter 4. However, these definitions are necessary in order to correctly interpret Table 2.4. Therefore, for ready reference, each of the important symbols is defined as follows:

> *ELEMENT NAME:* Denotes the name given to each component (including model circuit designations) of a circuit (e.g., RA, LLX, E17). No more than five alphanumeric characters may be used to name an element. These five characters must also include model circuit designations, if any.
>
> *NODE:* Denotes the designation assigned to each node of a circuit. No more than six alphanumeric characters may be used to name a node.
>
> *NUMBER:* A numerical constant that may be written as a signed quantity in either integer or decimal form and with or without an exponent. Up to 13 characters may be used to represent a number. For example, numbers may be

written in the following forms: 10, 10., 10.0, -.1, -0.1, +1.4, 6.4E9, -74.3E-7, 7E+11, -176.6667E5.

CONSTANT: Same as NUMBER except a decimal point must be included in the specification of the numerical constant.

VALUE: Will be used to denote any of the following: NUMBER, DEFINED PARAMETER, TABLE, EQUATION or EXPRESSION.

SPECIAL VALUE: Will be used to denote any of the following: VALUE, CONSTANT * Resistor Current, CONSTANT * Resistor Voltage, VALUE * Current Source, DIODE TABLE or DIODE EQUATION (X1, X2).

The second and third designations in the SPECIAL VALUE list are intended to accommodate, among other things, the class of linearly dependent sources that are encountered in small signal transistor equivalent circuits. The fourth designation was designed for the class of secondary dependent current sources that always appear in the large signal Ebers-Moll transistor equivalent circuit. The capability of specially processing these sources has been built into the program and should always be entered directly in the ELEMENTS section without parenthesis. Examples of these appear in Sec. 2.5.1.

The last two designations in the SPECIAL VALUE list are reserved for primary dependent current sources that represent diodes or transistor junctions. DIODE EQUATION (X1, X2) would be used when any diode or transistor junction has been entered, and the user wishes to employ the conventional closed form representation $(J = I_s(e^{\Theta V_s} - 1))$ The value of X1 must correspond to Is in the diode equation, and the value of X2 must correspond to Θ. The program will automatically use the voltage across that particular current generator as the independent variable, and for that reason this voltage need not be specified. For the Diode Table, the independent variable is also automatically set as the voltage across the current generator.

VARIABLE: Denotes any of the following:

1. The voltage or current associated with any element as VR1, VJ7, IE4, ILM, etc.
2. Any source or source derivative as J17, DJ17, E4, DE4, etc.
3. Any defined parameters as P7, DP7, etc.
4. Any element value as R17, CA, M12, etc.
5. Time as TIME.
6. Any internal parameter (see Sec. 2.20).

V ELEMENT NAME OR I ELEMENT NAME: Denotes the element voltage or current of ELEMENT NAME. For example, the voltage across capacitor CAB1 would be referred to as VCAB1, and the current through inductor LCHOK would be referred to by ILCHOK.

TABLE NAME (INDEPENDENT VARIABLE): Used when a variable circuit quantity is given in tabular form. The table used must be given a unique name prefixed by TABLE or simply T, and followed by a single independent variable in parenthesis. The name may consist of up to five alphanumeric characters. The independent variable may be any of the quantities defined under VARI-

ABLE, such as, TABLE 1A(VC1). If an independent variable, including the enclosing parenthesis is not supplied, then TIME is automatically chosen.

EQUATION NAME (ARGUMENT LIST): Used when a variable circuit quantity is given in closed form. The equation must be given a name prefixed by EQUATION or simply Q and followed by one or more arguments separated by commas and enclosed in a parenthesis. The EQUATION name may consist of up to five alphanumeric characters. The argument list may consist of any VARIABLE, CONSTANT and TABLE and its independent variable. For example, EQUATION 39(CX, J2, TIME, TABLE 2(VC7)). The detailed mathematical form of the equation must be given as a separate entry under the FUNCTIONS subheading.

EXPRESSION NAME (Math definition): This form is an alternative to the EQUATION entry that may be used to describe a variable quantity. It is somewhat more complex than the EQUATION form, but it has the virtue of needing no further description under FUNCTIONS. The expression must be given a name prefixed by EXPRESSION or simply X and followed by the mathematical definition. The EXPRESSION name may consist of up to five alphanumeric characters. It is suggested that numerical designations be used to avoid any possible confusion with some internal parameters. For example, X14 (10. * SIN (.628 * TIME)).

Examples of variable element entries using EQUATION, TABLE or EXPRESSION description are as follows:

```
E1, 1-2 = TABLE 3 (TIME)

E2, 1-6 = EQUATION 47 (VC1, TABLE 2 (ILO), 36.)

JM, 1-4 = EQUATION 47 (VC1, P5, 15.)

RA, 3-4 = TABLE 5 (TIME)

LZ, 1-5 = EXPRESSION 7 (10. * ILZ + 20.)
```

The tabular entries for TABLE 3 and TABLE 5 as well as the analytical expression for EQUATION 47 must be entered under the FUNCTIONS subheading. A good general rule to follow throughout the program is that <u>all constants</u> inside parentheses <u>must</u> include decimal points. A more convenient form that always may be used is to replace the word EQUATION by Q, the word TABLE by T and the word EXPRESSION by X.

```
E1, 1-2 = T3 (TIME)

JM, 1-4 = Q 47 (VC1, P5, 15.)

LZ, 1-5 = X7 (10. * ILZ + 20.)
```

2.4.2 Character Length

The alphanumeric character limits are listed in Table 2.3. The circuit designation used for models is always appended to any names used in a model description and the combined total characters must not exceed the circuit description limits. For example, element CB in a model designated as T15, in the circuit description becomes CBT15.

Table 2.3 Maximum alphanumeric character lengths allowed

Item	Circuit description maximum number	Model description maximum number[1]
Node names	6	3
Element names	5	2
Defined parameter names	6	3
Table names	5	2
Equation names	5	2
Model names	18	18
Output labels	6	3
Circuit designation (for calling models)	3	—

1. Recommended

This completes the general discussion and the definition of terms most often used. The new user will find it necessary to refer often to this section to be sure of correct applications. The following sections discuss the format for each subheading.

2.5 *ELEMENTS*

All elements (resistances, capacitances, inductances including mutual, voltage and current sources, source derivatives and model circuit designations) that are to be component parts of the network under analysis must be introduced under this subheading. Entries allowed under the ELEMENTS subheading are summarized in Table 2.4. Although any combination of alphanumeric characters (maximum of 4) can be used for model circuit designations, names unique from other element names are recommended. Alphanumeric character lengths are listed in Table 2.3. The general form for entries under the ELEMENTS subheading is:

```
ELEMENT NAME, NODE-NODE = SPECIAL VALUE
```

Each network element is defined by stating the ELEMENT NAME, the NODES between which the branch is connected, and the component value. The connection nodes are specified in a "from-to" order corresponding to the assumed direction of current flow. The actual tabular values or mathematical definitions of elements that are implicitly defined as variable are specified under FUNCTIONS (see Sec. 2.9). More than one element may be described on a card if the elements are separated by commas.

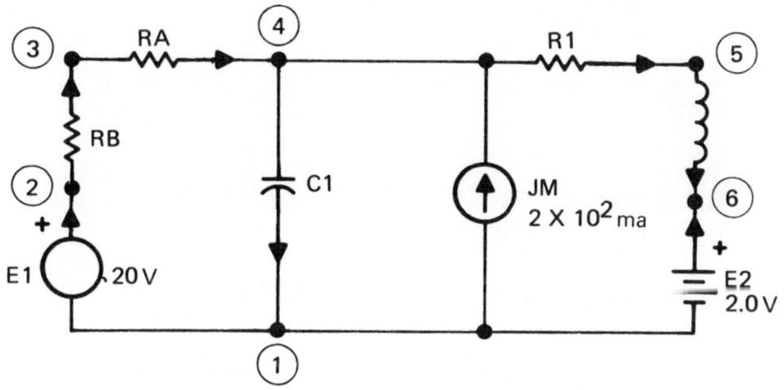

Fig. 2.6. Circuit in SCEPTRE form.

Some examples that illustrate proper entries under ELEMENTS for the *constant valued* elements of the network of Fig. 2.6 are:

```
E1, 1-2 = 20.
E2, 1-6 = 20
JM, 1-4 = 2E2
```

Note that any of these constant elements may be entered with or without decimal points or by use of the E format (Power of Ten). Note also that the proper reference direction for voltage sources correspond to the direction of current or positive charge movement within the voltage source.

2.5.1 More Element Examples

Examples of typical component descriptions that appear in the ELEMENTS group are shown below:

```
J18, 10-3 = DIODE TABLE 1
```

The tabular entries for DIODE TABLE 1 must be entered under the FUNCTIONS sub-heading.

```
R7, 4-5 = 11.5, E1, GND-1 = 6, R9, 3-1 = 17
```

```
JA, 0-4 = .98* J18
```

```
JK, 6-B = 50. * IRE
```

```
E12, C-7 = .0005 * VRC
```

JA is a secondary dependent current source (see Table 2.4).
JK and E12 are linearly dependent sources.

```
LX3, 9-3 = EQUATION 15X (ILX3, TIME)
```

```
T11, 2-3-7 = MODEL 2N7479AA
```

Reference to the FUNCTIONS section is never required for a given entry when the EXPRESSION format is used. The rules for the mathematical definition become specialized only in the case when a table is to be used as an argument. For example, if it is desired to enter capacitor C1 as 10 + (80) (Table 7), where Table 7 is a function of VC1, an appropriate entry would be:

```
C1, 7-8 = X314 (10. +80. * XTABLE (T7, VC1))
```

Note that in this special case the word TABLE is preceded by X and that both the table name, in this case T7, and the independent variable of the table, VC1, must be included in parenthesis. Decimal points must always be given with all constants used in an EXPRESSION.

2.5.2 Mutual Inductance

Mutual inductance is entered according to the general format:

```
M NAME, L NAME - L NAME = VALUE
```

If coupling exists between inductors L1 and L2, the appropriate entry must include these elements in place of the node identification as:

```
MX, L1-L2 = TABLE 1(IL1)
```

or

```
MX, L1-L2 = 32.4
```

In addition, a physical limitation of the principle of mutual inductance must be observed in order to have a physically realizable circuit. That is, since coefficient of coupling, k, is always less than unity and by definition,

$$k = \frac{M}{\sqrt{L1 \quad L2}} < 1,$$

the user should be certain that $M < \sqrt{L1\ L2}$. Stated in words, the mutual inductance between any two inductors *must be less* than the square root of the product of self-inductances of the components between which the mutual inductance exists. The sign of M is positive if, in a given winding, the induced voltage of mutual inductance acts in the same direction as the induced voltage of self-inductance. If the induced voltage of mutual inductance opposes the induced voltage of self-inductance in a given winding, M is negative. The proper sign for M for the assumed current directions is illustrated in Fig. 2.7.

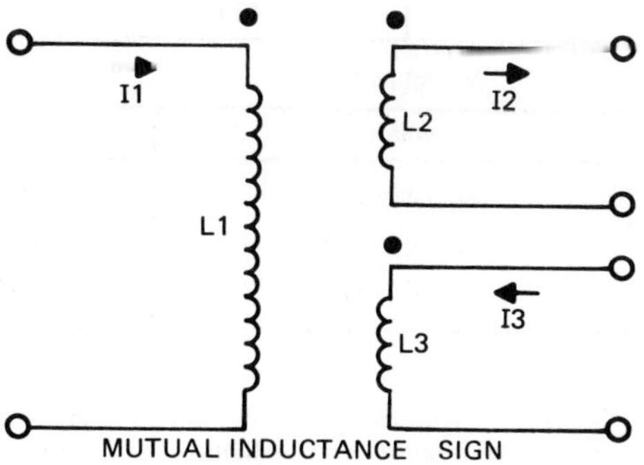

MUTUAL INDUCTANCE	SIGN
M12	−
M13	+
M23	−

Fig. 2.7. Mutual inductance polarities.

One of the more important tables in this book is Table 2.4. If the user pays close attention to the rules set up in this table, most of the common problems will not occur.

Table 2.4 Entries under elements

Name		Nodes			Value specification
R C L E J	, , , ,	NODE-NODE	=	VALUE	NUMBER TABLE NAME (Independent Variable)[1] Defined parameter EQUATION NAME (Argument list)[1]
M	,	L NAME-L NAME			EXPRESSION NAME (Mathematical definition)[1] EXTERNAL FUNCTION (Argument list)[1]
Linearly dependent sources					
E	,	NODE-NODE	=		CONSTANT * VR
J	,	NODE-NODE	=		CONSTANT * IR
Primary dependent current sources[2]					
J	,	NODE-NODE	=		DIODE TABLE NAME DIODE EQUATION (X1, X2)
Secondary dependent current sources[2]					
J	,	NODE-NODE	=		VALUE * J (J is a primary dependent current source)
Voltage and current source derivatives					
DE			=		VALUE
DJ			=		VALUE

Model calls[3]

Circuit designation		Nodes		Model name
D	,	NODE-NODE	=	MODEL 1NXXX (PERM)
T	,	NODE-NODE-NODE	=	MODEL 2NXXX (TEMP)
AX1	,	NODE-NODE-...NODE	=	MODEL XYZ (TEMP)

Notes:

1. No INITIAL CONDITIONS solution will be computed if any resistor current is used as an independent variable or argument in a table, equation, or expression.
2. Primary and secondary dependent current source specifications should always be used to represent semiconductor junctions when requesting INITIAL CONDITIONS solutions.
3. See Sec. 2.3.
4. Model circuit designation names can be any combination of no more than four alphanumeric characters. Names unique from other element names are recommended.

2.6 DEFINED PARAMETERS

This subheading permits the user considerable flexibility in the use of SCEPTRE. Any variable that can be described in terms of any network variable and/or any NUMBER may be defined, and this quantity may be used as an ELEMENT value, argument in an equation, expression, table, or an output at each time step of the problem, in the same manner as any conventional output. Examples of the use of this feature are given in Chapters 4, 5, and 6. More than one defined parameter may be entered on a card if they are separated by commas.

The input format for defined parameters requires that the first letter be P followed by no more than five alphanumeric characters. The general form for entries under DEFINED PARAMETERS is:

```
P NAME = VALUE
```

Possible combinations are:

```
PWR = X2 (IE3 * E3)
P1X72 = NUMBER
P2 = TABLE 1(VC7)
PX7 = EQUATION 2(VC7, VR1)
```

For the special case in which the derivative of a quantity is supplied, the first two letters must be DP followed by no more than four alphanumeric characters, or in general:

```
DPXXX = VALUE
```

To appreciate the usefulness of this feature see especially the example given in Sec. 5.2.2.

2.7 OUTPUTS

Any output must consist of some dependent variable which is a function of some independent variable. SCEPTRE outputs consist of printed tabular listings of requested dependent variables as functions of time and/or plots of the dependent variables as functions of time or some other independent variable. In SCEPTRE, the following general quantities may serve as either dependent or independent variables:

The voltage or current associated with any passive element as VR1, IL6.

The voltage or current associated with any source as E1, IE1, J2, VJ2.

Any internal parameter as defined in Table VII.

Any element value as C17.

Any defined parameter as P12.

Any Defined Parameter derivative as DP12, if the user has supplied one.

Any transient state variable derivative, as DC4, DL3 (which is actually, $\frac{dVC4}{dt}$ and $\frac{dIL3}{dt}$).

The last two entries are especially useful in control system problems when it is useful to output the phase-plane plot.

All requested outputs in SECPTRE are supplied in printed tabular form. In addition, the user has the option of requesting plotted outputs for any or all quantities. The general format for requesting printed outputs is:

```
VARIABLE, VARIABLE, VARIABLE
```

and/or

```
VARIABLE
```

```
VARIABLE
```

```
VARIABLE
```

and/or

```
VARIABLE, PLOT
```

```
VARIABLE, PLOT
```

Some typical output requests follow:

```
VR3, IR3, VR2
```

```
VR5, VC29
```

```
VRY, VC1, ESUP, PLOT
```

```
IC8
```

Note that more than one output can be requested on a single card. The third card indicates that three quantities are required and that all three are to be plotted as well as printed. The quantity, IC8, (the current through capacitor C8) would be output in printed form only. If the word PLOT is used, *no* other dependent variables may follow it on that card.

All indicated variables in the above example use time as the independent variable. If a different independent variable is desired for the plotted form, the following format is used:

```
IC14, PLOT(VC14)
```

In this case, the current through capacitor C14 would be plotted as a function of the voltage across it; however, the printed tabular output is IC14 as a function of time.

2.7.1 Output Label Change

Additional flexibility is available to permit the user to attach a different label to any output quantity except TIME. Consider that elements C1 and R7 exist in a given network and that the voltage across both (VC1 and VR7) are of interest.

If the user decides to rename them as VIN and VOUT, the outputs may be requested as VC1 (VIN), VR7 (VOUT), PLOT. All renames must be limited to six alphanumeric characters and the first character may be any alphanumeric character. The plot, rename, and choice of independent variable options can be presented in terms of the general format as:

```
YQTY(YLABEL),...., PLOT(XQTY (XLABEL))
```

If no rename is desired, the general format reduces to

```
YQTY,...., PLOT (XQTY)
```

If, in addition, only time is desired as the independent variable, this is further reduced to:

```
YQTY,...., PLOT
```

And if no plotted information is desired, the simplest form arises as:

```
YQTY,....
```

2.7.2 Changes in Outputs

Assume a stored transistor model called for the output of three quantities (VCE, VCC, J1). Normally, any run that uses this stored model produces these three outputs without further instruction from the user. If some other quantity in the stored model is desired as output; e.g., J2, it must be requested under the main circuit OUTPUTS subheading, e.g., J2T5, if the circuit designation T5 were used to call the model under the main circuit ELEMENTS subheading.

Output requests in the stored model may be inhibited as well. If any quantity specified as OUTPUT in the model listing is not desired as output for a given run; e.g., J1, it may be inhibited in the statement that locates the stored model in the main circuit. The proper format is:

```
T1, 7-8-12 = MODEL 2N1734B (PERM, SUPPRESS J1)
```

If it is desired to suppress more than one ouput the format is:

```
T1, 7-8-12 = MODEL 2N1734B (PERM, SUPPRESS JL, VCE)
```

All output associated with a stored model may be inhibited by entering:

```
T1, 7-8-12 = MODEL 2N1734B (PERM, SUPPRESS ALL)
```

The normal output routine will produce a listing of the entire circuit without a detailed printout of any stored models unless the model has been stored with that particular run. The proper language for a detailed printout of any stored model is:

```
T1, 7-8-12 = MODEL 2N1734B (PERM, PRINT)
```

Models which have a temporary designation are always printed out in detail.

2.8 *INITIAL CONDITIONS*

The complete solution of the general transient analysis problem requires that all independent initial conditions be supplied. The set of all capacitor voltages and inductor currents that exist at the start of the problem are sufficient for this purpose. This section is usually superfluous if the run is being made in either the initial conditions only or the automatic initial conditions mode since the initial conditions will be computed by the program in these situations.

However, when initial conditions are supplied by the user, the format is:

```
VCXXX = NUMBER,....., ILXXX = NUMBER
```

Initial capacitor voltages and inductor currents are supplied by simply listing the desired values. Any initial conditions not specified are taken as zero. If all initial conditions are zero, neither the heading card nor the data are required. Care must be taken to establish the proper polarities for initial conditions. Initial inductor currents are positive if they flow in the same direction as the assumed current direction for the inductor. Also, the initial capacitor voltages are positive when they are consistent with the assumed voltage polarity for the capacitor.

The assumed current direction through inductor L5 of Fig. 2.8 is from node 7 to node 8. If the initial inductor current is in this direction, it is entered under the INITIAL CONDITIONS subheading as a positive quantity. If, however, the current flows in the other direction, it is preceded by a negative sign. The same convention applies to the capacitor, where the assumed positive sense of the voltage is associated with the tail of the reference arrow (node 4 in Fig. 2.8). If the actual initial voltage polarity agrees with the arbitrarily chosen reference direction, that initial voltage is entered as positive; if not, it is entered as negative.

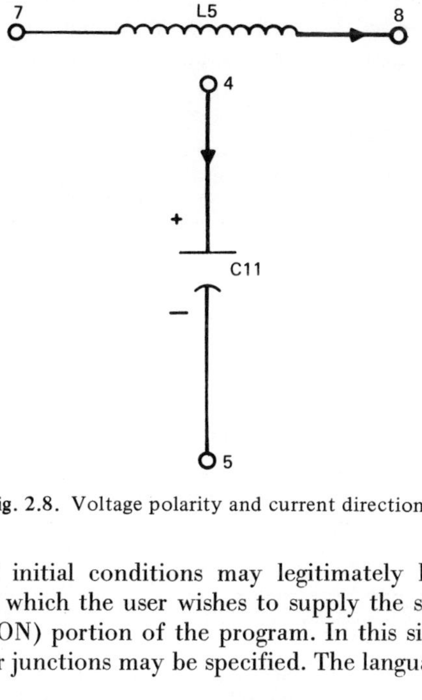

Fig. 2.8. Voltage polarity and current direction.

One other class of initial conditions may legitimately be entered. This class is intended for the case in which the user wishes to supply the starting values for the DC (i.e., INITIAL CONDITION) portion of the program. In this situation the initial voltage across diodes or transistor junctions may be specified. The language is:

```
VJ———    = NUMBER
```

If the DC portion of the program is used, the iterative process will begin with any VJ entries that are supplied. One practical application of this type of input would be to bias the ON and OFF sides of symmetric circuits such as flip flops in the desired state. Approximate values could be supplied and the DC solution would then provide the correct voltages for the desired state. The result of the DC solution would then carry over to the beginning of the transient solution if one is specified (see Sec. 4.8).

2.8.1 Initial Conditions for a Stored Model

Users rarely store initial conditions with stored models, since that would make the model circuit dependent. However, to supply a set of initial conditions for a stored model, the user need only supply entries of the form:

```
VC2T6 = NUMBER
```

for a circuit designation of T6 for the stored model. These entries would be made under the INITIAL CONDITIONS subheading of the main circuit.

2.9 FUNCTIONS

In this data group, each of the tables and equations referred to under "ELEMENTS" and "DEFINED PARAMETERS" (Sec. 2.5 and 2.6, respectively) is defined in detail. If no such references have been made, neither the FUNCTIONS subheading card nor data need be supplied. The equation definition sequence is discussed first.

2.9.1 Equation Definition Sequence

Each unique equation (used to define the variation of an element of defined parameter) is defined by giving the equation name, a dummy variable list, and the mathematical definition. The general format is:

EQUATION NAME (Dummy Variable List) = (Mathematical Definition)

or

Q NAME (Dummy Variable List) = (Mathematical Definition)

The dummy variable list must contain the same number of entries as does the argument list in the original equation reference. Each dummy variable may contain up to six alphanumeric characters, the first of which must *not* be a number or the letters I through N inclusive. For example, if an equation has been referenced under ELEMENTS as:

```
LX3, 9-3 = EQUATION 15 (ILX3, TIME, VC1)
```

then this equation could be explicitly defined under FUNCTIONS as:

```
EQUATION 15 (A,B,C) =(MATHEMATICAL DEFINITION)
```

or

```
Q 15 (A,B,C) = (MATHEMATICAL DEFINITION)
```

The dummy variables in this case are A, B, and C which replace ILX3, TIME, and VC1, respectively. The mathematical definition itself must be included in parentheses and must be written in terms of A, B, and C, along with any constants and allowable subprogram functions that apply. It is important to mention that there would be no need for the user to reserve quantities A, B, and C for Eq. (15) alone. These dummy variables may be freely used in other equations to represent other circuit quantities.

As another example, consider the equation where it is desired to enter capacitor C1 as 10 + (80) (TABLE 7) where TABLE 7 is a function of VC1. If under ELEMENTS, the user enters C1, 7-8 = EQUATION 2 (TABLE 7 (VC1)), then EQUATION 2 must be explicitly defined under FUNCTIONS. An appropriate entry would be:

```
EQUATION 2 (A) = (10. + 80. *A)
```

At each solution step, the ordinate value of TABLE 7 would replace the dummy variable A and the computation 10. + 80. *A would be carried out. Note that *decimal points are required* for the constants 10 and 80 because they appear with an EQUATION designation.

Still another method can be used that is particularly efficient when more than one equation of the same general form is used in a given run. Let it be desired to enter C1 as in the above paragraph in addition to C2 as 5 + (120) (TABLE 4) where TABLE 4 is a function of VC2. Under ELEMENTS the user may enter two cards as:

```
C1, 7-8 = EQUATION 2 (10., 80., TABLE 7 (VC1))
```

```
C2, 6-1 = EQUATION 2 (5., 120., TABLE 4 (VC2))
```

and under FUNCTIONS, Eq. (2) is explicitly described as:

```
EQUATION 2 (A,B,C) = (A+B*C)
```

At each solution step, C1 is evaluated in the program by replacing dummy variables A, B, and C by 10., 80., and the ordinate value of TABLE 7, respectively. C2 is then evaluated by replacing A, B, and C by 5., 120., and the ordinate value of TABLE 4, respectively. Note that two quantities of the same mathematical form have been accommodated by one equation.

The mathematical definition may be any combination of the allowable operations, functions, or variables. The following mathematical operations and corresponding symbols are included in SCEPTRE:

Operation	Symbol
Exponentiation	**
Multiplication	*
Division	/
Addition	+
Subtraction	-

The order in which operations are performed is indicated by the order in which the operators are listed. The use of parentheses, to denote clearly the intended mathematical combination, is suggested to avoid ambiguity. For example, X+Y*Z should be written as (X+Y)*Z if X+(Y*Z) is not intended.

2.9.2 Fortran Library Subprograms

Any program that is available in the Fortran IV Subprogram Library may be used in an equation or expression. A few of the most common are listed below.

Function	Symbol*
Square root	SQRT (argument)
Sine function	SIN (argument in radians)
Cosine function	COS (argument in radians)
Exponential function	EXP (argument)
Arctangent function	ATAN (argument in radians)
Hyperbolic tangent function	TANH (argument in radians)
Natural logarithm function	ALOG (argument)
Common logarithm function	ALOG10 (argument)

The argument of any function may be any allowable mathematical definition. In addition, the user may supply subprogram functions that he has written himself (see Sec. 2.17). When these functions are replaced by an equation, all variables that appear as the arguments of the function must be given in terms of dummy variables. Mathematical definitions are not limited to 72 characters (one card) and may be continued on subsequent cards, using as many as necessary.

2.9.3 Tabular Definition Sequence

Every TABLE NAME that has been referenced under ELEMENTS or DEFINED PARAMETERS is explicitly defined under the FUNCTIONS subheading. The general format is:

```
TABLE NAME, or DIODE TABLE NAME
NUMBER, NUMBER, NUMBER, NUMBER, etc.
```

Acceptable variations of this format are shown in the following examples.

Note that the designations TABLE 1 and DIODE TABLE 1 refer to the same table; therefore, these may not be used to name two different tables in any run. The tabular data itself is represented by a series of numbers in pairs, separated by commas, in which the number representing the independent variable point comes first with the dependent variable point following. Note that as Fig. 2.9 suggests, any number of point pairs can be supplied per card. The data pair points must be supplied such that the independent variable is in increasing algebraic order. No specific limit to the number of point pairs that may constitute any table is defined. Only single valued functions are allowed. It is permissible to supply two consecutive independent-variable values that are equal, but which have different dependent variable values. This may be done to produce step functions as illustrated in Fig. 2.9. When the table values are updated at each solution time step, linear interpolation is used between the points supplied. For independent variable values falling outside the range of values supplied in a table, linear extrapolation is performed to

*Some of the symbols are slightly different, depending upon the computer. For example, in the S/360 version, the Sine is entered as DSIN.

determine the correct table value and, therefore, proper termination may be necessary. For example, in Fig. 2.9, for any value of TIME in excess of 3 the dependent variable ERIN is assigned the value zero because of the use of the point pair 4,0.

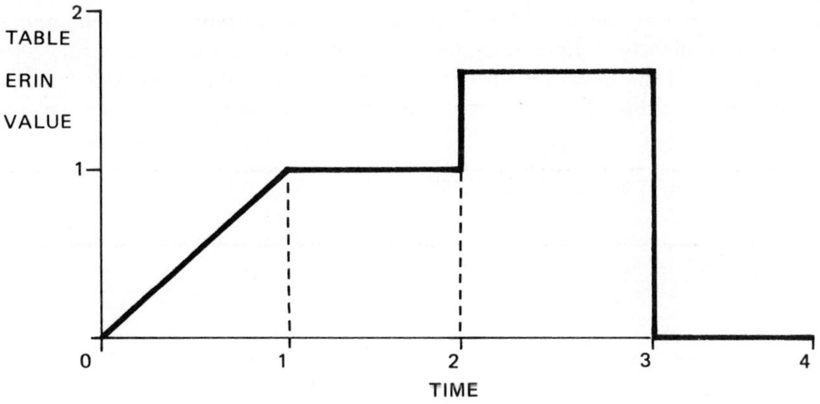

Fig. 2.9. TABLE ERIN values as a function of time.

```
TABLE ERIN
0,0
1,1
2,1
2,1.5
3,1.5
3,0
4,0
```

or

```
   TABLE ERIN
0,0,1,1,2,1,2,1.5,3,1.5
3,0,4,0
```

or

```
TERIN = 0,0,1,1,2,1,2,1.5,3,1.5,3,0,4,0
```

In situations where the same table will serve to define more than one quantity, the user need only explicitly define the table once. A common situation occurs when two current generators in the network are defined by the same tabular data, differing only in the independent variable. In this case the user may define both generators under ELEMENTS as:

```
J1, 1-2 = TABLE 1 (VC1)
```

```
J2, 7-8 = TABLE 1 (VC2)
```

Under FUNCTIONS, the user would define TABLE 1 only once.

2.10 RUN CONTROLS

This subheading contains all the auxiliary information needed to control the run. The appropriate listings are made under the RUN CONTROLS subheading. The information does not directly affect the network. Most of these quantities have automatically preset entries that hold unless specific entries are supplied by the user. The preset entries are given in Table 2.5. All possible entries under the RUN CONTROLS subheading are given in the following subsections. These entries may be made in any order, and as many may be placed on a card as will fit, if they are separated by commas.

Table 2.5 Preset run control quantities in SCEPTRE

Quantity	Preset information
STOP TIME	None
COMPUTER TIME LIMIT	None
MAXIMUM INTEGRATION PASSES	20,000
START TIME	0
INTEGRATION ROUTINE	XPO
MINIMUM STEP SIZE	1×10^{-5} *(STOP TIME)
MAXIMUM STEP SIZE	1×10^{-2} *(STOP TIME)
STARTING STEP SIZE	1×10^{-3} *(STOP TIME)
MINIMUM ABSOLUTE ERROR	0.0001
MAXIMUM ABSOLUTE ERROR	0.005
MINIMUM RELATIVE ERROR	0.0002
MAXIMUM RELATIVE ERROR	0.005
Mode of Analysis	Transient Only
NEWTON-RAPHSON PASS LIMIT	100
RELATIVE CONVERGENCE	0.001
ABSOLUTE CONVERGENCE	0.0001
MAXIMUM PRINT POINTS	1000
COMPUTER SAVE INTERVAL	15

2.10.1 Run Limits

All *transient runs must* have a problem duration in the time units consistent with those used to describe the circuit. The form is simply:

```
STOP TIME = NUMBER
```

Since the user never knows in advance the computer time required for the solution of a given run, it may sometimes be desirable to enter a limit that automatically terminates the run if that limit is exceeded. The limit in *minutes* is entered as:

```
COMPUTER TIME LIMIT = NUMBER
```

Another type of limit is available that operates on the number of passes made by the integration routine. In the absence of any instruction from the user, the program imposes a limit of 20,000 passes. The user may change this limit by entering:

```
MAXIMUM INTEGRATION PASSES = NUMBER
```

2.10.2 Start Time

Most transient problems start at time equal to zero, and this entry may be omitted. Otherwise the appropriate entry is:

```
START TIME = NUMBER
```

2.10.3 Integration Routine

Three integration routines are available for use with SCEPTRE. These routines are XPO, TRAP, and RUK. Tests indicate that XPO gives the best compromise between speed and accuracy for most circuits. This routine is used unless the user requests one of the others. For cases where the use of TRAP or RUK is preferred, the format is:

```
INTEGRATION ROUTINE = TRAP
```

or

```
INTEGRATION ROUTINE = RUK
```

2.10.4 Minimum Step Size

Any transient run automatically terminates whenever a time step size is required that is smaller than the minimum step size. If this quantity is not supplied, the program automatically computes a minimum limit equal to 10^{-5} times the problem duration (STOP

TIME). If a specific minimum step size independent of the problem duration is desired, the format is,

```
MINIMUM STEP SIZE = NUMBER
```

For example, if:

```
STOP TIME = 5000
```

is specified the minimum size would automatically be 0.05. To change this to 0.01 use:

```
MINIMUM STEP SIZE = 0.01
```

2.10.5 Maximum Step Size

This quantity sets an upper limit to the time solution increment that can be used to prevent a complete transient solution composed of few solution points. If this quantity is not supplied, the program automatically computes a maximum limit equal to 10^{-2} times the problem duration (STOP TIME). If a specific maximum step size independent of the problem duration is desired, the format is:

```
MAXIMUM STEP SIZE = NUMBER
```

2.10.6 Starting Step Size

This quantity is the size of the first time solution increment that is taken. Subsequent increments are automatically chosen by the integration routine. If this quantity is not supplied, the program automatically computes a starting step equal to 10^{-3} times the problem duration (STOP TIME). If a specific starting step size independent of the problem duration is desired. the format is:

```
STARTING STEP SIZE = NUMBER
```

2.10.7 Error Criteria

Unless otherwise specified, any of the three integration routines will operate with preset relative error criteria.

If it is desirable to modify these criteria, the proper formats would be:

```
MINIMUM ABSOLUTE ERROR = NUMBER
MAXIMUM ABSOLUTE ERROR = NUMBER
MINIMUM RELATIVE ERROR = NUMBER
MAXIMUM RELATIVE ERROR = NUMBER
```

2.10.8 Mode of Analysis

As indicated previously, SCEPTRE can compute transient and/or d-c steady state initial condition solutions. Unless otherwise specified by the user, only the transient response is computed. If it is desired to compute the initial conditions of a network and then to inspect the results, the user must enter RUN INITIAL CONDITIONS ONLY. If both the initial conditions and transient solutions are desired in one run, the appropriate entry is RUN INITIAL CONDITIONS.

2.10.9 Initial Condition Pass Limit

The Newton-Raphson iteration method should produce convergence, and hence, d-c solutions in less than 30 passes for most networks. If convergence is slow or does not occur, some realistic limit should be placed on the number of passes that are made. Unless otherwise specified, a limit of one hundred passes is built into SCEPTRE. To change this limit, the proper entry is:

```
NEWTON-RAPHSON PASS LIMIT = NUMBER
```

2.10.10 Convergence Criteria of the Newton–Raphson Routine

Convergence is considered to have occurred if:

$$\left| Y_{(n)} - Y_{(n-1)} \right| \leq A * \left| Y_{(n)} \right| + B$$

where Y refers to the vector of initial condition state variables, n is the number of completed passes, A is the relative criteria, and B is the absolute criteria. The values A = 0.001, B = 0.0001 are built into SCEPTRE. The user probably will not want to change these values, but if either quantity is to be changed, the proper entries are:

```
RELATIVE CONVERGENCE = NUMBER
```

```
ABSOLUTE CONVERGENCE = NUMBER
```

2.10.11 Output Control

Some transient runs may require thousands of solution points and it is usually unnecessary to have every solution printed out. SCEPTRE is processed to output a maximum of 1000 spaced points regardless of the number of steps taken. The maximum number of printed output points may be changed for any particular run by entering (See Sec. 4.3),

```
MAXIMUM PRINT POINTS = NUMBER
```

2.10.12 Automatic Termination

The user may often desire to monitor certain voltages or currents in a network to determine their relation with some predetermined quantity. If the relations are satisfied, there may be no further interest in continuing the run. The run is terminated at that point by the entry:

```
TERMINATE IF (XXXXXX RELATIONAL OPERATION XXXXXX)
```

The X quantities refer to any variable or constant (see Sec. 2.4.1). The user frequently supplies some element voltage or current for this purpose. The following relational operators may be used:

.LE.	(less than or equal to)
.LT.	(less than)
.GT.	(greater than)
.GE.	(greater than or equal to)
.EQ.	(equal to)
.NE.	(not equal to)

Note: decimal points are required before and after an operator.

A typical entry would be:

```
TERMINATE IF (VCC .GT. 0.)
```

which would automatically terminate that particular run whenever the voltage across capacitor CC became positive.

The termination feature may be extended to cover AND/OR logical situations by using the logical operators .AND. and .OR. A legal entry could be:

```
TERMINATE IF (VCC .GT. 0. .OR. IL 17 .GE. 7.)
```

This would terminate the run if the voltage across capacitor CC became positive or the current through inductor L17 exceeded 7 units of current. The other logical possibility would be simply:

```
TERMINATE IF (VCC .GT. 0..AND. IL17 .GE. 7.)
```

All automatic Termination conditions apply to all reruns that are associated with a given master run. One variation is available. If it is desired that all remaining reruns be cancelled if a termination condition is met, the word STOP is used instead of TERMINATE. Then a typical entry is:

```
STOP IF (VCC .GT. 0.)
```

2.10.13 Computer Save Interval

Particularly long computer runs should be protected against complete loss arising from improper termination caused by electrical malfunction, operator error, etc. This is done by periodically recording the status of the run on tape such that it can always be continued from the last saved point if improper termination does occur. A 15 minute save interval is preset in the program which ensures that no more than 15 minutes of computer solution time is ever lost in this way. The user may change the save interval to any number of minutes of computer solution time by the entry:

```
COMPUTER SAVE INTERVAL = NUMBER
```

2.11 RERUN DESCRIPTION

The preceding eight sections detailed the various subheading uses and rules. These rules are applicable for the subheadings under both MODEL DESCRIPTION and CIRCUIT DESCRIPTION. The uses of the same subheadings under the RERUN DESCRIPTION heading require some obvious variations due to the application involved.

The rerun feature permits the user to run multiple versions of a master run which differ from the master in one or more ways. The user need supply only the quantities that have been changed from the master run. An unlimited number of reruns may be made from a single master run, but each rerun requires approximately as much computer solution time as did the master. Each individual rerun utilizes all information from the master run except that which is specifically modified for that particular rerun. Any intermediate reruns that may have been made have no effect upon a subsequent rerun.

2.11.1 General Usage

As indicated in the general sequence of Table 2.2, the RERUN DESCRIPTION heading card follows the RUN CONTROLS subheading card of the main CIRCUIT DESCRIPTION. The RERUN DESCRIPTION heading card is followed by any of the five possible subheadings that are listed. The format of the card is:

```
RERUN DESCRIPTION (N)
```

where N is the number of reruns desired. The (N) designation may be omitted if just one rerun is desired.

2.11.2 Elements Under Rerun

The subheading ELEMENTS is used if any constant-valued element is to be changed for any of the reruns. If two reruns are desired in which resistor R1 is to assume constant values 9.3 and 9.5, then, under ELEMENTS:

```
R1 = 9.3, 9.5
```

regardless of the constant value R1 had in the master run.

2.11.3 Defined Parameters Under Rerun

This subheading is used if any constant defined parameter is to be changed for any of the reruns. If a defined parameter PX was used as constant in the Master run, a legal entry for two reruns under DEFINED PARAMETERS is:

```
PX = 5.1,6
```

2.11.4 Initial Conditions Under Rerun

This subheading is used to rerun any number of transient runs that start at different operating points. For two reruns, a sample form under INITIAL CONDITIONS is:

```
VC1 = 1.5, 3
VC2 = 7, 2.5
IL = 0, 1.5
```

2.11.5 Functions Under Rerun

This subheading is used to change any part of a table that describes an element or defined parameter of the original run. Changes in equations are discussed in section 2.12.2. If TABLE 7 was used to describe an element or group of elements in the master run, it may be modified under FUNCTIONS in the rerun section as:

```
TABLE 7 = 0,0,0
          1,3,5
          3,4,8
          5,7,4.9
```

These entries reflect the situation shown in Fig. 2.10. The independent variable values of the original table (0,1,2,4) are replaced by 0,1,3,5 for both reruns. The original corresponding table values (0,10,27.1,47) are replaced by 0,3,4,7 for the first rerun and 0,5,8,4.9 for the second rerun.

Master run		First rerun		Second rerun	
Independent variable	Table value	Independent variable	Table value	Independent variable	Table value
0	0	0	0	0	0
1	10	1	3	1	5
2	27.1	3	4	3	8
4	47	5	7	5	4.9

Fig. 2.10. Tables in rerun.

If both reruns are intended to use tables in which the independent variables differ from each other, another procedure must be used. Instead of one RERUN DESCRIPTION card with N = 2, two such cards are used (each represents one rerun). The sequence of subheadings could appear as:

```
RERUN DESCRIPTION
FUNCTIONS
RERUN DESCRIPTION
FUNCTIONS
```

Each new table is given separately under one of the FUNCTIONS subheadings. The length of any table in a rerun must be equal to its counterpart in the master run.

2.11.6 Run Controls Under Rerun

Any of the quantities except Automatic Termination messages that are defined under the RUN CONTROLS of a master run may be changed in any rerun. The Automatic Termination messages of the master run apply to all reruns. If three reruns are to be made in which the problem duration differs for each rerun, an appropriate sequence would be:

```
RUN CONTROLS
STOP TIME = 500, 600, 700
```

The problem durations of the three reruns would be 500, 600, and 700 units of time, regardless of the duration of the master run.

2.12 LIMITATIONS OF THE RERUN FEATURE

There are certain features of the master run which cannot be changed in any of the corresponding reruns. These are in the areas of output, element form, and operational mode.

2.12.1 Output

There can be no difference between the quantities that are output in the master run and those that are outputs in any of the reruns. Therefore, no OUTPUTS subheading is allowed under RERUN.

2.12.2 Element Form

No change in element form is allowed between the master run and the reruns. If an element originally appears as a constant, its rerun version must also be a constant. An element originally appearing as a table, must also be a table in the rerun version. The new table would be defined under FUNCTIONS. An element originally appearing as an equation, must also be an equation in the rerun. The only parts of an equation that may be changed are those that are defined under DEFINED PARAMETERS. To illustrate consider that the master run had an element:

```
JE, 1-8 = DIODE EQUATION (PX1, PX2)
```

where PX1 and PX2 are defined under DEFINED PARAMETERS. A legal rerun is:

```
RERUN DESCRIPTION
DEFINED PARAMETERS
PX1 = NUMBER
PX2 = NUMBER
```

2.12.3 Mode of Analysis

A master run may consist of a transient solution only, initial condition solution only, or initial condition plus transient solution. A general rule is that no rerun may contain more than its master run did. If the master run was a transient solution only, all associated reruns must also be transient solutions. If the master run was an initial condition only, all associated reruns must also be initial condition solutions. If the master run was an initial condition solution plus a transient solution, there is some leeway. The user should determine whether the changes in the network affect the initial conditions. (Usually only changes in resistors or batteries will have an effect.) If it is decided that the initial conditions do not change, the initial condition run that was made for the master will suffice for all reruns. In this case, no special entry is made for the reruns. If, however, it is decided that a new initial condition run is required for each rerun, the user must request this with the entry under RUN CONTROLS under RERUN DESCRIPTION as:

```
RUN INITIAL CONDITIONS
```

The single entry will allow for recomputation of initial conditions for all reruns in that group. The user may ask for new initial condition solutions for each rerun even if the master initial conditions are valid. However, this action is wasteful of computer time and should be avoided.

2.12.4 Rerun With Stored Models

Many users will probably apply the rerun option to networks in which stored models appear. If network changes leading to rerun occur in the body of the stored model, a slightly different format is required. To illustrate, assume that a model has been stored with the following card sequence:

```
MODEL DESCRIPTION
MODEL XXXX (PERM) (1-7-9-0)
ELEMENTS
R1, 1-2 = 1
C1, 2-3 = TABLE 1 (VC1)
E1, 9-1 = 20
J1, 8-7 = EQUATION 1 (VC1, P1)
DEFINED PARAMETERS
P1 = 2.7
OUTPUTS
VR1, VC1, IC1, PLOT
FUNCTIONS
TABLE 1
-5, 1, 0, 2, 1, 5, 2, 10
EQUATION 1 (A,B) = (3. *A + B)
```

Assume that this stored model is designated as T8 under CIRCUIT DESCRIPTION. To change the constant element R1 under the rerun option, the entry should be R1T8 = NUMBER. Note that the circuit designation for the model has been appended to the element name. In the same way, changes to P1 and TABLE 1 may be affected by the entries P1T8 = NUMBER and TABLE 1 T8 = NUMBER,...,NUMBER. All changes must be made under the appropriate subheading. For rerun, changes may be made to EQUATION 1 of the stored model only by changing the defined parameter which appears in the equation. Therefore, the user is advised to use the DEFINED PARAMETERS section liberally if frequent recourse is expected to both stored models and reruns.

2.13 CONTINUE FEATURE

The purpose of this feature is to allow the user to continue transient runs that have been run previously and have been properly terminated. Without this feature, the user is forced to start the problem again. This would obviously waste computation.

2.13.1. General Usage

This continue feature is intended primarily to extend the time duration of transient runs. The only changes permitted under this heading are most of the changes allowed under RUN CONTROLS. No changes are allowed to the network under this heading, so the subheadings ELEMENTS, DEFINED PARAMETERS, OUTPUTS, INITIAL CONDITIONS, and FUNCTIONS can never appear under CONTINUE.

The following representative sequence is appropriate for the case where an original run is carried to its specified duration and the user subsequently wishes to continue the run to a new duration of 2000 units of time:

```
CONTINUE
RUN CONTROLS
STOP TIME = 2000
END
```

Other changes that are permissible under continued RUN CONTROLS are those described in Secs. 2.10.1, 3, 4, 5, 6, 7, and 11.

2.13.2 Limitations of the Continue Feature

This feature is intended for transient runs and has no control over initial condition computations. For this reason, the entries described in Secs. 2.10.8, 9, and 10 cannot be applied under CONTINUE. When the original run contains a number of reruns, the continue option applies only to the last run that was processed, which is usually the last rerun.

2.14 RE-OUTPUT FEATURE

The user sometimes desires additional copies of output from previous runs without the necessity of repeating the runs. This can be done in SCEPTRE through the use of the Re-output feature if the original run has been preserved on tape. In addition to repeating the original output, the Re-output feature permits some degree of modification. Any quantity that was originally output in printed form may be obtained in both printed and plotted form by the use of this feature.

If the printed output results of a previous run are desired without modification, the appropriate entries are simply:

```
RE-OUTPUT
END
```

If, in addition to the printed results, any outputs are desired in plotted form (regardless of how many were plotted in the original run), appropriate entries would be:

```
RE-OUTPUT
VR1, VR2, VC10, PLOT
END
```

Re-output generally does not permit the printed output to exceed that which appeared in the original run. The RE-OUTPUT heading cannot be used in the same run with any other heading or subheading except END.

2.15 END

The END heading is used to terminate every input deck submitted to SCEPTRE. Each of the headings and subheadings listed in Table 2.2 has now been discussed.

2.16 TOPOLOGICAL RESTRICTIONS ON SCEPTRE

The purpose of this section is to illustrate circuit topologies that cannot be accommodated by SCEPTRE. Most of these situations are rather impractical and those that are not, may readily be remedied by the user. It should be mentioned here that, in general, the transient portion of the program is more versatile than that used for initial condition solutions. In other words, there are some circuit configurations that cannot be accommodated in the initial condition solution that could be in the transient solution.

2.16.1 Restrictions on Both Transient and Initial Condition Solutions

No run may contain a loop composed exclusively of voltage sources (Fig. 2.11 and Fig. 2.12) or a cut set* composed exclusively of current sources (Fig. 2.13 and Fig. 2.14). If either configuration is presented to the program, it is rejected with an appropriate diagnostic.

*An all-current source cut set is recognized by the presence of one or more nodes which cannot be connected to all other network nodes without traversing current sources.

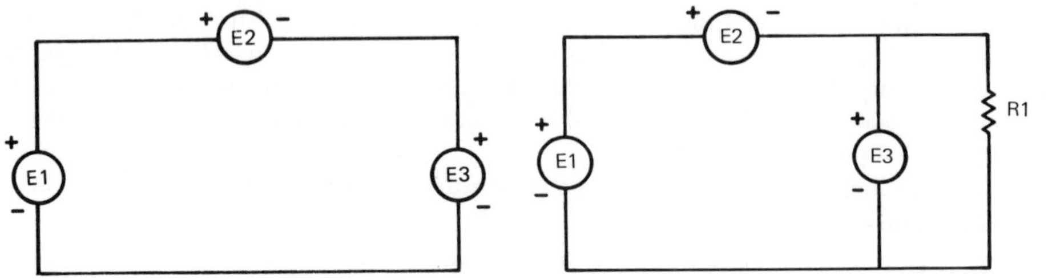

Fig. 2.11 and Fig. 2.12 Voltage source loop.

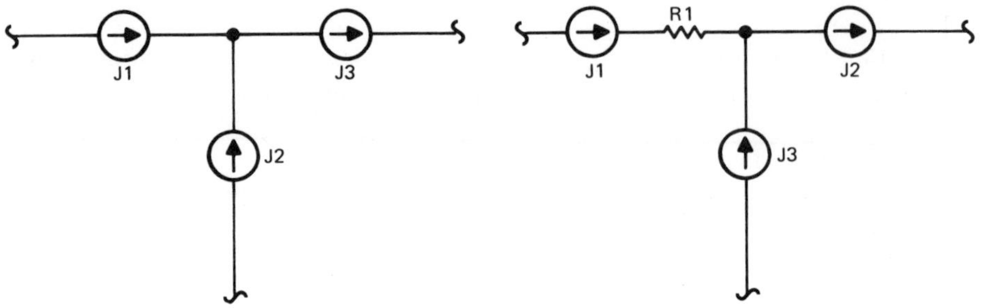

Fig. 2.13 and Fig. 2.14 Current source cut sets.

2.16.2 Restrictions on Initial Condition Solutions

No initial condition solution is performed if any loop exists that is composed solely of voltage sources and inductors (Fig. 2.15). Configurations which contain cut sets composed entirely of current sources and capacitors are also prohibited (Fig. 2.16). If either configuration is presented to the initial condition formulation, it is rejected with an appropriate diagnostic. It is worth noting that any J or J–C cut set may be easily broken up by the insertion of a resistor in the proper position. For example, the JC cut set in Fig. 2.16 may be eliminated by the insertion of a resistor between nodes X and ground.

In addition, no initial condition solution is computed if any network resistor current is used as an independent variable in an equation or table, also quantities which are functions of capacitors and inductors, since these are non-existent for initial conditions.

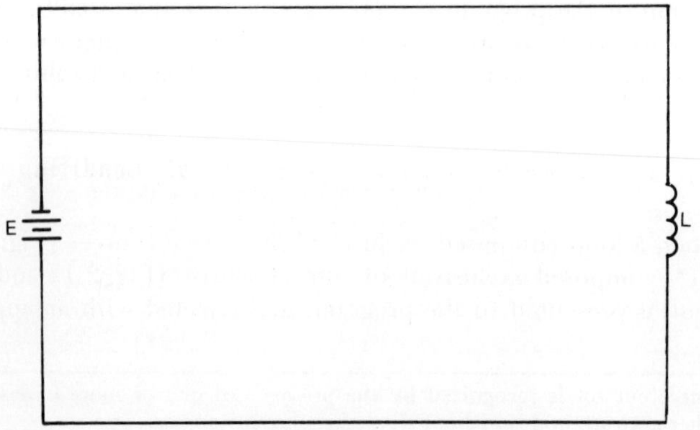

Fig. 2.15. Voltage source – inductor loop.

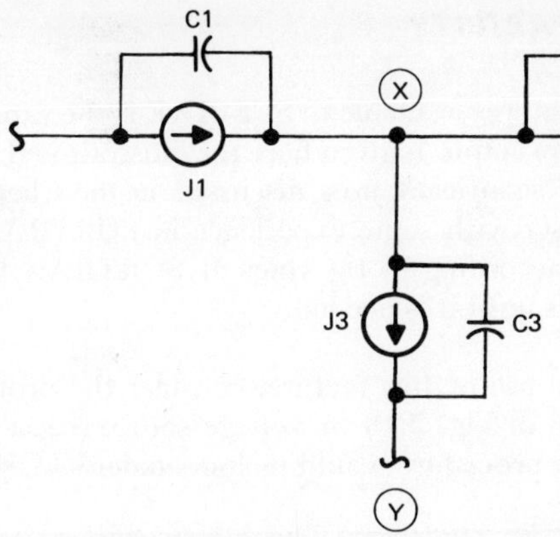

Fig. 2.16. Current source – capacitor cut set.

2.16.3 Source Derivatives

The time derivatives of sources must be supplied as input data when certain network configurations are encountered. These situations occur whenever a variable voltage source is connected in a loop containing only capacitors and other voltage sources, and whenever a variable current source is connected in a cut set containing only inductors and other current sources (see Fig. 2.17). If the sources in question are constant, the zero derivative will automatically be supplied and the user need not be concerned. If the user fails to supply a source derivative when one is required, the run terminates with an appropriate diagnostic message. The general form for a source derivative entry under ELEMENTS is:

```
DE NAME = VALUE
DJ NAME = VALUE
```

where the name is that of the appropriate E or J source. An example would be:

```
DERIVATIVE E7 = TABLE 2
```

or more simply:

```
DE7 = T2
```

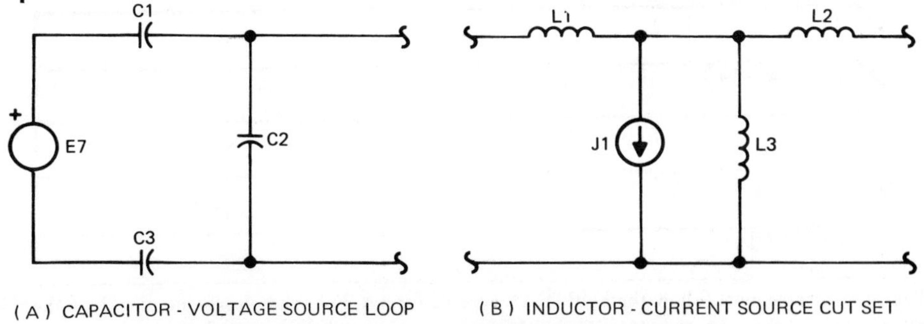

Fig. 2.17 Configurations that require source derivatives

2.17 SUBPROGRAM CAPABILITY

One of the most powerful extra features available in SCEPTRE is the capability to readily handle Fortran Subprograms. An example is given here for illustration; however, the user should also study carefully other significant uses discussed in the Chapter 4 examples, This feature is intended for the user with some experience in FORTRAN programming. Any subprogram may be written according to the rules of FORTRAN IV Function Programs and retained on tape or cards until it is needed.

As an example of a practical use of this feature, consider the problem of creating a periodic train of pulses as shown in Fig. 2.18. A voltage source (e.g.,E1) may generate a wave of this type. An appropriate procedure would include under ELEMENTS:

```
E1, NODE-NODE = FNAME (X,........, X, TIME)
```

where the X quantities refer to constants or network variables and FNAME refers to the chosen subprogram name. The subprogram name must be unique and begin with the letter F. It may not exceed six alphanumeric characters.

With specific reference to Fig. 2.18, the user could enter under ELEMENTS:

```
E1, 1-2 = FGEN (0., 6., 20., 40., 100., 100.,
300., 2., TIME)
```

which references the following subprogram: *

```
FUNCTION FGEN (VO, V1, TO, TR, TD, TF, TP, ZO,
TIME)
DATA Z/1./
1 IF (Z-ZO) 3, 3, 4
3 IF (TIME. GE. (Z*TP+TO)) GO TO 2
X = TIME-(Z-1.)*TP-TO
IF (TIME-TO) 4, 4, 5
2 Z = Z+1.
IF (Z-ZO-1.) 1, 4, 4
4  FGEN = VO
RETURN
```

* The subprograms must follow the FORTRAN type of format. For example, the number 1 on the third card must be in columns 1–5 and the rest of the text cannot begin before column 7.

```
5 IF(X.LE.TR) GO TO 6
IF (X.GT.TR.AND.X.LE.(TR+TD))GO TO 7
IF (X.GT.(TR+TD).AND.X.LE.(TR+TD+TF)) GO TO 8
IF (X.GT.(TR+TD+TF)) GO TO 4
6 FGEN = VO + X*(V1-VO)/TR
RETURN
7 FGEN = V1
RETURN
8 FGEN = V1-(X-TR-TD)*(V1-VO)/TF
RETURN
END
```

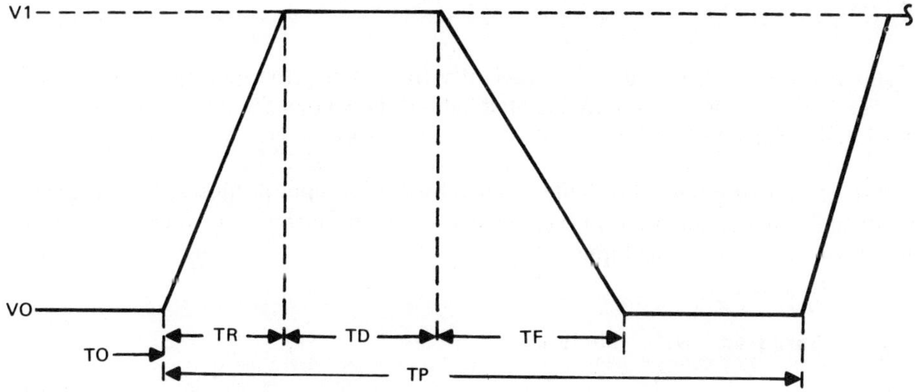

VO –INITIAL VOLTAGE
V1 –PEAK VOLTAGE
TO –PULSE DELAY
TR –TOTAL RISE TIME
TD –PEAK DURATION TIME
TF –TOTAL FALL TIME
TP –PERIOD
ZO –NUMBER OF COMPLETE PULSES
Z –INTERNAL VARIABLE THAT IS INDEPENDENT OF THE DESIRED WAVESHAPE

Fig. 2.18 Pulse train

The value of +1 will be computed in the function subprogram FGEN at each time step of the transient problem. The user may change any of the parameters given in Fig. 2.18 to create variations in waveshape. It is the user's responsibility to avoid format errors in the body of any included subprogram since they cause the entire run to be aborted.

2.17.1 Subprogram Insertion

Once the subprogram cards are generated, they are inserted, in the IBM 7094 version, in the second half of the SCEPTRE Program Control Deck behind the $IBJOB card PGMC 1940. Assuming that a source deck is to be submitted, the proper sequence of cards is:

```
$IBJOB              ALT10, NOFLOW
$IBFTC   FGEN.   FULIST, DECK
FUNCTION FGEN (VO, V1, TO, TR, TD, TF, TP, ZO,
   TIME)
END
```

This feature may be used more efficiently by inserting the subprogram in compiled form. The compiled deck is simply inserted behind the second $IBJOB card, PGMC 1940 as illustrated in Chapter 7.

This example is intended to build a general wave for input purposes. The imaginative user will find other applications in the course of practical analysis which cannot be implemented through conventional input.

2.17.2 Subprogram with Models

There is no difficulty in combining the subprogram feature with stored models. The models may be stored (either permanently or temporarily) with appropriate reference to the user's subprograms. To illustrate:

```
MODEL DESCRIPTION
MODEL XXX (PERM) (1-6-4-A)
ELEMENTS
L2, 1-2 = FCOR (IL1, IL2, IL3, 25., 100.)
```

Here the user would ensure that the model includes elements L1, L2, and L3 because the currents through these elements have been used as arguments in the subprograms. The subprogram itself should be supplied in either source or object form as in section 2.17.2.

2.18 ADDITIONAL OUTPUT AND CONTROL

Some additional output and control is available to the interested user. This information is probably not desired for most runs that are made but can be useful in special situations.

Normally, the program outputs the results of initial conditions computations once after convergence has been achieved regardless of the number of iterations required. However, all resistor and primary dependent current source voltages may be output after each iteration of the DC program. In addition, all the formatted data that is used by the FORTRAN program (SIMUL8) written by SCEPTRE is available. This information may be obtained by the request WRITE SIMUL8 DATA in the RUN CONTROLS section.

In the absence of any user direction, the program forms the network tree by giving priority to the smaller elements within any passive element type. Variable elements are considered to have zero value for this purpose. If the instruction NO ELEMENT SORT is inserted in the RUN CONTROLS section, tree priority is given within passive element types according to orders of appearance under ELEMENTS. This instruction of course does not change the basic element type preference order of E, C, R, and L.

The network F matrix, which gives the topological relation between all network links and tree branches, is available in labeled and constructed form. The required entry is BTFORM = 1 to obtain the matrix for the transient program. The corresponding matrix for the DC program is called by BTFORM = 2. A card indicating the desired option must be placed immediately after the $DATA card that normally precedes MODEL DESCRIPTION or CIRCUIT DESCRIPTION.

2.19 ERROR DIAGNOSTICS

The free form input data format of SCEPTRE is intended to minimize formatting errors. Other types of errors such as those of omission, ambiguity, inconsistency, and violation (of syntax, program limits, etc.) must be detected and diagnosed and the user alerted. For this purpose, the program possesses a comprehensive input data diagnostic capability. As the input data cards are read-in by the input processor, each card is printed-out and then scanned for errors. If an error is found, an error message stating the trouble is printed immediately following the card in error. The severity of errors detected in this manner is also indicated. There are three levels of severity:

Level 1: (warning only) Errors of this type are not critical and will be repaired by the input processor. The error scan is continued and the analysis phase executed providing that errors of a higher level are not detected elsewhere in the input data.

Level 2: (simulation deleted) Errors of this type are critical and thus prohibit execution of the analysis phase. However, the error scan is continued in order to detect, diagnose and alert the user of any remaining errors in the input data.

Level 3: (execution terminated) This type of error will not permit the proper execution of the input processor and, thence successful continuation of either the error scan or execution of the analysis phase.

Most input data errors fall into Level 2 as illustrated by the following example of an initial condition specification which should contain an "I" prefix designating the initial current through inductor L3:

```
INITIAL CONDITIONS
VC1 = 100
L3 = 0.01
```

***Error Message 1 – level 2 – simulation deleted – ERROR SCAN CONTINUES
INITIAL CONDITION LACKS A 'V' OR 'I' PREFIX

The program generator portion of the program also contains a diagnostic capability to inform the user of errors that can only be detected after the network topology is analyzed. Such errors may or may not cause the execution of the analysis phase to be aborted as indicated by the error messages which occur.

2.20 INTERNAL PARAMETERS

A number of parameters are carried internally in the program. The user is permitted access to these parameters for use as equation arguments, independent variables for tables, and automatic termination quantities. The nomenclature and definitions of these parameters are given in Table 2.7.

2.21 PROGRAM LIMITS

The standard core memory capacity available on the IBM 7090/94 data processing system is 32,000 words which imposes certain program limits on the capacity of the SCEPTRE program. Two factors that establish these limitations are program storage and data storage.

The program storage requirements have priority over data storage and must be satisfied first. The entire SCEPTRE program would require approximately 64,000 words of core storage, but this capacity is not necessary if the overlay feature of the IBSYS monitor system is used. This feature breaks up the program into several segments called "links" which are executed sequentially. Each segment overlays or occupies the portion of core storage used by the preceding segment. Therefore, the amount of program storage required is equal to the largest program segment.

The storage capacity that remains after the program storage requirements are satisfied is available for data. In the SCEPTRE program, there are about 10,000 words of storage available for this purpose. Within this area, the program data are limited by FORTRAN dimension statements which allocate specific amounts of storage. Although these limits are usually considered fixed, they can be changed somewhat by a programmer. However, increasing limits in this manner usually is accomplished by decreasing one

Table 2.6. Program data limits.

Description of data	Maximum number
Heading cards	11
Elements	300
Nodes	301
Source derivatives	50
Defined parameters	100
Defined parameter differential equations	100
Mutual inductances	50
Arguments in equation value specification	50
Model table changes	15
Model equation changes	15
Model output suppressions	10
Output requests	100
Supplied initial conditions	100
Equation functions (1 equation per card)	80
Cards per equation function	20
Table functions	80
Optional termination conditions	10
Models on library tape (combined)	250
Characters in model name	18
Model terminals (external nodes)	25
Model internal nodes	301

Table 2.7. Internal parameter table.

Internal parameter name	Definition
XSTOPT	Transient solution duration (STOP TIME)
XIR	Integration routine (TRAP = 1, RUK = 2, XPO = 3)
XTISSS	Starting step size
XMNISS	Minimum step size allowed
XMXISS	Maximum step size allowed
XMNAIE	Minimum absolute error
XMXAIE	Maximum absolute error
XRERNO	Rerun number
XMXPAS	Maximum pass limit
XMNRIE	Minimum relative error
XMXRIE	Maximum relative error
XMXICP	Newton-Raphson pass limit
XICRER	Newton-Raphson relative convergence
XICAER	Newton-Raphson absolute convergence
XMXOTP	Maximum number of print points
XICPAS	Newton-Raphson pass number
XNOPRQ	Number of output requests
XNDFEQ	Number of differential equations
XERT	Elapsed computer time
XSTPNO	Transient solution step number
XPASNO	Transient solution pass number
XRUNNO	Run number (including reruns)
XSTPSZ	Transient solution step size
XSAVE	Save internal
XTMON	Elapsed computer time at start of transient solution

or more of the other program data limits. Also, any change which affects the total amount of data storage will necessitate a change to the program storage; and therefore, the program operation.

Table 2.6 gives the program data limits which have been established in the program as it is distributed. These limits are considered adequate for most normal circuit analysis work. Approximate limits are also given for the number of TABLE and EQUATION functions. If only one-half of an input card is used to write the equation specification, 160 equation functions could be used. Conversely, if each equation specification took two full data cards, only 40 equations would be allowed. Defining table functions involves a similar process, which considers the number of x–y coordinates and the number of tables. Forty tables, each containing 20 point pairs, or eight tables each containing 100 point pairs would be allowed.

The limits shown in Table 2.6 are checked by SCEPTRE as the data is read into storage. If any limits are exceeded, a message stating the limit is produced, and the computer run terminated.

When large networks are encountered, SCEPTRE automatically reverts to an internal renaming feature (VECTORIZED NOTATION) which effectively enlarges the capacity of the Fortran Compiler, "Large" networks, in this case, refer to those containing 70 or more elements if a DC solution is requested, or 100 or more elements if only a transient solution is to be made. This feature does not affect the user operationally. It does however, cause SIMUL8 to be written in terms of the renamed quantities. A complete listing of all circuit Elements and Defined Parameters along with their corresponding internal names is provided for every run in which renaming occurs.

PROBLEMS

2.1 Draw the input waveform given by the following TABLE under the FUNCTIONS subheading. What is the independent variable?

Table A1 = 0, 0, 10, 5, 30, 5, 30, 1, 40, 1, 50, 5, 70, 5, 70, 0, 80, 0

2.2 Give the sign of the mutual inductance, M, for each of the following.

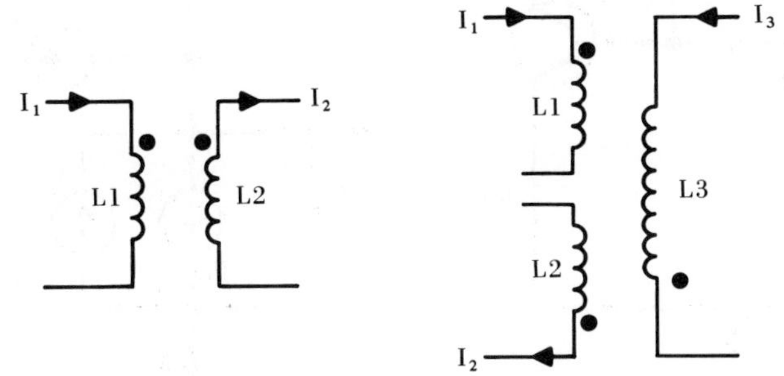

Fig. p2.1

2.3 For any two inductors, L1 and L2, what are the relationships between L1, L2 and M that are necessary for a physically realizable circuit?

2.4 Which of the following changes to a stored Model are incorrect? Why?

a. T1, 1-2-3 = MODEL 2NNPN (PERM, CHANGE RB = .01, P1 = .98)

b. T2, 8-9-10 = MODEL TRAN (PERM, CHANGE TABLE 1 = TABLE 2 (IR1))

c. T3, 5-6-7 = MODEL 2NPNP (PERM, SUPPRESS ALL, CC = 25)

d. T4, 1-9-6 = MODEL 2N608 (PERM, CHANGE TABLE 1 = Q4)

e. T5, 2-5-A = MODEL ABC (CHANGE JB = TABLE 2, RE = .05, PRINT)

2.5 What must be added to this circuit in order to compute initial conditions? Why?

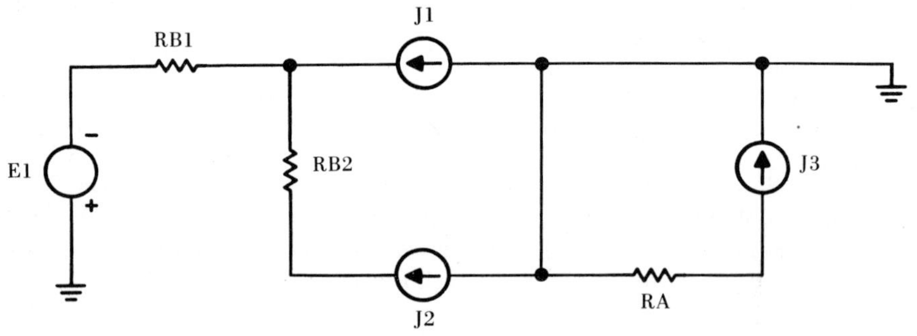

Fig. p2.2

2.6 Observing the transistor model Fig. p2.3 and the circuit diagram, determine what problem is present. How can the trouble be corrected?

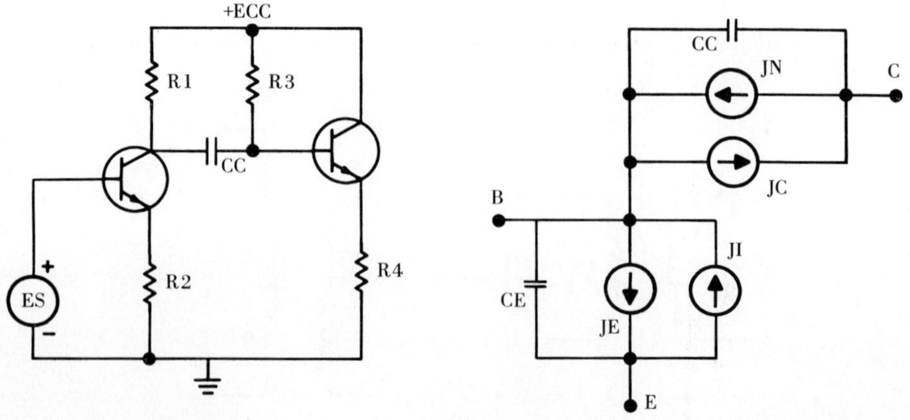

Fig. p2.3

2.7 Write a SCEPTRE program using Fig. p2.4 to find the amplitude of signal voltage, ES = Esin 377t, to give a voltage of 5 volts peak across R3.

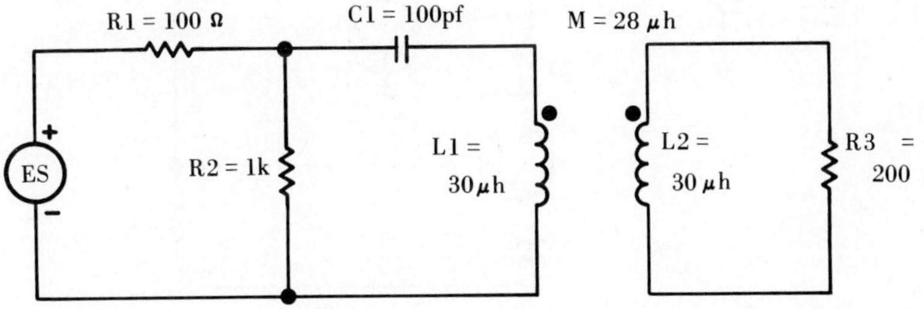

Fig. p2.4

2.8 For two reruns, a TABLE value is to be changed as shown below. Write the FUNCTIONS subheading under RERUN DESCRIPTION required.

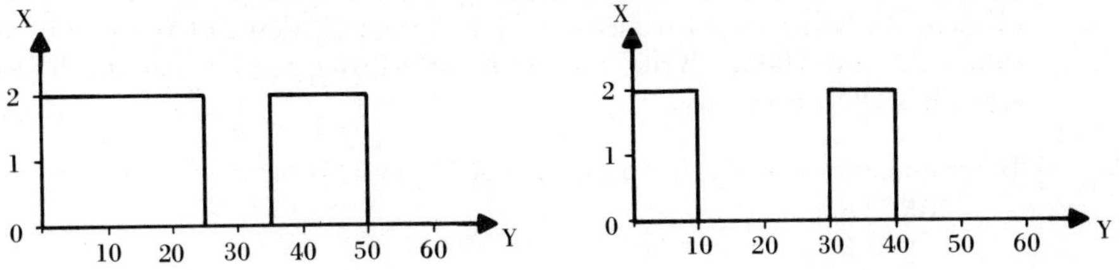

Fig. p2.5

2.9 A master run contained the elements,

```
RC, 1-2 = 5.3
J3, 2-B = DIODE Q (.98, 3.E-10)
```

Below is given a rerun description for this master run. Find and discuss the errors in the listing.

```
RERUN DESCRIPTION
ELEMENTS
RC, 1-2 = TABLE 9
J3, 2-B = DIODE Q (.90, 4.5E-5)
OUTPUTS
VRC, PLOT
FUNCTIONS
TABLE 9 = 0, 0, 20, 2, 40, 3, 50, 3
RUN CONTROLS
STOP TIME = 300
END
```

2.10 A run made with the circuit of Fig. p2.6 terminated with the diagnostic, THE DERIVATIVE OF THE VARIABLE SOURCE ES HAS NOT BEEN SUPPLIED. What is the cause and what must be done to correct the error? The transistor model is the same as shown in Fig. p2.3.

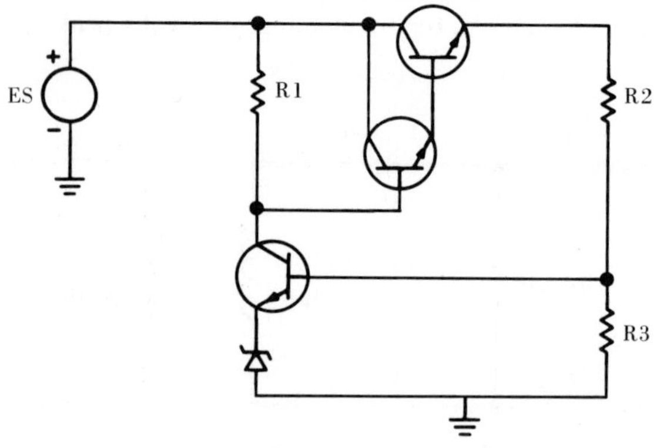

Fig. p2.6

2.11 The rerun option is to be used to simulate a circuit changing one of the independent variables, W, of the equation, Q1 (W,T) = SIN (W * T). Assume an element ES is connected from node 1 to 2 and the values of W are to be 100, 200, 500 and 1000. Write the master run listing for ES and the RERUN DESCRIPTION necessary.

2.12 Determine the errors in the following portion of a listing.

```
CONTINUE
ELEMENTS
R10, 2-3 = 5
RUN CONTROLS
STOP TIME = 1.E4
RUN INITIAL CONDITIONS
MINIMUM STEP SIZE = .001
END
```

2.13 In the following problems assume the appropriate transistor and diode models have been permanently stored in the computer.

a. Write a correct listing for the circuit shown in Fig. p2.7.

b. Write a correct listing for the circuit shown in Fig. p2.8.

c. Write a correct listing for the circuit shown in Fig. p2.9.

d. Write a correct listing for the circuit shown in Fig. p2.10.

e. Write a correct listing for the circuit shown in Fig. p2.11.

f. Write a correct listing for the circuit shown in Fig. p2.12.

DIODE CLAMP

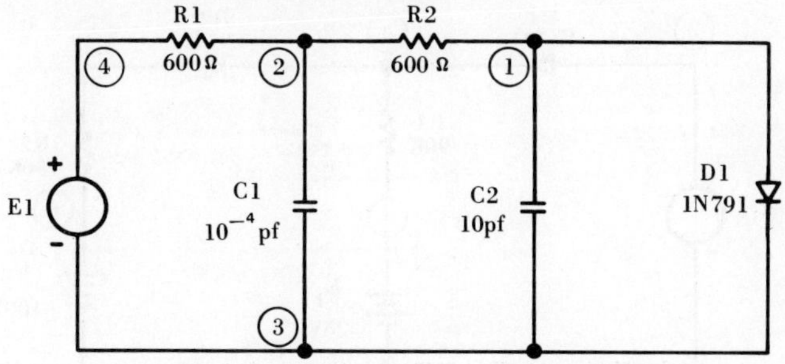

Fig. p2.7

POSITIVE–NEGATIVE FEEDBACK AMPLIFIER

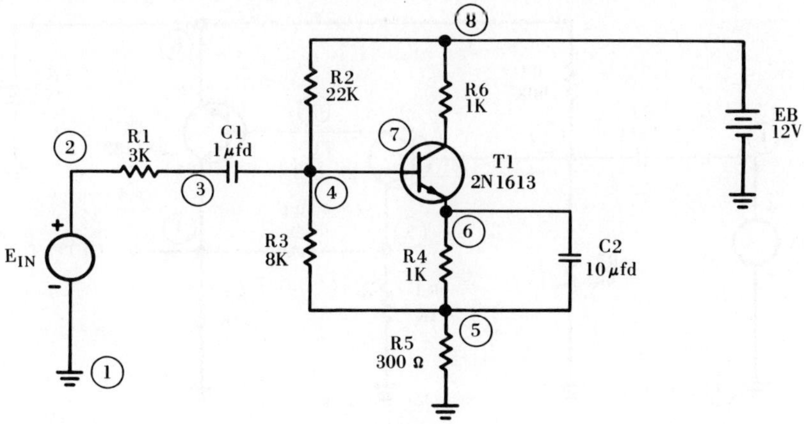

Fig. p2.8

LC FILTER

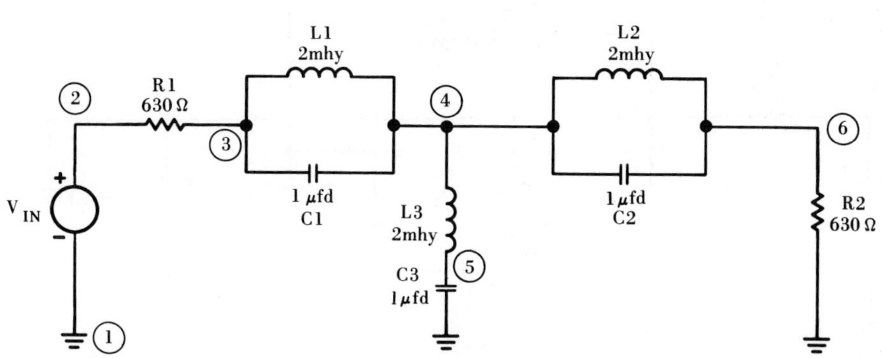

Fig. p2.9

TWO LEVEL CLIPPER

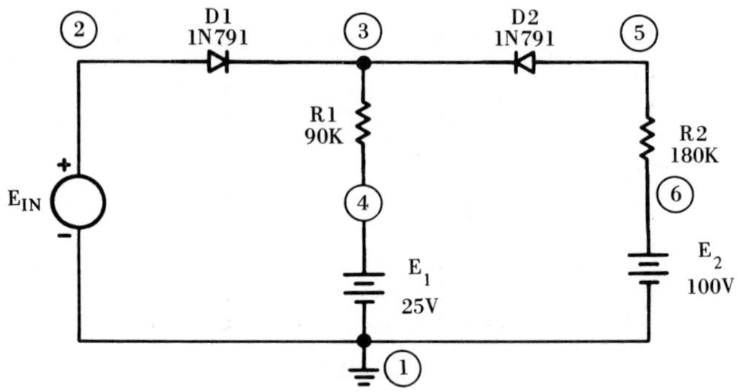

Fig. p2.10

SERIES FEEDBACK AMPLIFIER

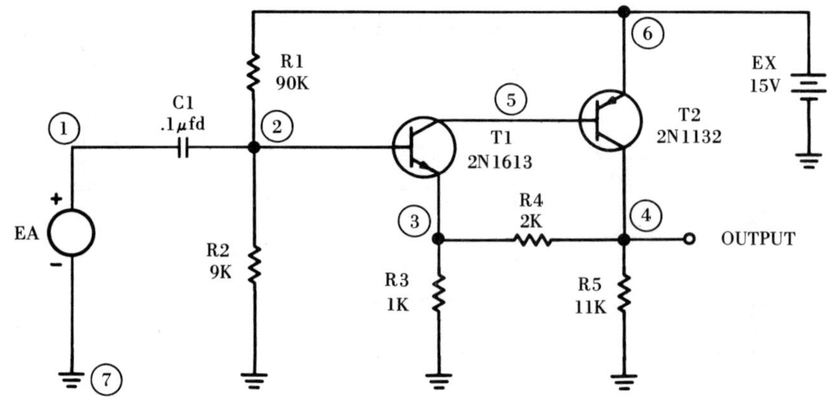

Fig. p2.11

COMPLIMENTARY EMITTER FOLLOWER

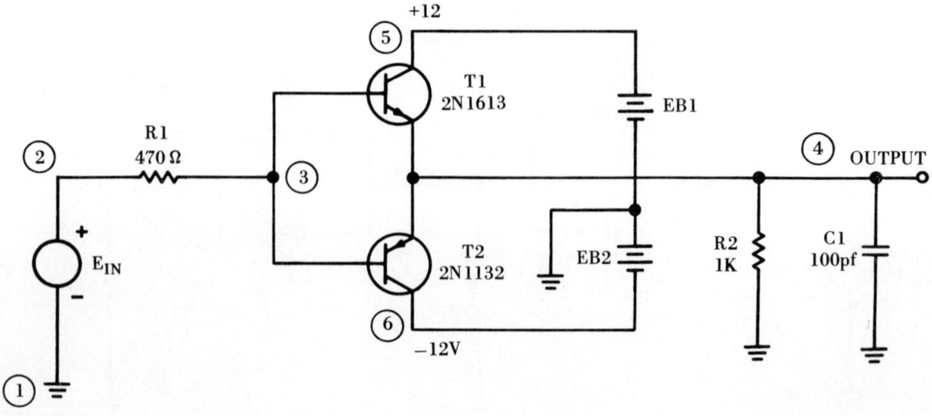

Fig. p2.12

REFERENCE

1. "SCEPTRE Support, Vol. I, Revised SCEPTRE User's Manual," *Air Force Weapons Laboratory Technical Report* TR–69–77.

3

SOLID-STATE MODELING

3.1 INTRODUCTION

The accuracy of the results in computer-aided-analysis is always proportional to the accuracy of the models. Given "perfect" mathematical descriptions for all components, the computer produces "perfect" results, within the machine's capability. It is unfortunate that the models of some devices are sometimes quite difficult to obtain and many times must be approximated. Fortunately, models which are reasonably accurate for a wide variety of applications are available and are presented in this chapter. There are two basic steps in device modeling. The first step is to develop the equivalent circuit which is to represent the particular device. The second step is to determine how to obtain the parameters, which are necessary for the equivalent circuit representation. Many equivalent circuits have been presented in the literature for the various devices. The most common one is probably the Ebers-Moll representation for the transistor. However, much less has been said about how one actually obtains the parameter values necessary for the Ebers-Moll and other models.

In this chapter equivalent circuits are presented for the six most common solid-state devices. In addition, a detailed step-by-step procedure is given for obtaining the parameters required for the equivalent circuits. The models presented are not, in all cases, the most sophisticated possible. In fact, any model may always be further improved! These equivalent circuits do, however, give excellent results in most applications. The measurement procedures are purposely developed using the minimum of complex equipment, in order to make the process available to the widest variety of users.

In addition to illustrating the model and showing the measurement procedures, each device is used in a SCEPTRE application, which verifies to some extent the model obtained. Obviously, for complete verification, many additional tests must be made, however those shown illustrate the proceedure. The device models presented in each section are not derived or developed in this text. References are given for the interested reader; however, the measurement techniques which are not generally available, are described in great detail.

3.2 SCEPTRE DIODE MODEL

Fig. 3.1 illustrates the selected diode model, along with the defining equations.

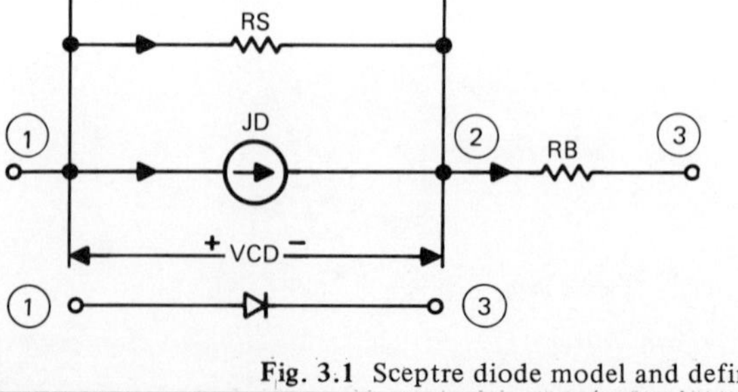

$$JD = IS(e^{\Theta \cdot VCD} - 1)$$

$$CD = \frac{CO}{(\phi - VCD)^n} + KD(JD+IS)$$

Fig. 3.1 Sceptre diode model and defining equations

3.2.1 Definition of Terms

All of the terms necessary for the diode derivation are defined as follows:

CD = The sum of the diode transition and diffusion capacitances. Where,

$\frac{CO}{(\Phi - VCD)}$ = Transition Capacitance = CJ, and KD(JD+IS) = Diffusion Capacitance (pf)

JD = Current generator representing the diode junction current. The generator is a function of the voltage VCD.

IS = Diode Saturation Current (ma)

RB = Diode Bulk Resistance (Kilohms)

θ = Constant of the diode equation (Volts^{-1})

CO = Constant of the transition capacitance equation

Φ = Junction Contact Potential (Volts)

n = Junction Grading Constant

KD = Diffusion Capacitance Constant = $\frac{\theta}{2\pi F}$ (pfd/ma)

F = Frequency Parameter (GHZ)

VCD = Voltage across capacitor CD, which is equal to the diode junction voltage (Volts)

3.2.2 Diode Selected

Parameters are derived in this section for the FD300 diode. The FD300 is a high conductance, low leakage device. Pulse measurement techniques are employed for measurements involving currents greater than 10 ma. It has been experimentally determined that junction heating effects occur for higher current levels, introducing significant error in the resulting parameters. The use of pulse techniques, with very low duty cycles, minimizes this problem. All measurements are made at an ambient temperature of 25°C or 298°K.

3.2.3 Model Parameter Determination for the FD300

The measurement technique for each diode parameter is discussed in the following sections.

3.2.4 θ, Diode Equation Constant

θ is determined from the forward region voltage-current characteristics of the diode. The test configuration used in obtaining the diode forward characteristic for low current

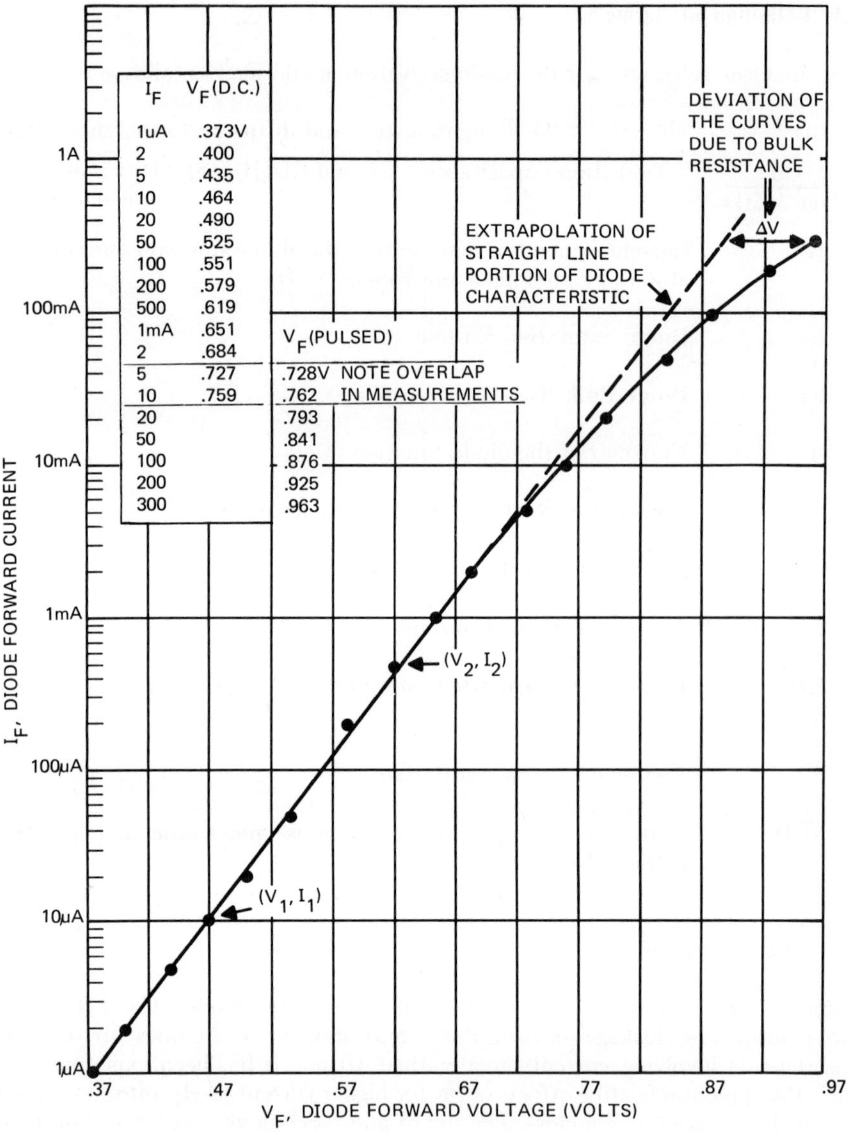

Fig. 3.3 FD300 forward region voltage-current data

levels is shown in Fig. 3.2. The desired current level is set up on the front panel dials of the current source. The source is then switched to the operate mode and the resulting diode voltage is read from the digital voltmeter. The diode voltage is measured for forward currents of 1ua to 10 ma. This data is tabulated and plotted in Fig. 3.3.

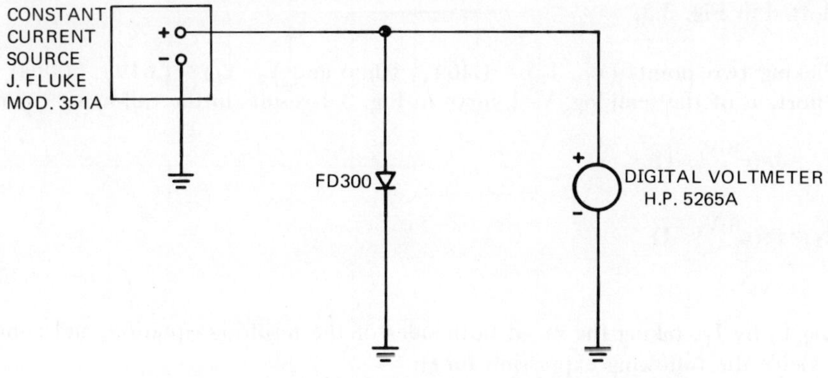

Fig. 3.2 Test configuration for low diode current levels

For current levels greater than 10 ma, the test configuration shown in Fig. 3.4 is used to obtain the V-I characteristics.

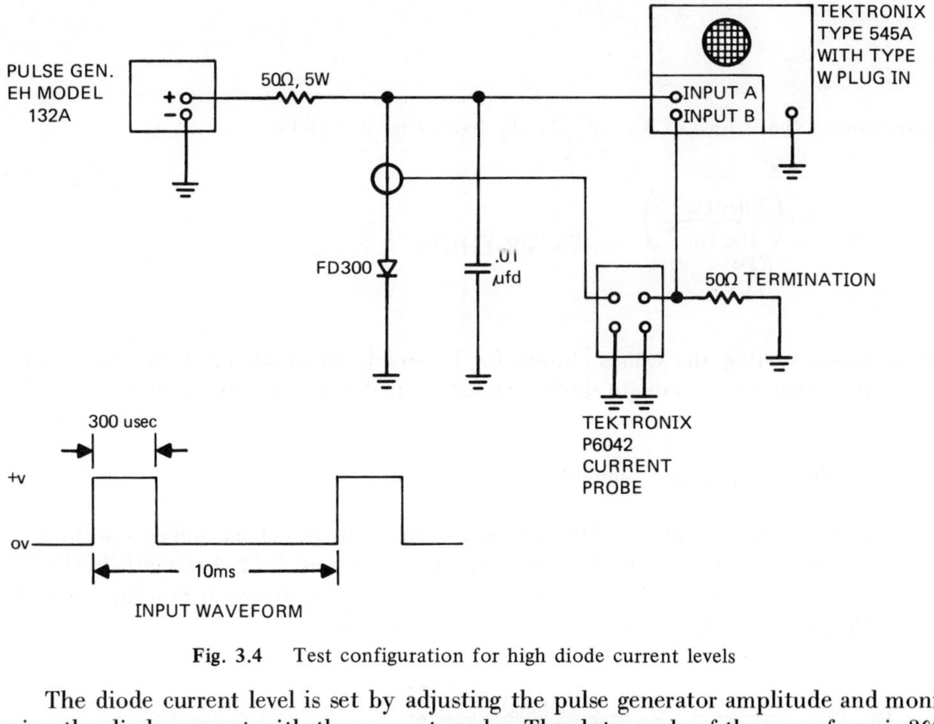

Fig. 3.4 Test configuration for high diode current levels

The diode current level is set by adjusting the pulse generator amplitude and monitoring the diode current with the current probe. The duty cycle of the wave form is 3%, minimizing heating of the junction. After the current is adjusted to the desired amplitude, the voltage is measured using the comparator on the type W plug in. It is desirable to overlap the pulse measurements with the D.C. measurements to assure the data taken by both methods fit the same diode curve. All instrumentation should be carefully calibrated before making these measurements. The diode voltage is measured using the pulse method for forward currents of 5 ma to 300 ma. The 5 ma and 10 ma measurements are chosen to overlap the D.C. measurements which terminate at 10 ma. The results are also shown and plotted in Fig. 3.3.

Picking two points $(V_1, I_1) = (.464_v, 10\mu a)$ and $(V_2, I_2) = (.619_v, 500\mu a)$ on the linear portion of the semi-log, V-I curve in Fig. 3.4 results in the following equations:

$$I_1 = IS(e^{\theta \cdot V_1} - 1)$$

$$I_2 = IS(e^{\theta \cdot V_2} - 1)$$

Dividing I_2 by I_1, taking the ℓn of both sides of the resulting equation, and combining terms yields the following expression for Θ:

$$\theta = \frac{\ell n \left(\frac{I_2}{I_1}\right)}{V_2 - V_1},$$ For $e^{\theta V} >> 1$, which is normally the case at these current levels

Substituting the values of V_1, V_2, I_1, I_2 from Fig. 3.3 yields:

$$\theta = \frac{\ell n \left(\frac{500 \times 10^{-6}}{10 \times 10^{-6}}\right)}{.619 - .464} = 25.239 \text{ VOLTS}^{-1}$$

It is necessary that the values chosen for I_1 and I_2 in obtaining θ are chosen in the microamp region to eliminate diode voltage drop due to bulk resistance.

3.2.5 IS, Diode Saturation Current

Saturation current is determined by substituting the value of voltage and current measured at some point on the straight-line portion of the forward V-I characteristic (Fig. 3.3) into the diode equation. This point is chosen in the low microamp region for best results. From the point $(V_1, I_1) = (.464v, 10\mu a)$, IS is calculated:

$$IS = \frac{I_1}{e^{\theta \cdot V_1} - 1} = \frac{10 \times 10^{-6}}{e^{(25.239)(.464)} - 1}$$

$$IS = 8.2045 \times 10^{-11} \text{ amps}$$

or

$$IS = 8.2045 \times 10^{-8} \text{ ma}$$

To obtain a check on the validity of IS and θ, the diode equation is re-arranged to the form:

$$V = \frac{1}{\theta} \ell n \left(\frac{I}{IS} + 1 \right)$$

Substituting in I = 1μa:

$$V = \frac{1}{25.239} \ell n \left(\frac{1 \times 10^{-6}}{8.2045 \times 10^{-11}} + 1 \right) = .3727\text{v}$$

The measured value at I = 1μa is, V = .373v.

Subsitituting in I = 500μa:

$$V = \frac{1}{25.239} \ell n \left(\frac{500 \times 10^{-6}}{8.2045 \times 10^{-11}} + 1 \right) = .619\text{v}$$

The measured value at I = 500μa is V = .619v

Other points may be checked.

3.2.6 RB, Bulk Resistance

A plot of the diode equation, I = IS ($e^{\theta \cdot V} - 1$), continues as a straight line if plotted on semi-log coordinates. The physical diode, however, has a series bulk resistance. At high forward currents, the voltage drop due to the bulk resistance becomes significant, resulting in deviation of the diode characteristic from the straight line expected on the semi-log plot. This effect is also seen in Fig. 3.3. The bulk resistance, RB, is calculated from:

$$RB = \frac{\Delta V}{I}$$

Where ΔV is the difference between the measured and the calculated value. For the FD300, the voltage drop due to RB becomes significant at 300ma. The value of V predicted from the diode equation at 300 ma is calculated from:

$$V = \frac{1}{\theta} \ell n \left(\frac{I}{IS} + 1 \right) = \frac{1}{25.239} \ell n \left(\frac{300 \times 10^{-3}}{8.2045 \times 10^{-11}} + 1 \right) = .872\text{v}$$

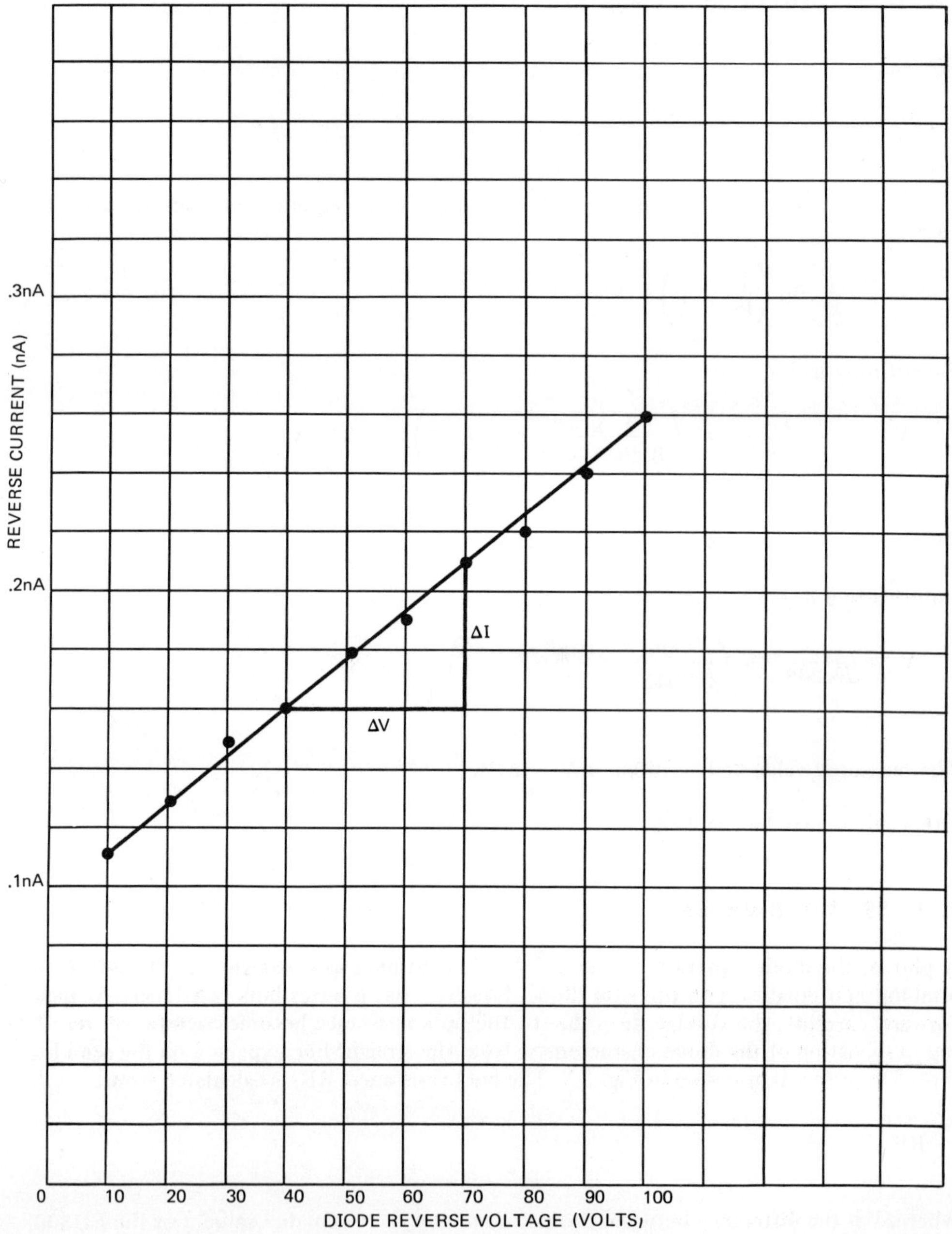

Fig. 3.5. Diode leakage resistance curve.

The value of V measured at I = 300 ma is .963v.

Therefore, $\Delta V\big|_{I = 300\ ma} = .963v - .872v = .091v$

Calculating RB:

$$RB = \frac{\Delta V}{I} = \frac{.091_v}{.3a} = .303\Omega$$

or

$$RB = .000303 \text{ kilohm}$$

3.2.7 RS, Leakage Resistance

Leakage resistance, RS, is determined from the slope of the reverse biased diode leakage current versus the reverse voltage plot shown in Fig. 3.5. The test circuit used to obtain this data is shown in Fig. 3.6.

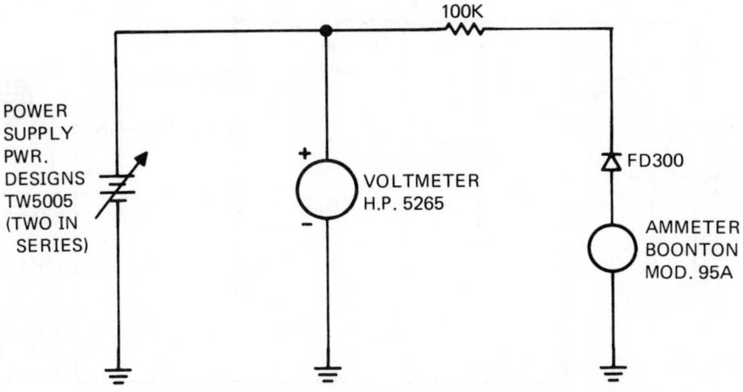

Fig. 3.6 Reverse leakage current test configuration

The reverse voltage is applied using the variable power supply and the leakage current measured for several values of reverse voltage. The data obtained is plotted in Fig. 3.5. Care must be exercised not to exceed the device breakdown voltage.

From Fig. 3.5, using a straight-line approximation to the curve, RS is calculated from:

$$RS = \frac{\Delta V \text{ reverse}}{\Delta I \text{ reverse}} = \frac{70v - 40v}{.21na - .16na} = \frac{30v}{.05 \times 10^{-9}a}$$

$$RS = 600 \times 10^9\Omega$$

or,

$$RS = 6 \times 10^8 \text{ Kilohms.}$$

3.2.8 KD, Diffusion Capacitance Constant

The diffusion capacitance constant is obtained from diode storage time measurements. The test configuration shown in Fig. 3.7 is used to determine the diode storage time.

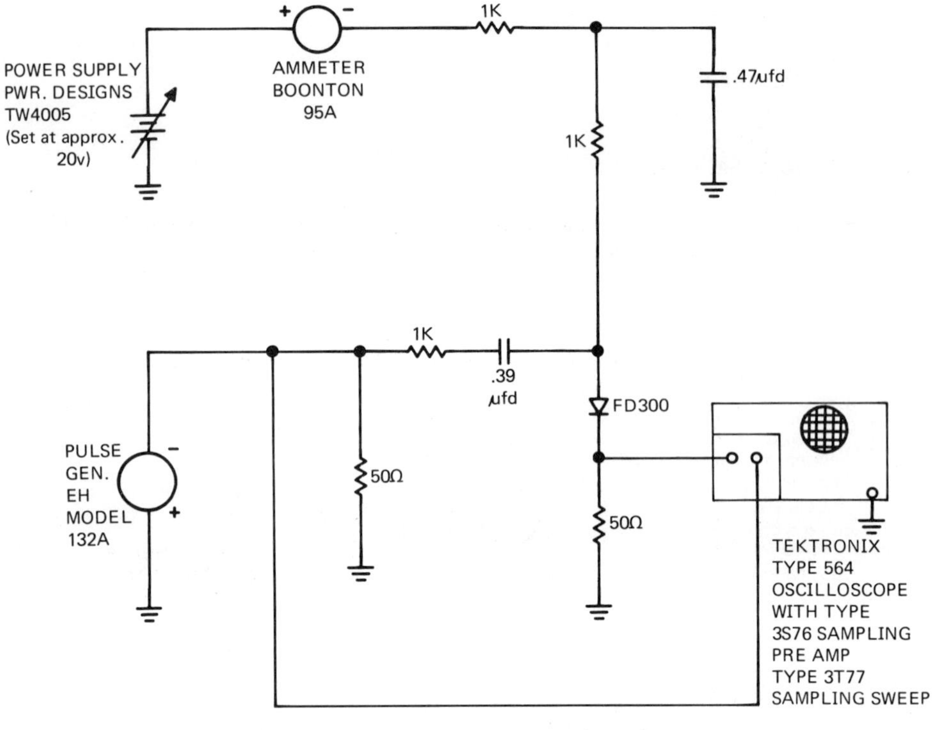

Fig. 3.7 Diode storage time test configuration

The power supply is adjusted to obtain the desired forward test current. A current of 10 ma is used (arbitrarily) for this device. The pulse generator is then adjusted to obtain the desired reverse recovery current of 10ma current that is again used for the FD300 (arbitrarily). The forward and reverse currents through the diode are seen in the top trace of Fig. 3.8. The horizontal axis is 100 ns/cm and the vertical axis is 500 mv/cm, which corresponds to 10 ma/cm. The storage time is the time from the beginning of the reverse current transition to the time when the reverse current begins to decay toward its D.C. value (nanoamps). For the FD300 with $I_F = 10$ ma, $I_R = 10$ ma, the storage time as determined from the top trace of Fig. 3.8 is 420 ns. The bottom trace of Fig. 3.8 is the pulse generator input with the vertical axis equal to 10v/cm and the horizontal axis equal to 100 ns/cm.

Knowing the storage time, the frequency parameter, F, required in the equation for KD, is calculated:

$$F = \frac{1}{2\pi} * \frac{\ln\left(1 + \frac{I_F}{I_R}\right)}{T_S}$$

Where:

I_F = DIODE FORWARD CURRENT = 10 ma
I_R = DIODE REVERSE CURRENT = 10 ma
T_S = STORAGE TIME = 420 ns

Substituting:

$$F = \frac{1}{2\pi} \frac{\ln\left(1 + \frac{10 \times 10^{-3}}{10 \times 10^{-3}}\right)}{420 \times 10^{-9}} = 2.627 \times 10^{5} \text{ HZ}$$

or

$$F = 2.627 \times 10^{-4} \text{GHZ}$$

KD, the diffusion capacitance constant, is now calculated from:

$$KD = \frac{\theta}{2\pi F} = 1.529 \times 10^{-4} \text{ pfd/ma}$$

TOP TRACE
VOLTAGE ACROSS 50Ω RESISTOR
(IN SERIES WITH DIODE)

VERTICAL 500mV/cm
HORIZONTAL 100nS/cm

BOTTOM TRACE
INPUT VOLTAGE

VERTICAL 10V/cm
HORIZONTAL 100nS/cm

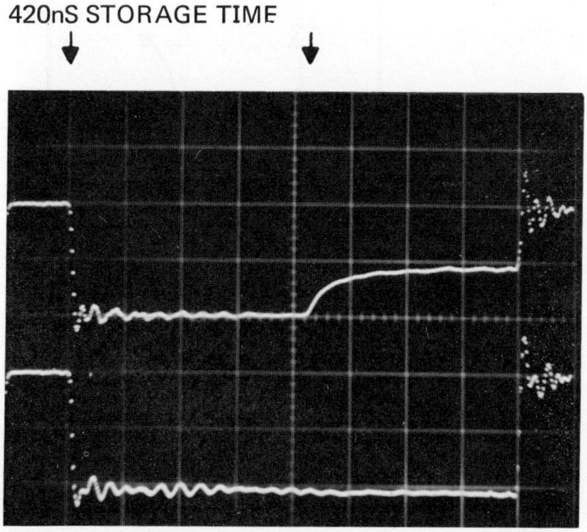

Fig. 3.8.
Diode storage time waveforms.

3.2.9 CO, Ø, n Transition Capacitance Equation Constants

The transition capacitance constants are obtained from measurements of diode transition capacitance as a function of reverse junction voltage. The test configuration used in this measurement is shown in Fig. 3.9.

The device is mounted between the high and low terminals of the bridge. The internal bias voltage is adjusted to the desired value and the capacitance measured. Transition capacitance for the FD300 is obtained for several values of reverse voltage. The resulting data is shown tabulated and plotted in Fig. 3.10. To determine the capacitance equation constants consider two points, (V_1, CJ_1) (V_2, CJ_2), on the measured transition capacitance versus reverse voltage plot in Fig. 3.10. From the transition capacitance equation:

$$CJ_1 = \frac{CO}{(\phi - V_1)^n}$$

$$CJ_2 = \frac{CO}{(\phi - V_2)^n}$$

Where V_1, V_2 are the reverse voltages corresponding to CJ_1, CJ_2. V_1 and V_2 are negative quantities.

Dividing the first equation by the second and taking the ℓn of both sides of the resulting expression eliminates CO, yielding:

$$\ell n\, CJ_1 - \ell n\, CJ_2 = n\, \ell n(\phi - V_2) - n\, \ell n(\phi - V_1)$$

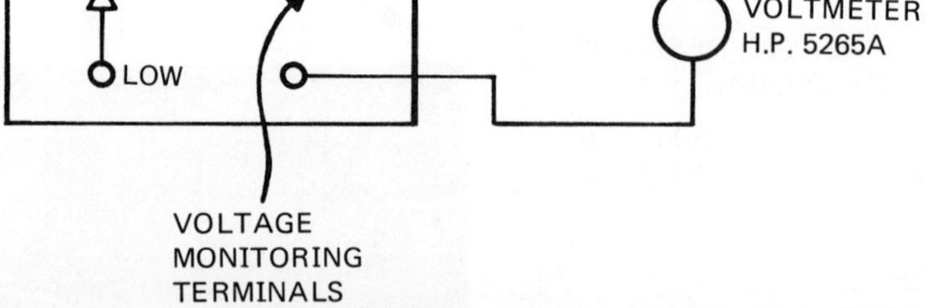

Fig. 3.9 Transition capacitance test configuration

The value for ϕ falls in the range of .6v to 1v for most devices. A value is assumed and is verified from the resulting capacitance equation. ϕ is assumed to equal .9v. With $\phi = .9$v, the measured values $CJ_1 = 4.86$ pfd at $V_1 = 0$v and $CJ_2 = 2.39$ pfd at $V_2 = -3$v,

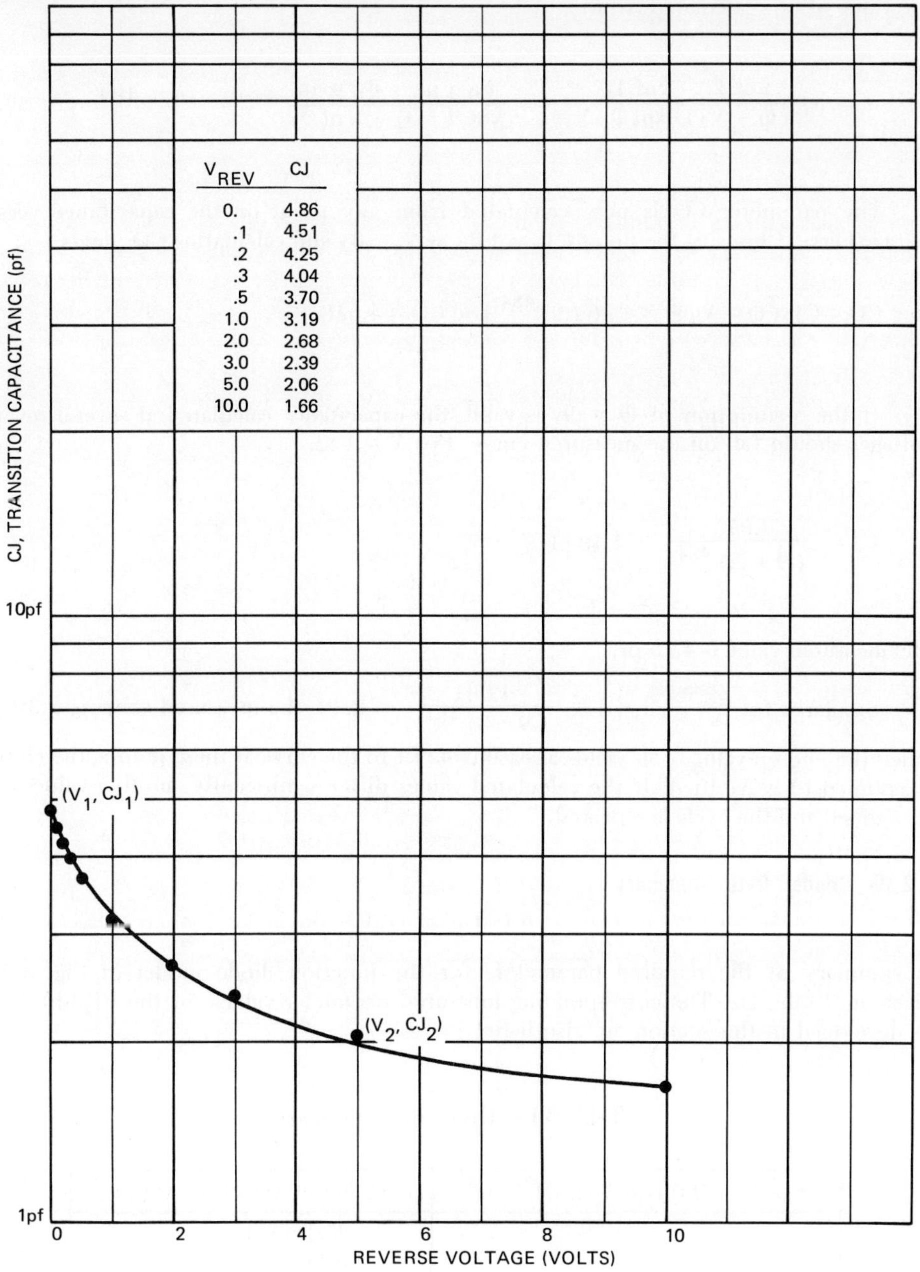

Fig. 3.10 Diode transition capacitance.

the value of n is calculated from:

$$n = \frac{\ell nCJ_1 - \ell nCJ_2}{\ell n(\phi - V_2) - \ell n(\phi - V_1)} = \frac{\ell n\, 4.86 - \ell n\, 2.39}{\ell n(.9+3) - \ell n(.9)}, \quad \text{or} \quad n = .484 .$$

The parameter CO is now calculated from any point on the capacitance versus voltage curve. Choosing the point $CJ_1 = 4.86$ at $V_1 = 0v$ and calculating CO yields:

$$CO = CJ_1(\phi - V_1)^n = 4.86\,(.9)^{.484} \text{ and } CO = 4.618$$

If the assumption of $\phi = .9v$ is valid, the capacitance calculated at several reverse voltages should fall on the measured curve. For $V = -.2$:

$$CJ = \frac{4.618}{(.9+.2)^{.484}} = 4.40 \text{ pf}$$

The measured value is 4.25 pf.

Similarly for $V = -3$, $CJ = \dfrac{4.681}{(.9+3)^{.484}} = 2.38$, the measured value is 2.39 pf.

Since the chosen value of ϕ yields a reasonable fit to the curve at these points, the choice is assumed to be verified. If the calculated values differ significantly, another value of ϕ is assumed and the cycle is repeated.

3.2.10 Diode Data Summary

A summary of the required parameters for the junction diode model of Fig. 3.2 is given in Table 3.1. The corresponding measured parameter values for the FD300 diode as developed in this section are also listed.

Table 3.1 Diode data summary

REQUIRED DIODE PARAMETERS	FD300 MEASURED VALUES
Θ	25.239 Volts^{-1}
IS	8.2045×10^{-8} ma
RB	.000303 Kilohms
RS	6×10^8 Kilohms
KD	1.529×10^4 pfd/ma
n	.484
ϕ	.9Volts
CO	4.618

3.2.11 SCEPTRE Diode Storage Time Test Circuit Example

In order to test the validity of the diode model, the test circuit shown in Fig. 3.7 is put into SCEPTRE form as shown in Fig. 3.11.

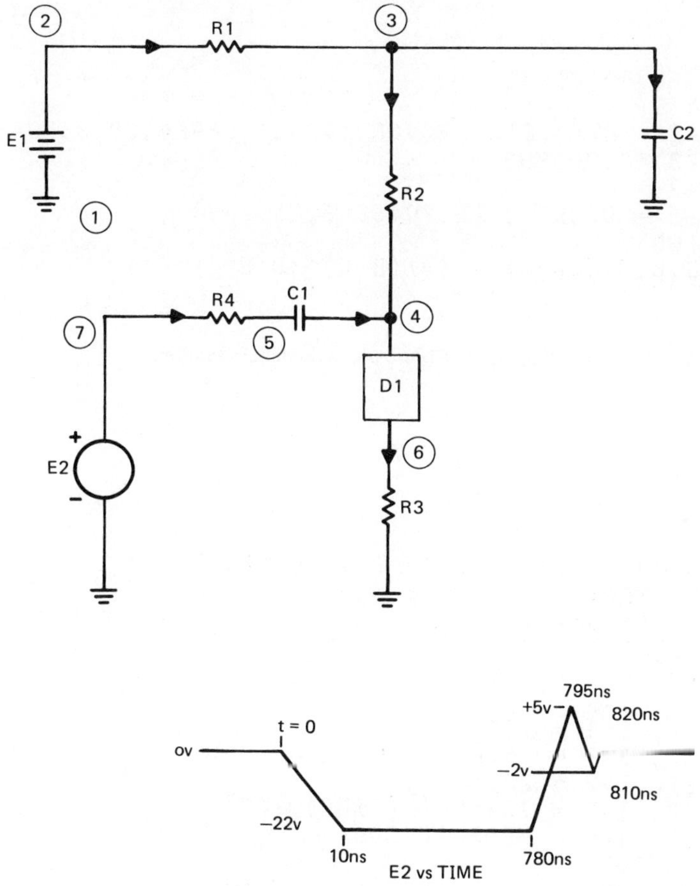

Fig. 3.11. Storage time test circuit (SCEPTRE form).

The input voltage E_2 is a tabular description of the wave form observed at the output of the pulse generator (see Fig. 3.8 bottom trace). SCEPTRE listings of the FD300 model and the test circuit are shown in Figs. 3.12 and 3.13 respectively. Outputs from the SCEPTRE run are shown in Figs. 3.14 and 3.15 and correspond to the input voltage E_2, and diode current IRBDI. Diode current IRBDI of Fig. 3.15 is seen to compare quite favorably with the wave form of Fig. 3.8a over the range 0 to 420 ns. The storage time is almost identical to that measured in the laboratory as shown in Fig. 3.8 (top trace). The decay of diode current which occurs in the actual test circuit is not adequately described by the model. However, experimental result exhibits a considerably longer delay time which is due to charge stored in remote regions of the device. This effect is usually negligible in low-current, high-speed devices in which the decay time is primarily due to transition capacitance and external circuit resistance.

```
MODEL FD300 (1-3)
ELEMENTS
CD, 1-2 = Q1(4.618,.9,VCD,.484,1.529E4,JD,8.2045E-8)
RB, 2-3 = .000303
RS, 1-2 = 6E8
JD, 1-2 = DIODE Q (8.2045E-8,25.239)
FUNCTIONS
Q1(A,B,C,D,E,F,G) = (A/(B-C)**D+E*(F+G))
```

Fig. 3.12. FD300 diode model description.

```
DIODE STORAGE TIME TEST CIRCUIT
ELEMENTS
R1, 2-3 = 1
R2, 3-4 = 1
R3, 6-1 = .05
R4, 7-5 = 1
C1, 5-4 = .39E6
C2, 3-1 = .47E6
E1, 1-2 = 20.5
E2, 1-7 = TABLE 1 (TIME)
D1, 4-6 = MODEL FD300
OUTPUTS
E2,IRBD1,VCDD1,JDD1,CDD1,PLOT
FUNCTIONS
TABLE 1
0,0,10,-22,780,-22,795,5,810,-2,820,0,850,0
RUN CONTROLS
RUN INITIAL CONDITIONS
STOP TIME = 850.
END
```

Fig. 3.13. Diode storage time test circuit description.

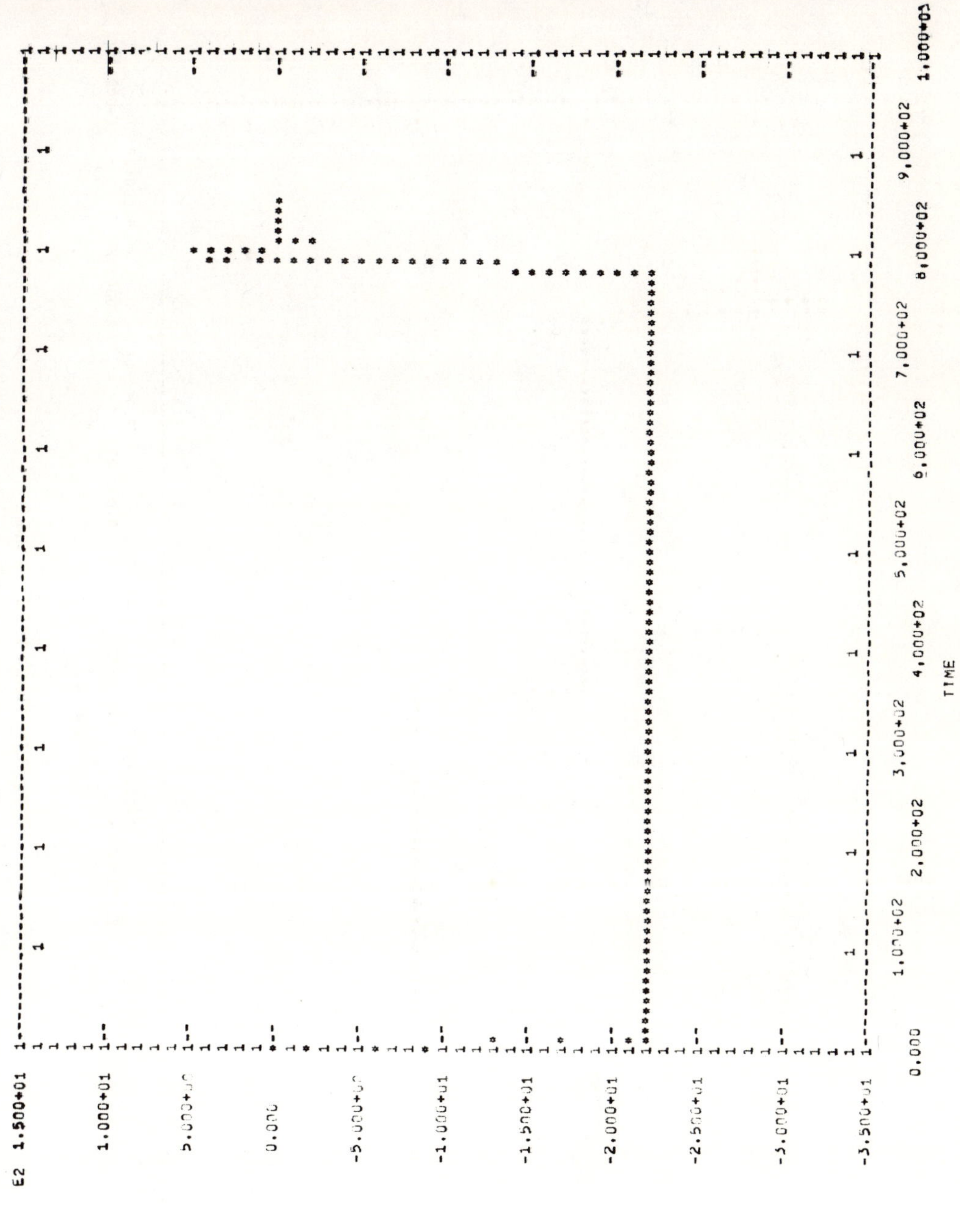

Fig. 3.14. Plot of E2 vs. time.

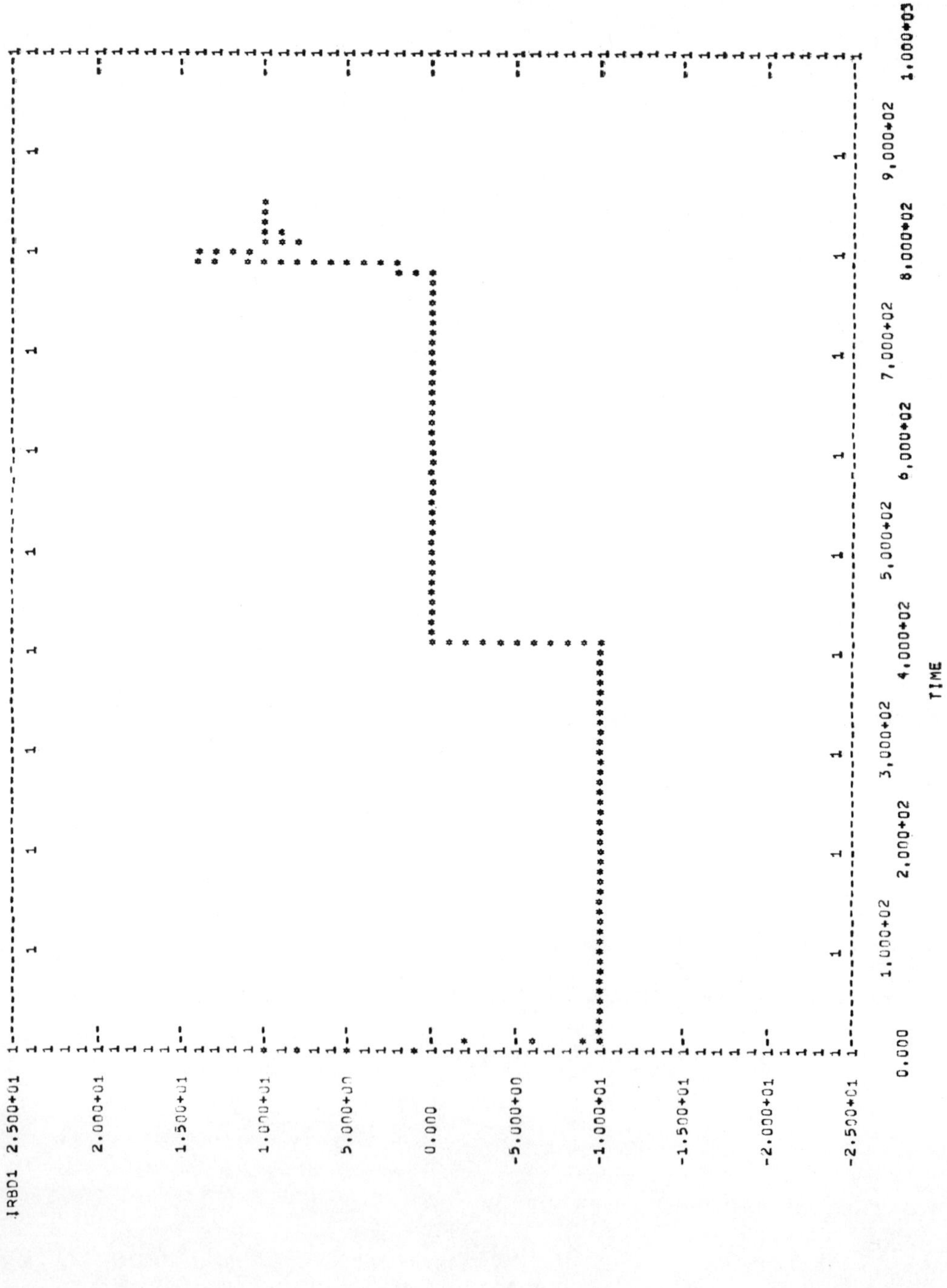

Fig. 3.15. Plot of 1RBD1 vs. time.

3.3 SCEPTRE BIPOLAR TRANSISTOR MODEL (NPN)[1,2,3,4]

Figure 3.16 illustrates the selected transistor model along with the defining equations.

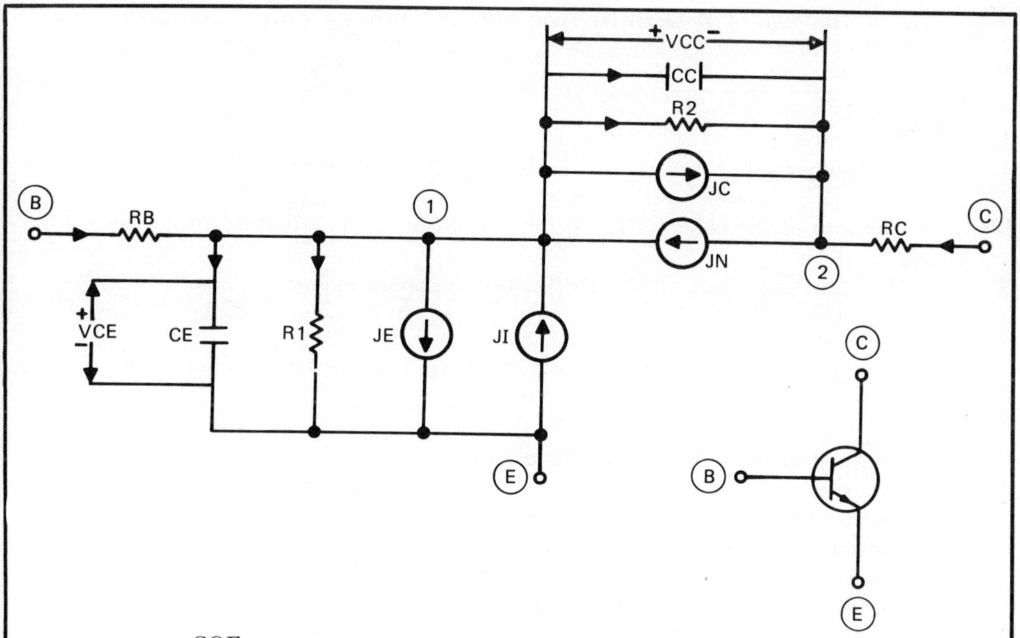

$$CE = \frac{COE}{[\phi_E - VCE]^{n_E}} + \theta_E T_E [JE + IES]$$

$$CC = \frac{COC}{[\phi_c - VCC]^{n_c}} + \theta_c T_s [JC + ICS]$$

$$JE = IES[e^{\theta_E \cdot VCE} - 1]$$

$$JC = ICS[e^{\theta_c \cdot VCC} - 1]$$

$$JN = \alpha_N JE$$

$$JI = \alpha_I JC$$

$$\alpha_N = f[JE]$$

$$\alpha_I = f[JC]$$

$$T_E = f[JE]$$

$$T_s = f[JC]$$

Fig. 3.16. SCEPTRE transistor model and defining equations.

3.3.1 Definition of Terms

All of the terms necessary for the transistor derivation are defined as follows:

CE = The sum of the emitter transition and diffusion capacitances (pfd) where:

$$\frac{COE}{[\phi_E - VCE]^{n_E}} = \text{Emitter Transition Capacitance}$$

and, $\Theta_E T_E [JE + IES]$ = Emitter Diffusion Capacitance

CC = The sum of the collector transition and diffusion capacitances (pfd) where:

$$\frac{COC}{[\phi_c - VCC]^{n_c}} = \text{Collector Transition Capacitance}$$

and, $\Theta_c T_S [JC + ICS]$ = Collector Diffusion Capacitance

COE = Constant of the emitter transition capacitance equation.
COC = Constant of the collector transition capacitance equation.
ϕ_E = Emitter base junction contact potential (volts).
ϕ_c = Collector base junction contact potential (volts).
JE = Current generator representing the emitter base junction.
JC = Current generator representing the collector base junction.
n_E = Emitter junction grading constant.
n_c = Collector junction grading constant.
JN = Current generator dependent on the emitter base junction current.
JI = Current generator dependent on the collector base junction current.
IES = Emitter base saturation current measured in the active region (ma).
ICS = Collector base saturation current measured in the active region (ma).
Θ_E = Constant of the emitter base junction equation (volts^{-1}).
Θ_c = Constant of the collector base junction equation (volts^{-1}).
T_E = Time constant of the emitter diffusion capacitance equation.

$$T_E = \frac{1}{2\pi F_N} \text{ (nsec)}$$

T_S = Time constant of the collector diffusion capacitance equation.

$$T_S = \frac{1}{2\pi F_I} = \alpha_I \, TCI$$

F_N = The average f_T normal.
F_I = The average f_T inverse.
TCI = Charge control parameter related to storage time.
α_N = Forward current gain = $\frac{I_c}{I_E}$
This parameter is entered as a function of JE in the SCEPTRE model.
α_I = Inverse current gain = $\frac{I_E}{I_c}$
This parameter is entered as a function of JC in the SCEPTRE model.

RB = Base bulk Resistance.
RC = Collector bulk resistance.
R1 = Emitter base junction leakage resistance.
R2 = Collector base junction leakage resistance.

3.3.2 Introduction

Parameters are determined in this section for the 2N2369 transistor. As in the diode model, D.C. measurement techniques are utilized only for low current measurements. Pulse measurements are employed for current levels greater than 10 ma. For the bipolar transistor model, four parameters:

$$\alpha_N, \alpha_I, T_E, T_s$$

are entered as functions of current. Considerable errors occur if these parameters are assumed constant when the device is operated over a wide current range.

The effective emitter bulk resistance is not included in this model. This simplification may lead to errors in saturation voltage. For most applications however, the magnitude of this error is not significant.

3.3.3 Model Parameter Determination for the 2N2369 Transistor

The measurement technique for each transistor parameter is now discussed in detail.

3.3.4 Θ_E, Emitter Base Junction Constant

Θ_E is determined from active region measurements of the base-emitter junction voltage as a function of emitter current. The test configuration used in obtaining the base emitter characteristic is shown in Fig. 3.17. This configuration is used for currents up to 10 ma.

The desired current level is set up on the front panel dials of the current source. The source is switched to the operate mode and the resulting base-emitter voltage is read from the digital voltmeter. The base-emitter is measured for $I_E = 1\mu a$ to $I_E = 10$ ma. This data is tabulated and plotted in Fig. 3.18.

For current levels greater than 10 ma, the test configuration shown in Fig. 3.19 is used.

The emitter current level is set by adjusting the pulse generator amplitude and monitoring the current with the current probe. The duty cycle is 3%, minimizing heating. After the current is adjusted to the desired amplitude, the base-emitter voltage is measured using the comparator on the type W plug-in. As in the junction diode measurement, the pulse measurements begin at $I_E = 5$ ma, to overlap the D.C. measurements. V_{BE} is measured using the pulse methods for $I_E = 5$ ma to $I_E = 100$ ma. The results are tabulated and plotted in Fig. 3.18.

Using the same equation as in the junction diode measurement, a value for Θ_E is calculated from two points, on the straight line portion of the V_{BE} versus I_E plot in Fig. 3.18.

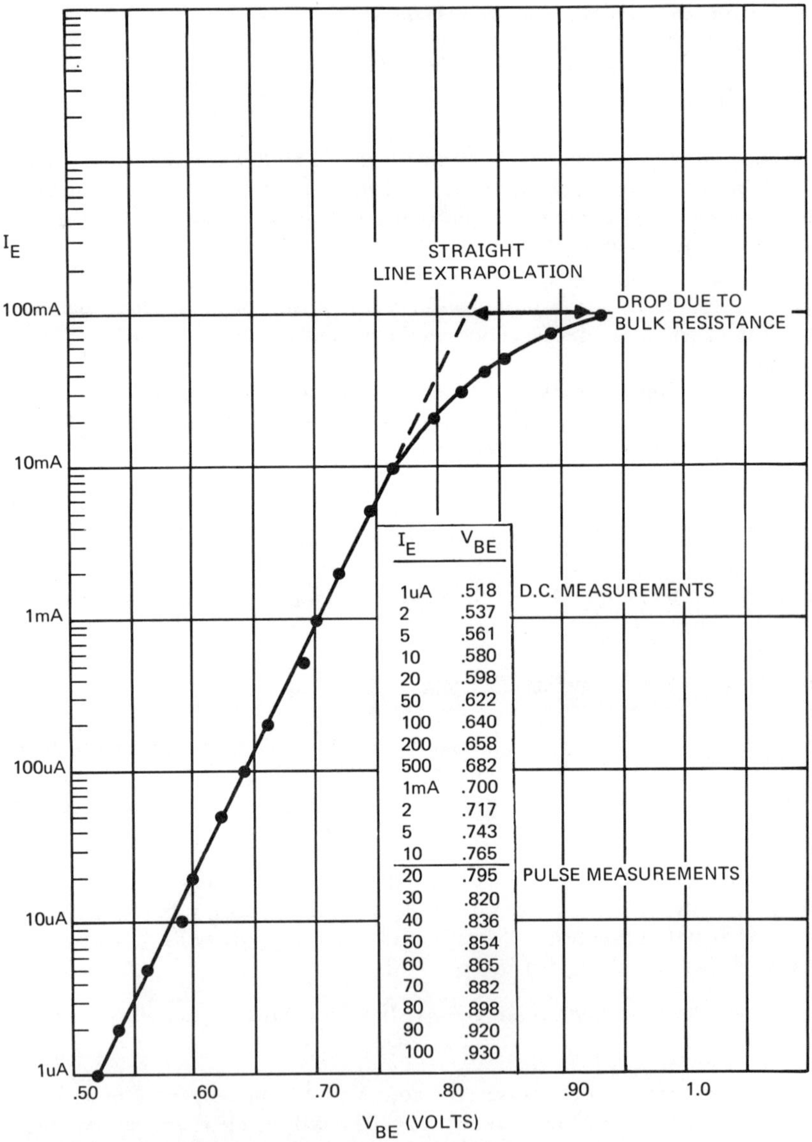

I_E	V_{BE}
1uA	.518
2	.537
5	.561
10	.580
20	.598
50	.622
100	.640
200	.658
500	.682
1mA	.700
2	.717
5	.743
10	.765
20	.795
30	.820
40	.836
50	.854
60	.865
70	.882
80	.898
90	.920
100	.930

Fig. 3.18 I_E vs. V_{BE} curve.

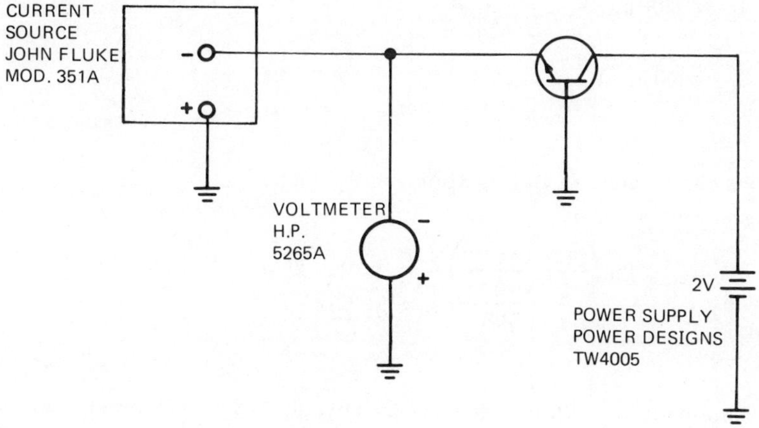

Fig. 3.17. Transistor base emitter characteristic test configuration for low currents.

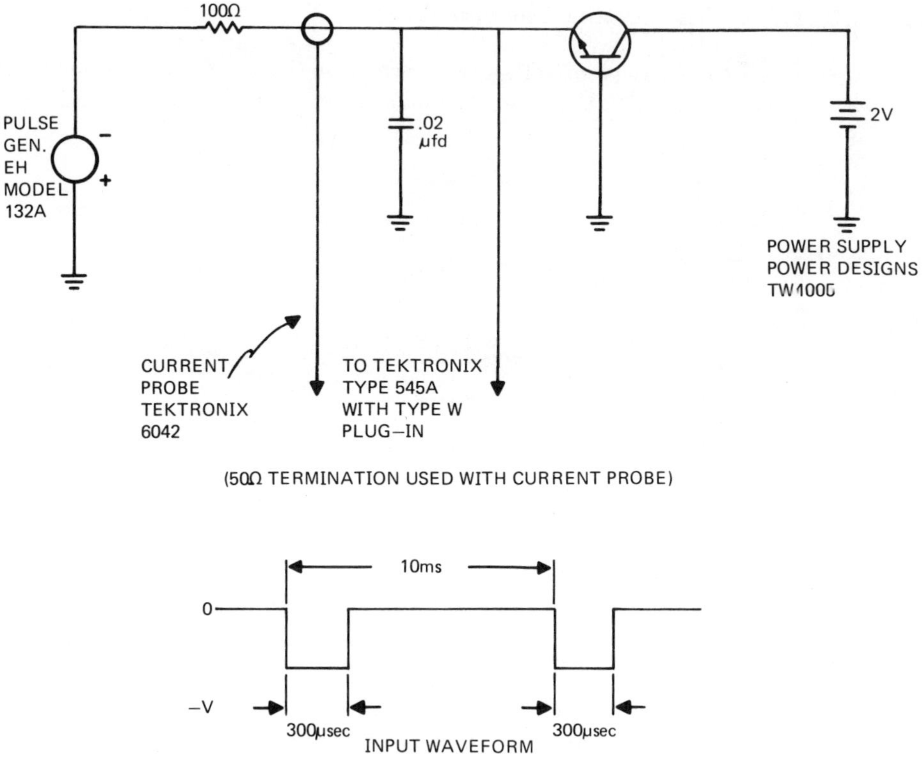

Fig. 3.19. Transistor base-emitter characteristic test configuration for high currents.

Choosing:

$$I_{E2} = 200\mu a, \ V_{BE_2} = .658$$

$$I_{E1} = 20\mu a, \ V_{BE_1} = .598$$

Substituting these values in the equation for θ_E yields:

$$\theta_E = \frac{\ln\left(\frac{200 \times 10^{-6}}{20 \times 10^{-6}}\right)}{.658 - .598} = 38.376$$

Low collector-emitter voltages, typically 2v, are utilized in determining the D.C. parameters for the transistor model. The reason for utilizing the low supply voltages is to obtain the best possible fit to the Ebers-Moll Equations which do not take into account variations in collector current with collector-emitter voltage.

3.3 5 IES, Emitter Base Junction Saturation Current

IES is determined from a value of voltage and current on the straight-line portion of the V_{BE} versus I_E curve, Fig. 3.18. Using the equation:

$$IE = IES\,(e^{\theta E\, V BE} - 1)$$

then,
$$IES = \frac{I_E}{e^{\theta E\, V BE} - 1}$$

Choosing the point $I_E = 20\mu a$, $V_{BE} = .598v$:

$$IES = \frac{20 \times 10^{-6}}{e^{(38.376)(.598)} - 1} = 2.1596 \times 10^{-15} a,$$

or,

$$IES = 2.1596 \quad 10^{-12} ma.$$

3.3.6 θc, Collector Base Junction Constant

θc is determined from active region measurements of the collector base junction voltage as a function of collector current. The test configuration used in obtaining the collector-base characteristies is shown in Fig. 3.20.

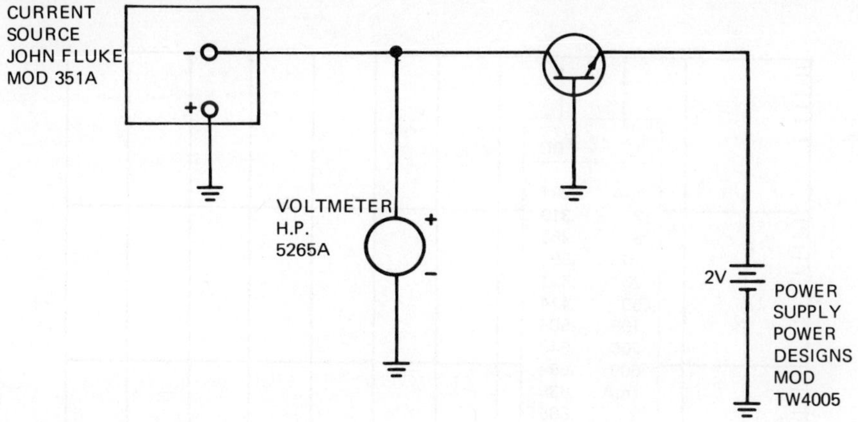

Fig. 3.20. Transistor base-collector characteristic test configuration for low currents.

The desired collector current level is set up on the front panel dials of the current source. The source is switched to operate and the resulting collector-base voltage is read from the digital voltmeter. The base-collector voltage is measured for Ic = 1 μa to Ic = 10 ma. The collector-base voltage is only measured from currents up to 10 ma for the 2N2369, since this junction is seldom forward biased past this level in normal applications. If higher current measurements are desired, the pulse measurement circuit in Fig. 3.19 is utilized with the collector and emitter leads interchanged. The results of the D.C., V_{BC} versus Ic measurements are shown in Fig. 3.21.

Using the same equation as in the junction diode measurements, a value for Θc is calculated from two points on the linear portion of the V_{BC} versus Ic plot in Fig. 3.21.

Choosing:

$I_{C2} = 200\ \mu a,\ V_{BC2} = .543$

and, $I_{C1} = 20\ \mu a,\ V_{BC1} = .431.$

Substituting these values in the equation for Θc yields:

$$\Theta c = \frac{\ln\left(\dfrac{200 \times 10^{-6}}{20 \times 10^{-6}}\right)}{.543 - .431} = 20.559$$

3.3.7 ICS Collector-Base Junction Saturation Current

ICS is determined from a value of voltage and current on the straight-line portion of the V_{BC} versus Ic curve, Fig. 3.21. Using the equation:

$$IC = ICS\,(e^{\Theta_C V_{BC}} - 1)$$

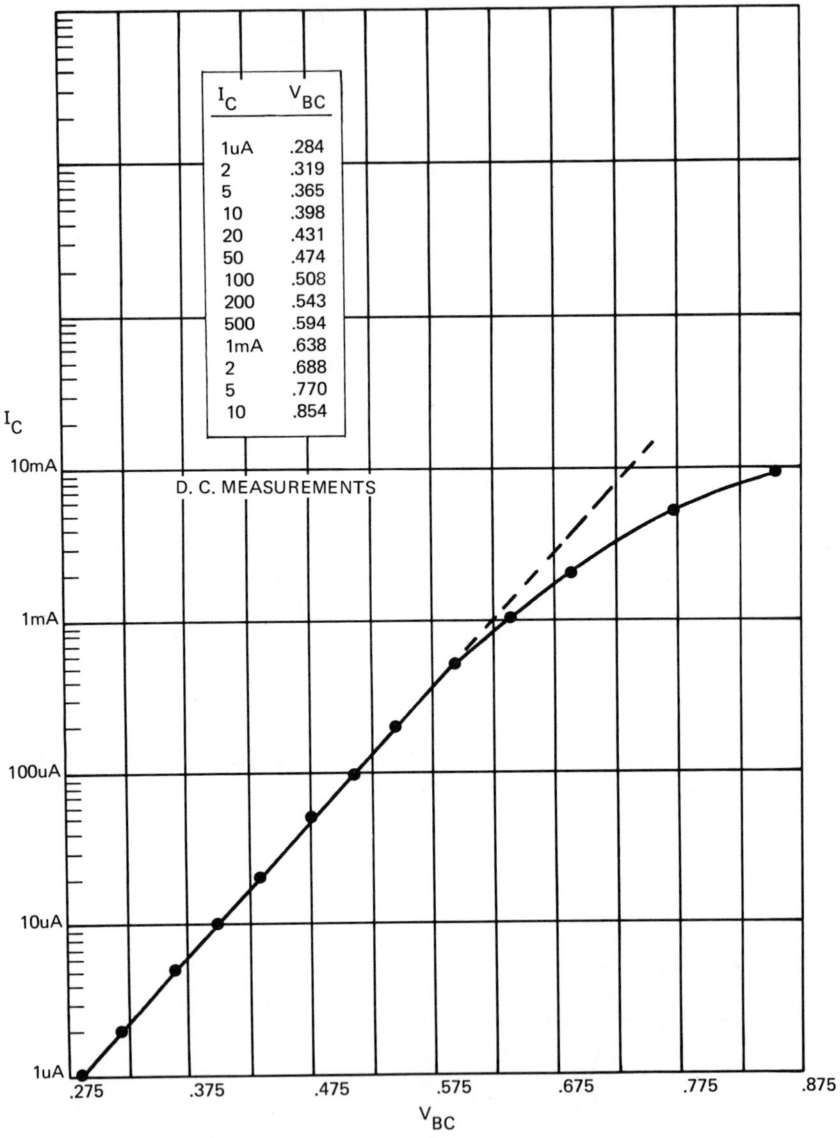

Fig. 3.21 I_C vs. V_{BC} curve.

then,

$$ICS = \frac{IC}{e^{\theta c\ V_{BC}} - 1}$$

Choosing the point Ic = 20 μa
$V_{BC} = .431$

$$ICS = \frac{20 \times 10^{-6}}{e^{(20.599)(.431)} - 1} = 2.8367 \quad 10^{-9} a.$$

or,

$$ICS = 2.8367 \quad 10^{-6} \text{ ma.}$$

3.3.8 α_N, Forward Current Gain

α_N is determined as a function of emitter current in the active region. To obtain α_N for low current levels, the test circuit of Fig. 3.22 is utilized.

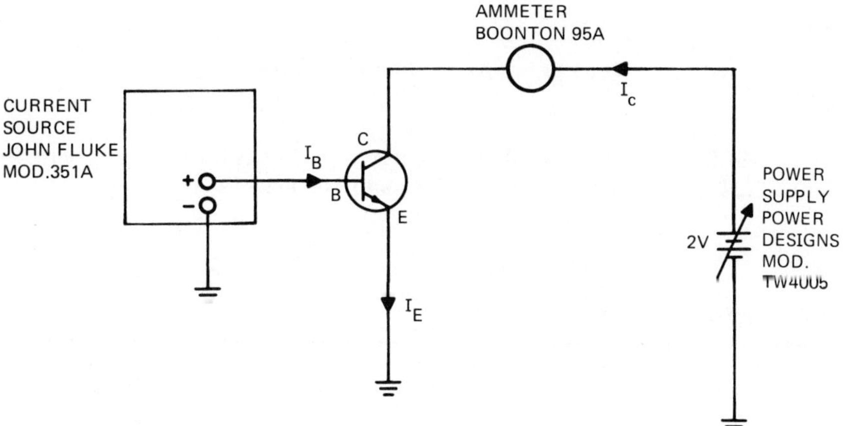

Fig. 3.22. α_N test circuit for low current levels.

The base current source is adjusted to obtain specific values of collector current from 50 μa to 10 ma. The base current corresponding to each value of collector current is recorded. Knowing the collector and base currents, the emitter current is calculated from $I_E = I_c + I_B$ [neglecting leakage currents]. α_N is then calculated from:

$$\alpha_N = \frac{I_C}{I_E}.$$

Values for I_C, I_B, and α_N are listed in Fig. 3.23, for the range $I_C = 50\ \mu a$ to $I_C = 10$ ma.

B_N is also calculated from $B_N = \frac{I_C}{I_B}$ and is utilized in rise time calculations. For collector current levels greater than 10 ma, the pulse test configuration of Fig. 3.25 is used.

The input pulse level is adjusted to obtain the desired collector current. The voltage V_A and V_B are then measured and the base current calculated from:

$$I_B = \frac{V_A - V_B}{IK}$$

IC	IB	$I_E = I_C + I_B$	$\alpha_N = \frac{I_C}{I_E}$	$B_N = \frac{I_C}{I_B}$
.05ma	.0014ma	.0514ma	.97276	35.71
.10	.0026	.1026	.97466	38.46
.20	.0047	.2047	.97704	42.55
.50	.0103	.5103	.97982	48.54
1.00	.0195	1.0195	.98087	51.28
2.00	.0353	2.0353	.98265	56.66
5.00	.0832	5.0832	.98363	60.10
10.00	.1650	10.1650	.98376	60.61

Fig. 3.23. α_N data for low current levels.

Knowing I_C and I_B, α_N is then calculated as before. The values for I_C, I_B, α_N and B_N are listed in Fig. 3.24, for collector currents of 20 ma to 100 ma.

I_C	V_A	V_B	I_B	$I_E = I_C + I_B$	α_N	B_N
20.00ma	1.120v	.795v	.325ma	20.325ma	.98400	61.53
30.00	1.330	.820	.510	30.510	.98328	58.82
40.00	1.535	.838	.697	40.697	.98287	57.38
50.00	1.775	.855	.920	50.920	.98193	54.34
60.00	1.980	.869	1.111	61.111	.98182	54.00
70.00	2.272	.886	1.386	71.386	.98058	50.50
80.00	2.550	.900	1.650	81.650	.97979	48.48
90.00	2.885	.922	1.963	91.963	.97865	45.84
100.00	3.215	.935	2.280	102.280	.97771	43.85

Fig. 3.24. α_N data for high current levels.

From the transistor model of Fig. 3.16, it is seen that $JE = I_E$ for the active region since the collector base diode is reversed [neglecting leakage current]. Each calculated value of α_N is then entered directly as a tabular function of JE. The value of JE at which a particular α_N is entered, is the value of emitter current, I_E, at which α_N is measured.

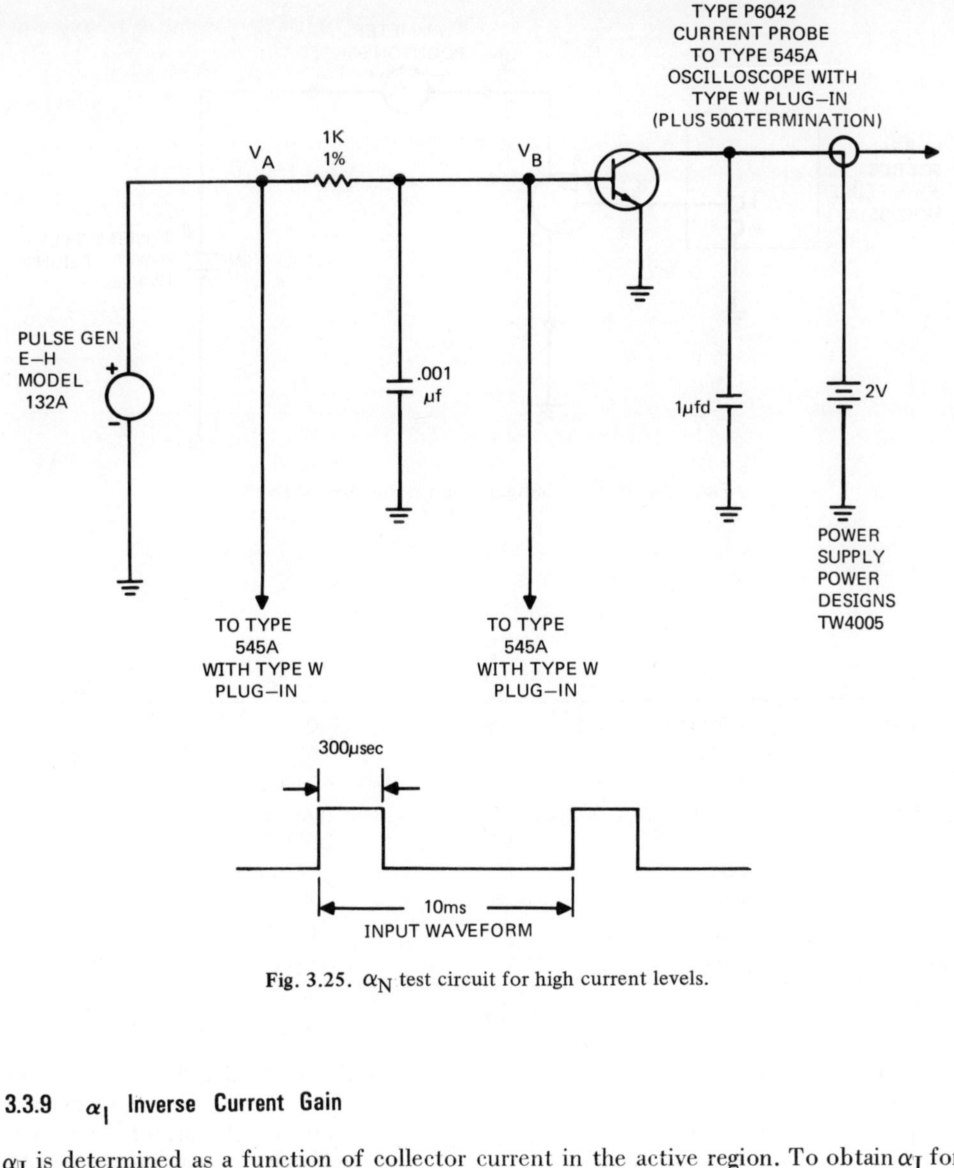

Fig. 3.25. α_N test circuit for high current levels.

3.3.9 α_I Inverse Current Gain

α_I is determined as a function of collector current in the active region. To obtain α_I for low current levels, the test circuit of Fig. 3.26 is utilized. The base current source is adjusted to obtain specific values of emitter current from 10μa to 2ma. The base current corresponding to each value of emitter current is recorded. Knowing the emitter and base currents, the collector is calculated from $I_C = I_E + I_B$ [neglecting leakage currents]. α_I is then calculated from:

$$\alpha_I = \frac{I_E}{I_C}$$

Values for I_E, I_B, and α_I are listed in Fig. 3.27 for the range I_E = 10μa to 2 ma. B_I is calculated from $B_I = \frac{I_E}{I_B}$.

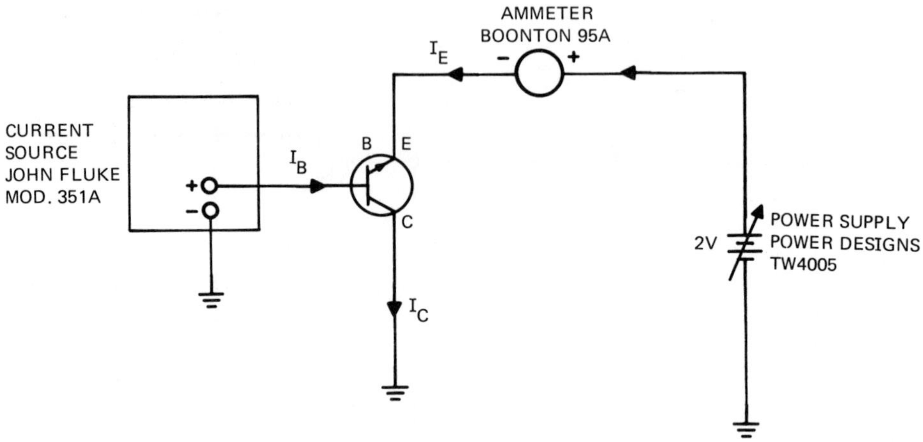

Fig. 3.26. α_I test configuration for low current levels.

I_E	I_B	$I_C = I_E + I_B$	$\alpha_I = \frac{I_E}{I_C}$	$B_I = \frac{I_E}{I_B}$
.010ma	.479ma	.489ma	.02045	.021
.020	.707	.727	.02751	.028
.050	1.20	1.25	.04000	.042
.100	1.82	1.92	.05208	.055
.200	2.75	2.95	.06779	.073
.300	3.58	3.88	.07731	.084
.500	4.88	5.38	.09293	.102
1.000	8.03	9.03	.11074	.125
2.000	13.0	15.0	.13333	.154

Fig. 3.27

α_I is not determined for higher current levels for this particular device since few applications require collector-base diode currents greater than 15 ma. This value occurs for $I_E = 2$ ma. For other devices requiring α_I measurements at high current levels, the test circuit of Fig. 3.24 is utilized with the collector and emitter leads interchanged.

From the transistor model of Fig. 3.16, it is seen that JC = IC for the active region, inverse connection, since the emitter base diode is reversed biased. Each calculated value of α_I is then entered directly as a tabular function of JC. The value of JC at which a particular α_I is entered is the value of collector current, Ic, at which α_I is measured.

3.3.10 n_E, COE, ϕ_E, Emitter Transition Capacitance Parameters

Transition capacitance parameters n_E, COE, and ϕ_E are obtained using the same techniques and test equipment utilized in the junction diode capacitance measurements. Fig. 3.28 shows the proper method of connecting the transistor to the capacitance bridge terminals.

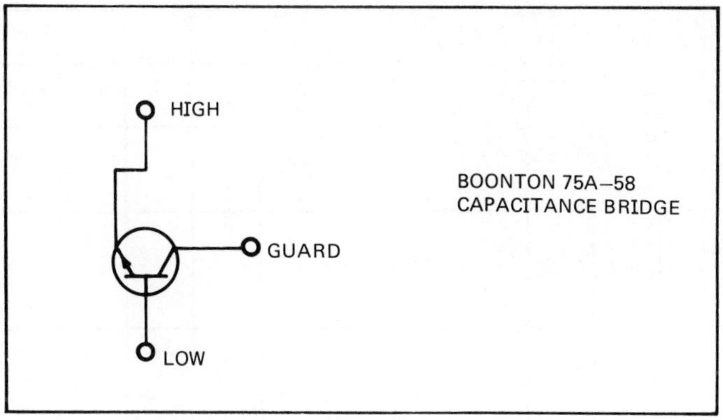

Fig. 3.28 Test configuration for C_{ib} measurement.

Tabular values of C_{ib} as a function of reverse base-emitter voltage appear in Fig. 3.29. The transition capacitance C_{Te} is obtained by subtracting the stray emitter-base capacitance, C_{se}, from each measured value of C_{ib}. C_{Te} is plotted in Fig. 3.29. C_{se} is obtained by destroying the internal connections between the transistor and case leads of another 2N2369 and measuring the capacitance on the bridge using the same test setup as shown in Fig. 3.28. Picking two points ($VR1$, C_{Te1}), ($VR2$, C_{Te2}) on the transition capacitance curve in Fig. 3.29, and assuming $\emptyset_E = .9v$, n_E is calculated:

$$n_E = \frac{\ell n\, C_{Te1} - \ell n\, C_{Te2}}{\ell n(\emptyset_E - VR_2) - \ell n(\emptyset_E - VR_1)} = \frac{\ell n\, 4.46 - \ell n\, 2.78}{\ell n(.9+2) - \ell n(.9)} = .3708$$

COE is obtained from:

$$COE = C_{Te}\,(\emptyset_E - VR)^{n_E}$$

Substituting the value of $C_{Te} = 4.46$ pf at $VR = 0v$, and $n_E = .3708$, COE is calculated:

$$COE = 4.46(.9)^{.3708} = 4.289.$$

Care must be taken not to apply more than a few volts reverse bias on the emitter base junction or breakdown will occur (typical breakdown voltage = 5v).

3.3.11 n_C, COC, $\emptyset_C$, Collector Transition Capacitance Parameters

Transition capacitance parameters n_C, COC, and $\emptyset_C$ are obtained using the same techniques and test equipment utilized in the junction diode capacitance measurements. Fig.

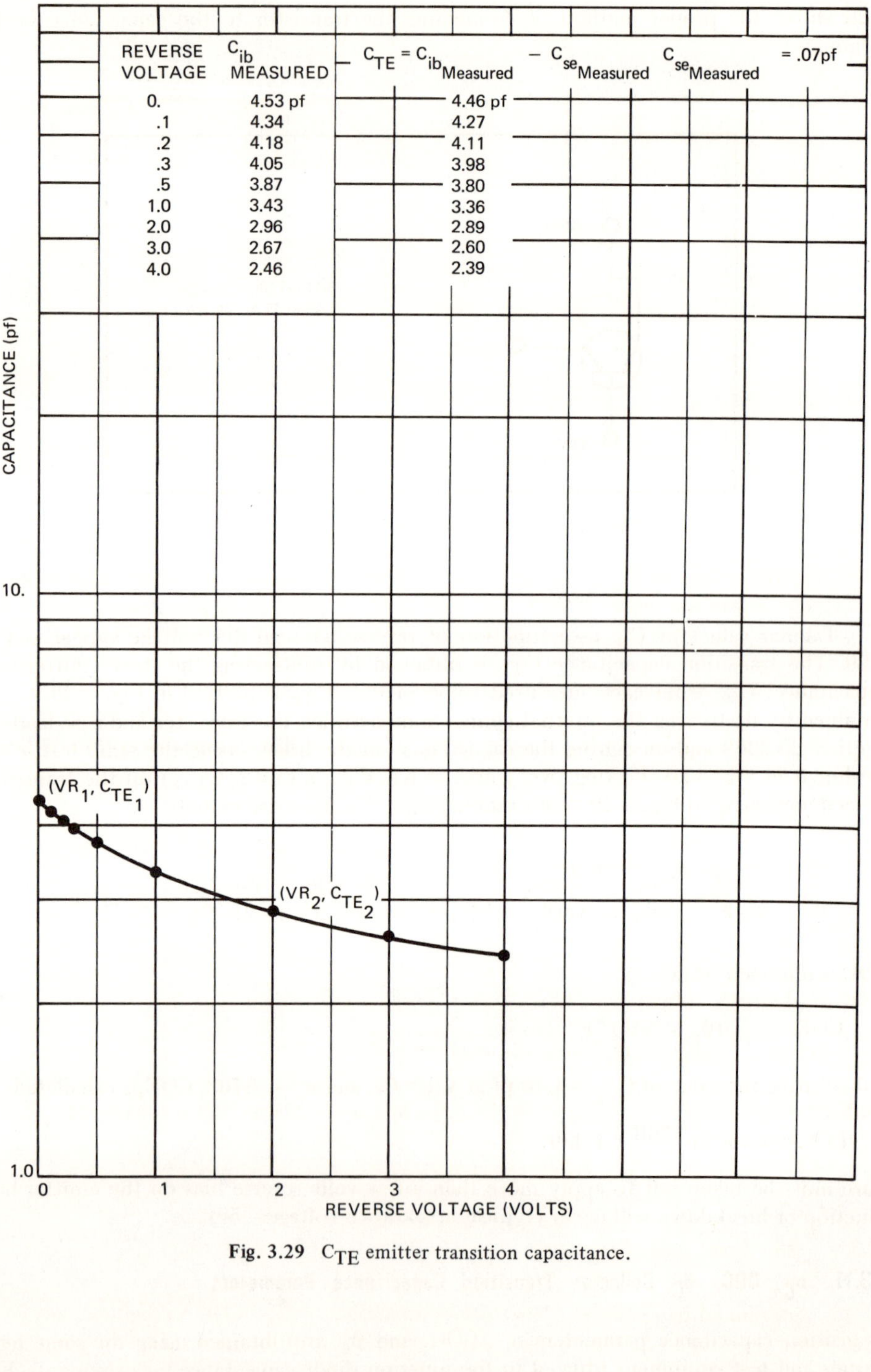

REVERSE VOLTAGE	C_{ib} MEASURED	C_{TE}
0.	4.53 pf	4.46 pf
.1	4.34	4.27
.2	4.18	4.11
.3	4.05	3.98
.5	3.87	3.80
1.0	3.43	3.36
2.0	2.96	2.89
3.0	2.67	2.60
4.0	2.46	2.39

Fig. 3.29 C_{TE} emitter transition capacitance.

3.30 shows the proper method of connecting the transistor to the capacitance bridge terminals.

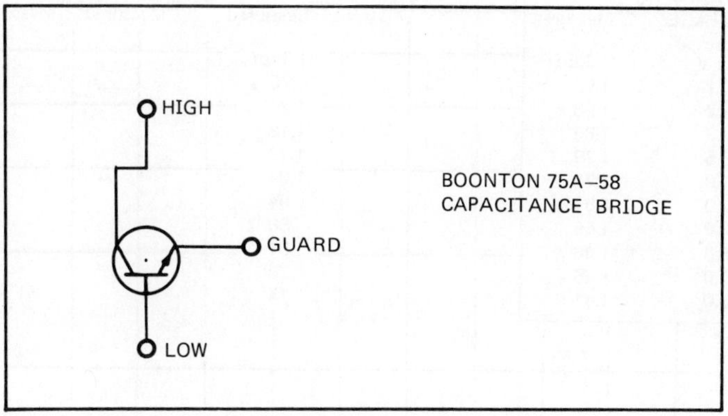

Fig. 3.30. Test configuration for C_{ob} measurement.

Tabular values of C_{ob} as a function of reverse base-collector voltage appear in Fig. 3.31. The transition capacitance C_{Tc} is obtained by subtracting the stray collector-base capacitance, C_{sc} from each measured value of C_{ob}. C_{Tc} is plotted in Fig. 3.31. C_{sc} is obtained by destroying the internal connections between the transistor and case leads of another 2N2369 and measuring this capacitance on the bridge.

Picking two points (VR_1, C_{Tc1}), (VR_2, C_{Tc2}) on the transition capacitance curve in Fig. 3.31, and assuming $\phi c = .9v$, n_c is calculated:

$$n_c = \frac{\ell n\, C_{Tc1} - \ell n\, C_{Tc2}}{\ell n(\phi_c - VR_2) - \ell n(\phi_c - VR_1)} = \frac{\ell n 1.31 - \ell n .81}{\ell n(5.9) - \ell n(.9)}$$

$$\therefore\ n_c = .2556$$

COC is obtained from:

$$COC = C_{Tc}\,(\phi c - VR)^{n_c}$$

Substituting the value $C_{Tc} = 1.31$ at $VR = 0v$, and $n_c = .2556$, COC is calculated:

$$COC = 1.31(.9)^{.2556} = 1.275$$

3.3.12 RB, Base Bulk Resistance

RB is determined from the V_{BE} versus I_E plot shown in Fig. 3.18. At high current levels, this curve deviates from the straight line extrapolation of the curve at low current levels,

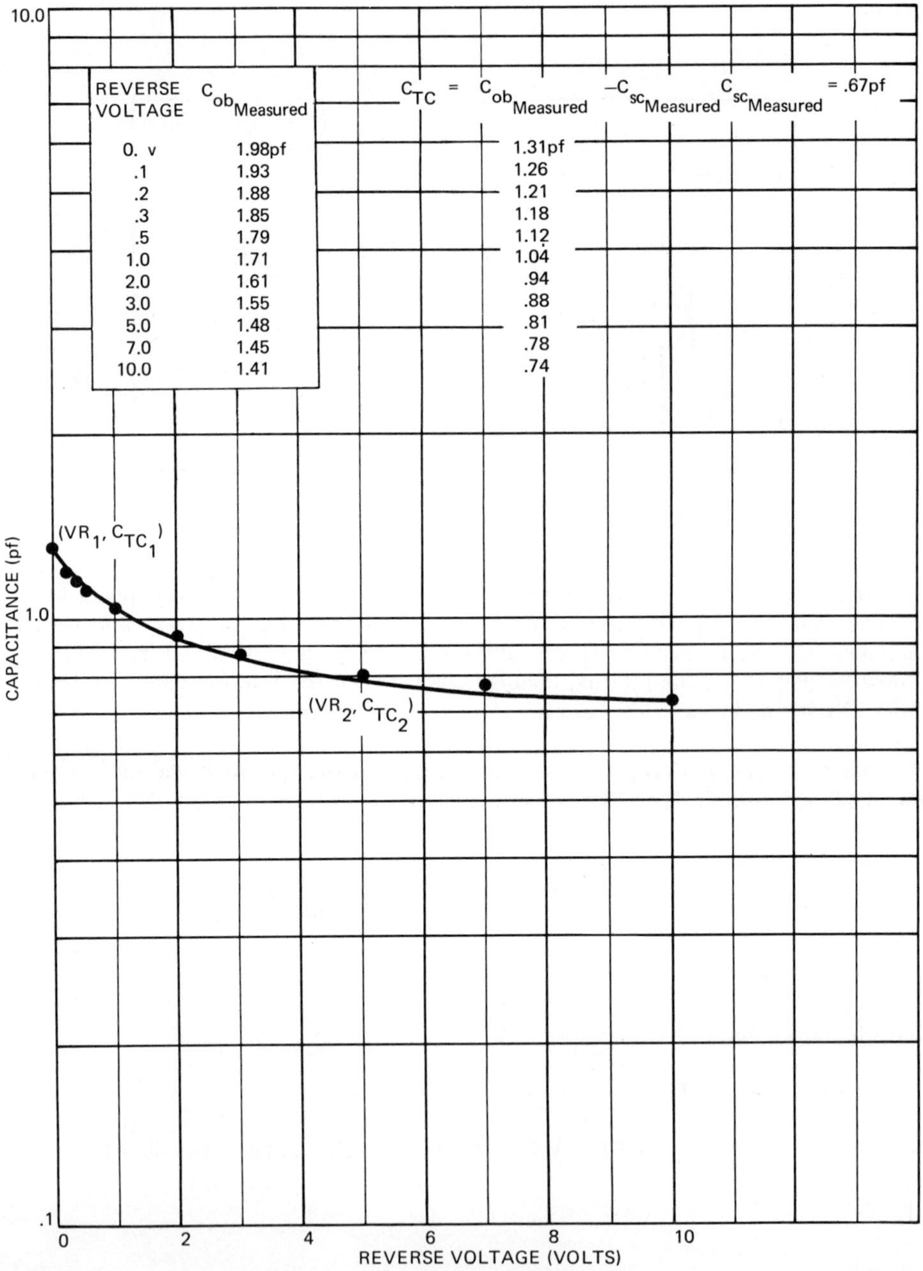

Fig. 3.31. C_{TC}, collector transition capacitance.

This deviation is due to the bulk resistance, RB. RB is calculated from:

$$RB = \frac{\Delta V}{I_B}$$

where,

I_B = The base current at which R_B is to be measured. $I_B = (1 - \alpha_N)I_E$.

and ΔV = The difference between the measured base-emitter drop and the value calculated from:

$$V_{BE} = \frac{1}{\Theta_E} \ln\left[\frac{I_E}{IES}\right]$$

A convenient point at which to obtain RB is the point I_E = 100 ma. At I_E = 100 ma, V_{BE} measured = .930v. I_B is calculated from I_E and the value of α_N at I_E = 100 ma. This value is interpolated from Fig. 3.24 to be $\alpha_N = .9779\big|_{I_E = 100\text{ ma.}}$ Solving for I_B, I_B = $(1 - \alpha_N)I_E = (1 - .9779)100$ ma. or, I_B = 2.21 ma.

At I_E = 100 ma, the calculated value for V_{BE} is $V_{BE} = \dfrac{1}{38.376} \ln\left[\dfrac{100 \times 10^{-3}}{2.1596 \times 10^{-12}}\right]$

$$= .8199.$$

ΔV is calculated from:

$$V = V_{BE_{\text{Measured}}} - V_{BE_{\text{Calculated}}} = .930 - .8199 = .1101\text{v}\ .$$

Solving for RB:

$$RB = \frac{.1101\text{ v}}{2.21\text{ ma}} = 49.8\ \Omega$$

or,

$$RB = .0498\ \text{Kilohm}$$

This method of obtaining RB is valid only when the assumption that R_E (emitter bulk resistance) equals zero is valid.

3.3.13 RC, Collector Bulk Resistance

RC is determined from collector-emitter saturation voltage measurements. The test circuit of Fig. 3.32 is utilized to determine $V_{CE_{SAT}}$

The base current is initially set to a convenient value. A current of 1 ma is chosen for this device. The collector current is then adjusted using the power supply to obtain a given forced gain, where the forced gain is defined as $B_F = \dfrac{I_C}{I_B}$ where $I_B = \dfrac{I_C}{B_N}$ (saturation).

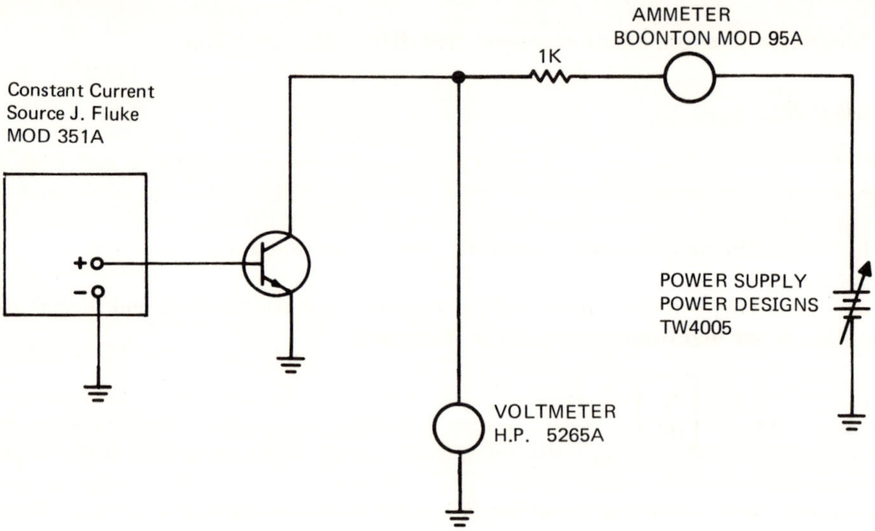

Fig. 3.32. RC measurement test configuration.

For this device a forced gain of 10 is utilized. This requires a collector current of 10 ma. For I_B = 1ma, I_C = 10ma, $V_{CE_{SAT}}$ is measured to be .176 volts.

RC is calculated from:

$$RC = \frac{V_{CE_{SAT(measured)}} - \left[\frac{1}{\theta_E} \ell n \left[\frac{I_E - \alpha_I I_C}{IES(1-\alpha_N \alpha_I)}\right] - \frac{1}{\theta_C} \ell n \left[\frac{\alpha_N I_E - I_C}{ICS(1-\alpha_N \alpha_I)}\right]\right]}{I_C}$$

The values for α_N, α_I are determined as follows: At $I_E = I_C + I_B$ = 11ma, α_N is interpolated to be .9847 from Fig. 3.24. For devices with low values of α_I, α_I is evaluated at a current level corresponding to the base current supplied in excess of that required to drive the device into saturation. The base current required for saturation is calculated from $I_B = (1-\alpha_N)\, I_E$ or I_B = (1 - .9847) (11ma) = .168 ma.

Since a base current of 1ma is supplied, the excess base current is 1 ma – .168 ma = .832 ma. α_I is found to be approximately .0315, at a collector junction current of .832 ma.

Substituting the values for α_N, α_I, Θ_E, Θ_c, IES, ICS, I_E, I_C in the equation for RC yields:

$$RC = \left[.176 - \left[\frac{1}{38.376} \ell n \left[\frac{11 \times 10^{-3} - (.0315)(10 \times 10^{-3})}{2.1596 \times 10^{-15} (1-(.9847)(.0315))}\right] - \frac{1}{20.559} \ell n \left[\frac{(.9847)(11) - 10}{2.9367 \times 10^{-9} (1-(.9847)(.0315))}\right]\right]\right] \Big/ 10 \times 10^{-3}$$

$\therefore$ RC = 2.74Ω
or,

RC = .00274 Kilohms.

This method for determining RC is valid only for devices with $\alpha_I << 1$ and with R_E effectively equal to zero. For other devices, a computer solution is required to obtain the bulk resistances, since the value for α_I is not known as a function of JC except in the case cited where JC approximately equals the effective base current.

3.3.14 R1, Emitter-Base Junction Leakage Resistance

The leakage resistance R1 is determined from emitter-base leakage current measurements. The test configuration in Fig. 3.33 is utilized. Reverse bias is applied to the emitter-base junction and the resulting data is shown in Fig. 3.34. R1 is determined from the slope of the reverse voltage versus reverse current plot. The value for R1 is determined from the region of the curve where the slope is constant. The region chosen for determining R1 should not be near the origin since the junction saturation current may be significant. From Fig. 3.34, R1 equals:

$$\frac{\Delta V_{Reverse}}{\Delta I_{Reverse}} = \frac{4v - 3v}{.15na - .03na} = 8.33 \times 10^9 \Omega$$

or, R1 = 8.33 x 10^6 Kilohms.

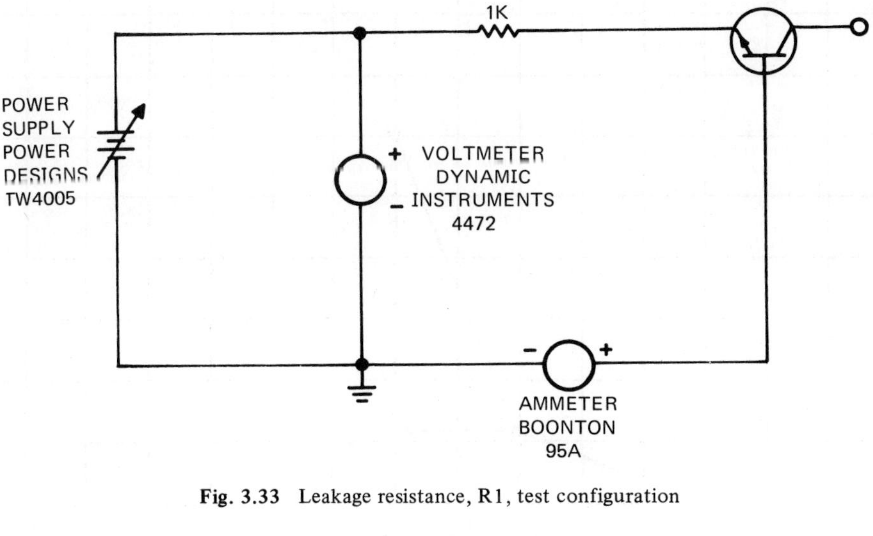

Fig. 3.33 Leakage resistance, R1, test configuration

3.3.15 R2, Collector-Base Junction Leakage Resistance

The leakage resistance R2 is determined from collector-base leakage current measurements. The test configuration in Fig. 3.35 is utilized. Reverse bias is applied to the collector base junction and the resulting leakage current is measured. Leakage current is

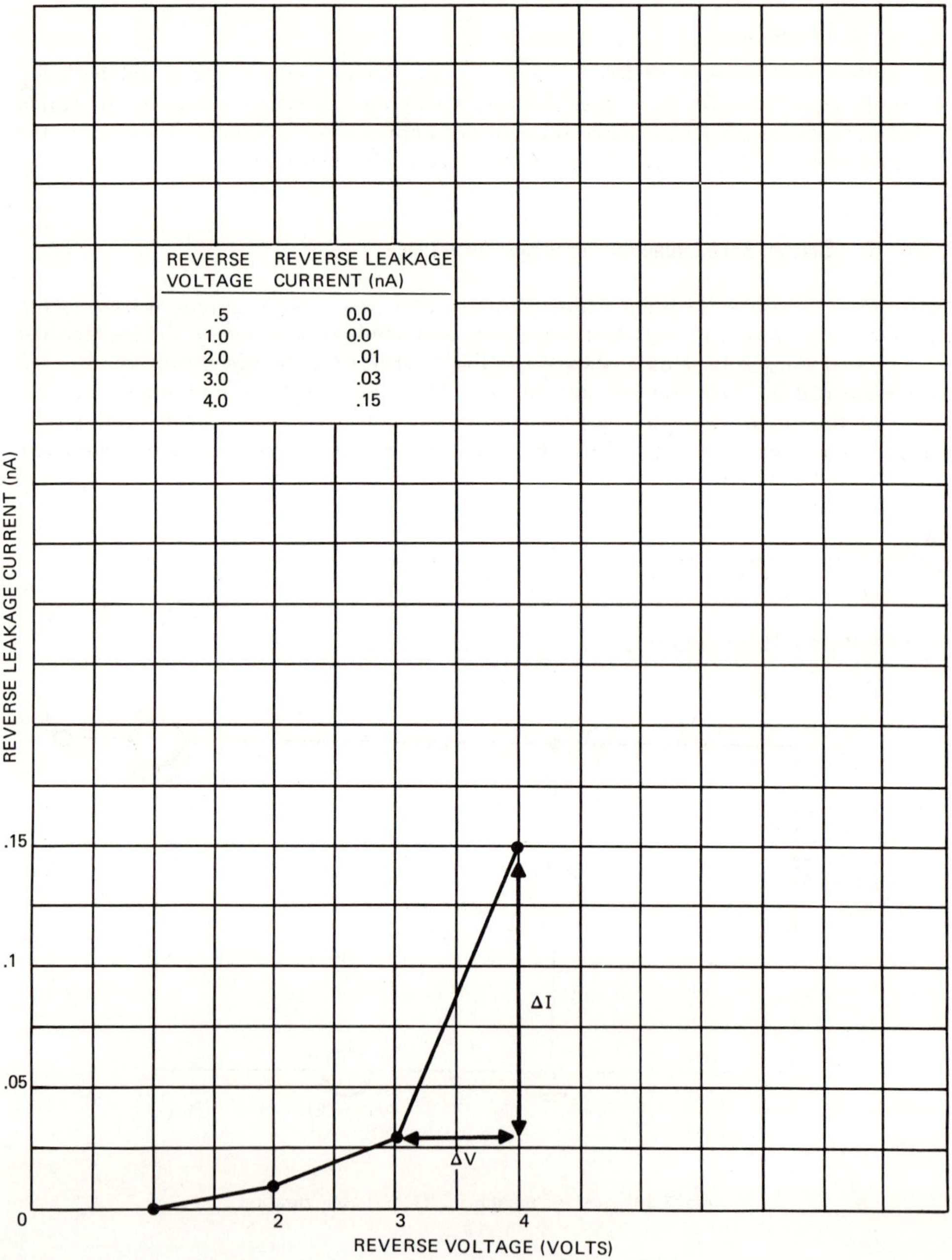

Fig. 3.34 Transistor base-emitter leakage resistance curve.

measured for reverse voltages of 1v to 20v. The resulting data is shown in Fig. 3.36. R2 is determined from the slope of the reverse voltage versus reverse current plot. The value of R2 is determined from the region of the curve where the slope is constant. The region chosen for determining R2 should not be near the origin since the junction saturation current may be significant. From Fig. 3.36:

$$R2 = \frac{\Delta V_{Reverse}}{\Delta I_{Reverse}} = \frac{20v - 10v}{10.4na - 8na} = 4.166 \times 10^9 \Omega$$

or,

$$R2 = 4.166 \times 10^6 \text{ Kilohms.}$$

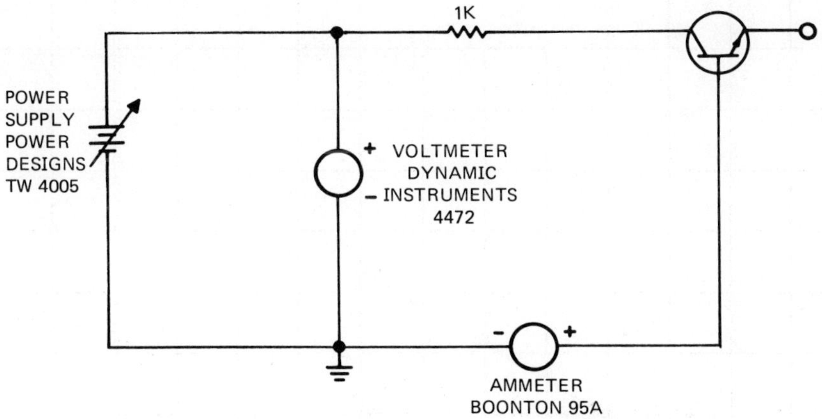

Fig. 3.35 Leakage resistance, R2, test configuration

3.3.16 T_E, Time Constant of the Emitter Diffusion Capacitance Equation

T_E is obtained from collector current rise time in the active region. This parameter is entered in the listing as a function of JE. The test circuit in Fig. 3.37 is used to obtain collector current rise time in the active region.

The pulse generator amplitude is adjusted to obtain the desired peak collector current. Rise time measurements are obtained for the 2N2369 for I_{Cpeak} = 2 ma to I_{Cpeak} = 100 ma. The rise time is measured from the 27% to 73% points on the collector current waveform. The collector current waveforms for I_C = 2, 5, 10, 20, 50, 100 ma are shown in Figs. 3.38, 3.39, 3.40, 3.41, 3.42, 3.43. The rise times for these waveforms are recorded in Fig. 3.44. T_E is defined for the 27% to 73% rise time measurements by:

$$T_E = \frac{tr}{(B_o + 1)} = \frac{1}{2\pi F_N}$$

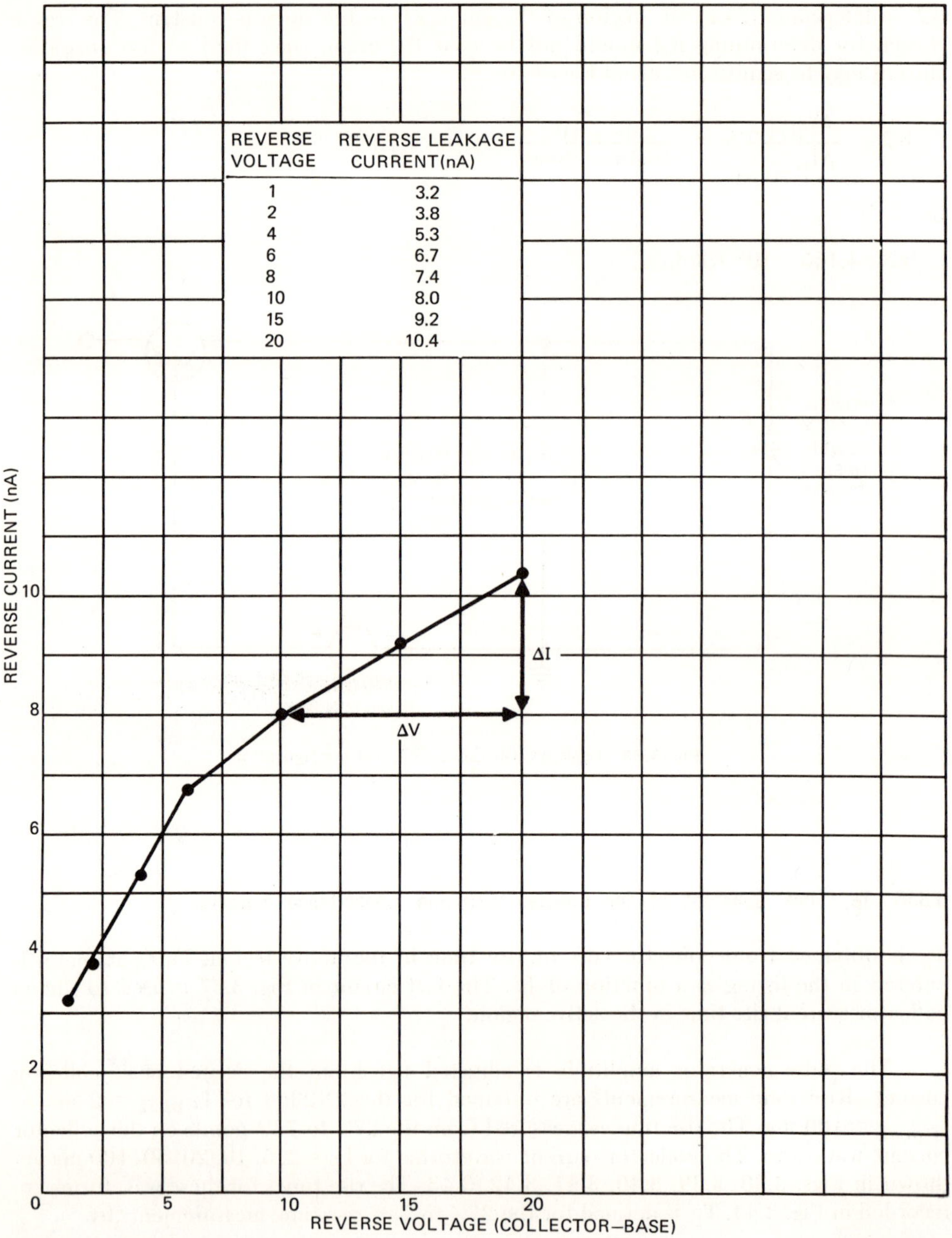

REVERSE VOLTAGE	REVERSE LEAKAGE CURRENT(nA)
1	3.2
2	3.8
4	5.3
6	6.7
8	7.4
10	8.0
15	9.2
20	10.4

Fig. 3.36 Transistor base-collector leakage resistance curve.

Where:

t_r = The time required for I_C to change from 27% of I_{Cpeak} to 73% of I_{Cpeak}

B_o = Average B_N over the 27% to 73% range of I_C

$$B_o = \frac{B_{N1} + B_{N2}}{2}$$

B_{N1} = B_N at I_C = 27% of I_{Cpeak}

B_{N2} = B_N at I_C = 73% of I_{Cpeak}

Values for B_{N1}, B_{N2} are obtained from interpolating B_N data given in Figs. 3.23 and 3.24. Knowing the rise time and B_o, T_e is calculated for each collector current transition. The calculated values are shown in Fig. 3.44.

T_E will be entered as a function of JE in the SCEPTRE program. The value of JE at which a particular value of T_E is entered is the average value of emitter current during the 27% to 73% transition of I_C. I_E average is calculated from,

$$I_{Eaverage} = \left(\frac{I_{Cpeak}}{2}\right) \frac{(B_o + 1)}{B_o}$$

Values of $I_{Eaverage}$ are given in Fig. 3.44 corresponding to the peak collector current under consideration.

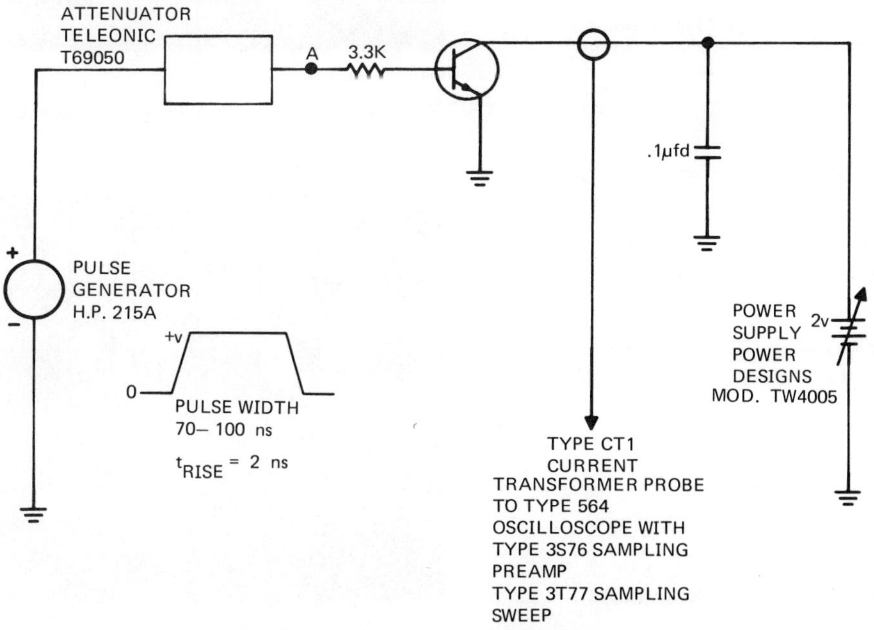

Fig. 3.37 Rise time test configuration

VERTICAL 1mA/cm
HORIZONTAL 10nS/cm

$I_{C_{PEAK}}$ = 2mA

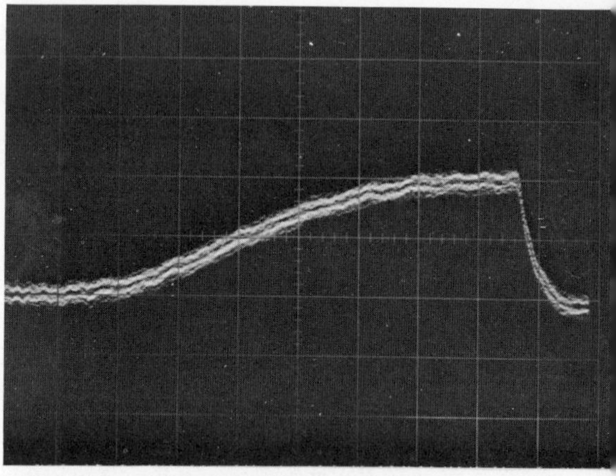

Fig. 3.38.
Collector current.

VERTICAL 1mA/cm
HORIZONTAL 10nS/cm

$I_{C_{PEAK}}$ = 5mA

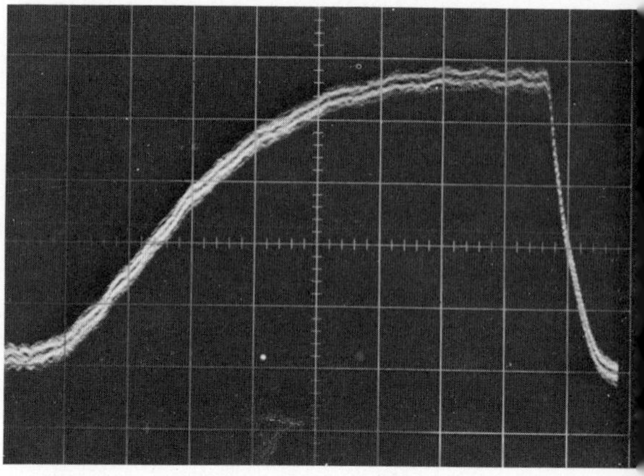

Fig. 3.39.
Collector current.

VERTICAL 2mA/cm
HORIZONTAL 10nS/cm

$I_{C_{PEAK}}$ = 10mA

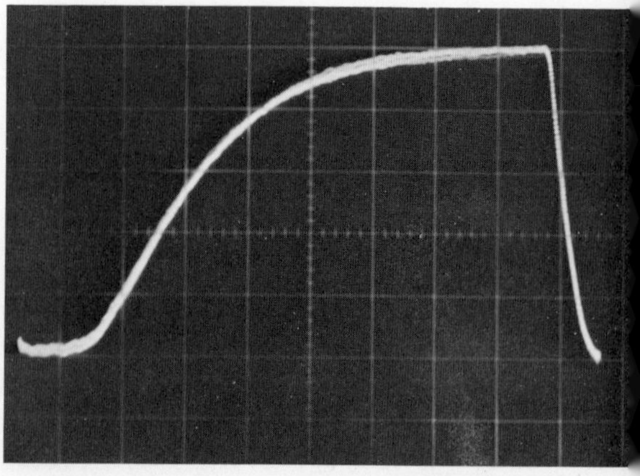

Fig. 3.40.
Collector current.

VERTICAL 4mA/cm
HORIZONTAL 10nS/cm

$I_{C_{PEAK}}$ = 20mA

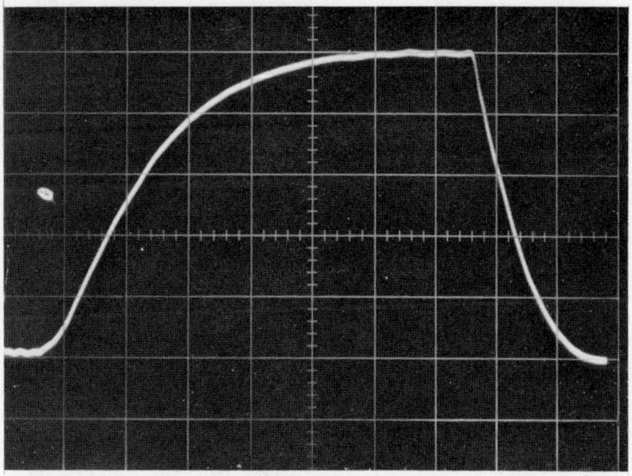

Fig. 3.41.
Collector current.

VERTICAL 10mA/cm
HORIZONTAL 10nS/cm

$I_{C_{PEAK}}$ = 50mA

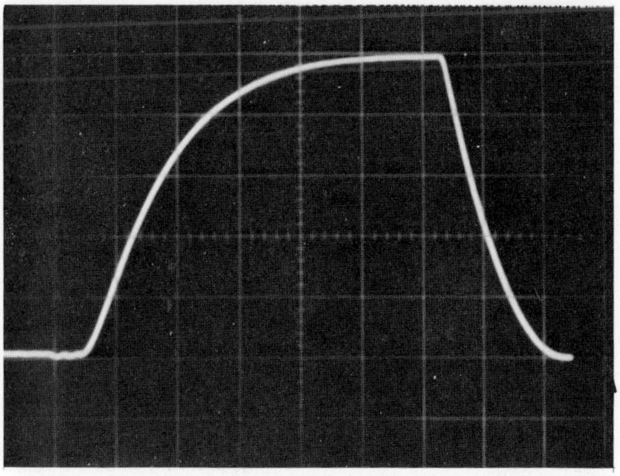

Fig. 3.42.
Collector current.

VERTICAL 20mA/cm
HORIZONTAL 10nS/cm

$I_{C_{PEAK}}$ = 100mA

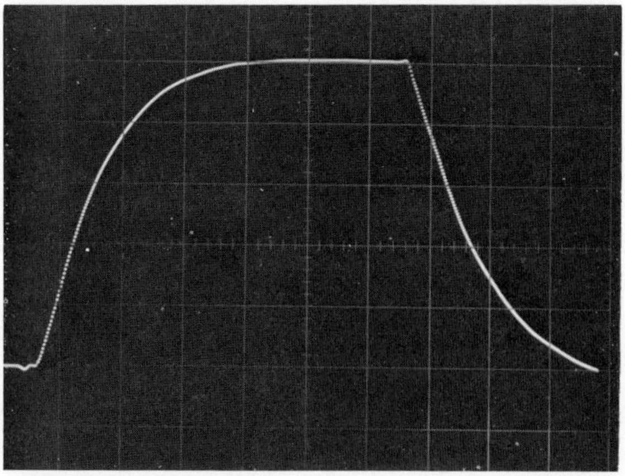

Fig. 3.43.
Collector current.

$I_{C_{Peak}}$	t_r 27% → 73%	B_{N_1}	B_{N_2}	B_0	$I_{C_{Average}}$	$I_{E_{Average}} = \left(\frac{I_{C_{Peak}}}{2}\right)\frac{(B_0+1)}{B_0}$	T_E
100 ma	8 ns	60	49	54.5	50 ma	50.92 ma	.1441 nsec
50 ma	11 ns	62	58	60	25 ma	25.41 ma	.1803 nsec
20 ma	13.5 ns	60	62	61	10 ma	10.16 ma	.2177 nsec
10 ma	17 ns	58	60.5	59.2	5 ma	5.08 ma	.2823 nsec
5 ma	20 ns	53.5	58	55.8	2.5 ma	2.54 ma	.3521 nsec
2 ma	24 ns	48.5	54	51.2	1 ma	1.02 ma	.4597 nsec

Fig. 3.44 Transistor current rise time data summary

3.3.17 T_S, Time Constant of the Collector Diffusion Capacitance Equation

T_S is calculated from storage time data obtained from the test circuit of Fig. 3.45.

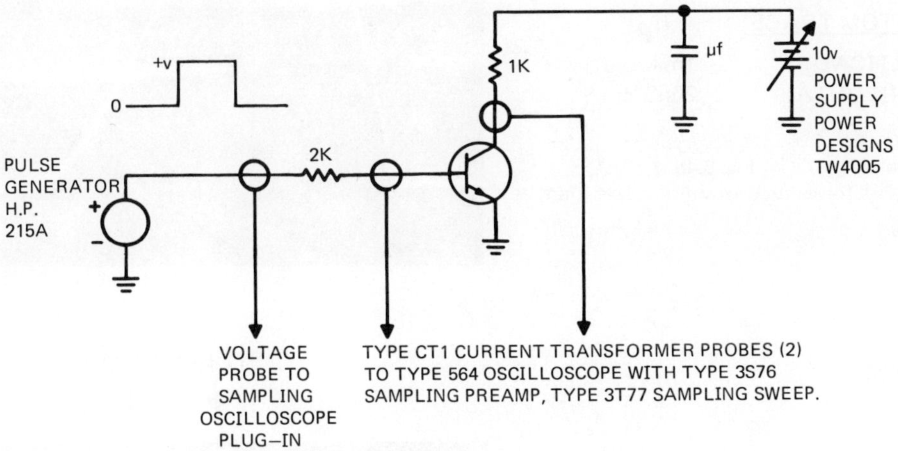

Fig. 3.45 Storage time test circuit

The amplitude of the pulse generator is adjusted to obtain several values of base current. Each base current level must be sufficient to drive the device into saturation, since storage time is to be measured. The pulse width is adjusted experimentally until storage time is not a function of pulse width.

Storage time is measured from the time the base current drops at the end of the input pulse to the time when the collector current falls to 90% of its saturation value. A slight error is introduced by using the $I_c = .9\ I_{cSAT}$ point as the end of the storage time interval. Actually the device is in its active region from the time the collector current begins to drop. The $I_c = .9\ I_{cSAT}$ point is used because it is very difficult to determine exactly when the collector current begins to decrease. The error introduced by this procedure is usually negligible. Collector and base current waveforms are shown for $I_B = 1$ ma, 2 ma, 3 ma, 4 ma, 5 ma, in Figures 3.46 through 3.55. The storage time, the test circuit input voltage, the base current (I_B), and negative base current (I_{B2}) are listed in Figure 3.56.

In order to calculate T_S and enter it as a function of JC, seven parameters are required: $t_{storage}$, α_I, h_{FE_I}, h_{FE_N}, I_B, I_{B2}, JC. Of these parameters, $t_{storage}$, I_B, I_{B2} are measured. Since the four remaining parameters are a function of operating point and cannot be measured in the test circuit, a preliminary SCEPTRE run is made to determine these parameters. The storage time test circuit is set up in SCEPTRE form in Figure 3.57. The input waveform (E1) is composed of a series of voltage steps, the amplitude of each step corresponding to the measured input voltage amplitude yielding, $I_B = 1$ ma, 2 ma, 3 ma, 4 ma, and 5 ma. The model capacitances used in this preliminary run are arbitrarily chosen as CE = 10pf, CC = 1pf, since the objective is to obtain steady state values for α_I, h_{FE_I}, h_{FE_N}, JC.

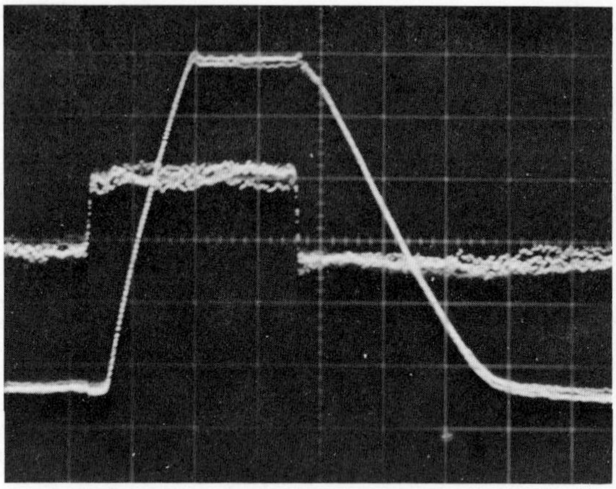

TOP TRACE (I_C)
VERTICAL 2mA/cm
HORIZONTAL 20nS/cm

BOTTOM TRACE (I_B)
VERTICAL 1mA/cm
HORIZONTAL 20nS/cm

Fig. 3.46.
Storage time waveforms, I_B = 1mA.

STORAGE TIME = 8nS

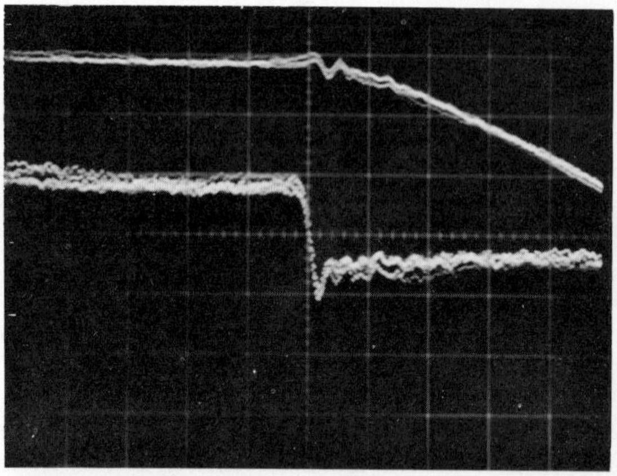

TOP TRACE (I_C)
VERTICAL 2mA/cm
HORIZONTAL 5nS/cm

BOTTOM TRACE (I_B)
VERTICAL 1mA/cm
HORIZONTAL 5nS/cm

Fig. 3.47
Storage time waveforms, I_B = 1mA.
Time Scale Expanded

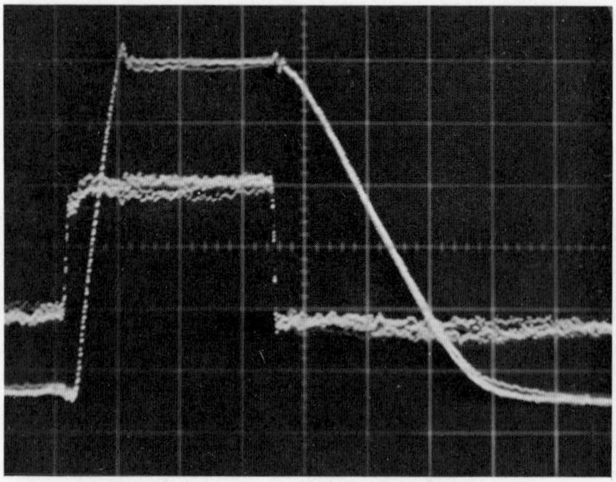

TOP TRACE (I_C)
VERTICAL 2mA/cm
HORIZONTAL 20nS/cm

BOTTOM TRACE (I_B)
VERTICAL 1mA/cm
HORIZONTAL 20nS/cm

Fig. 3.48
Storage time waveforms, I_B = 2mA.

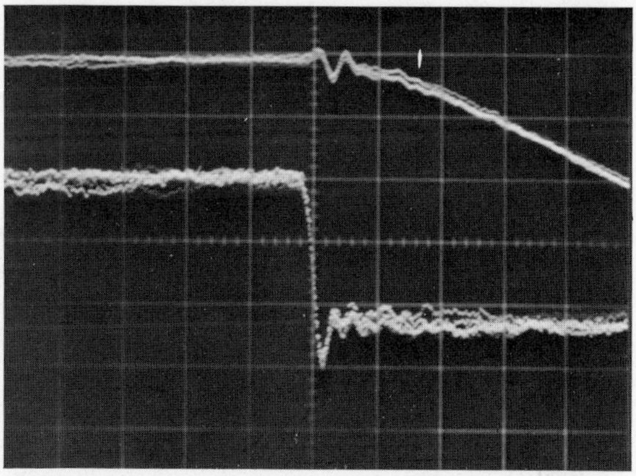

TOP TRACE (I_C)
VERTICAL 2mA/cm
HORIZONTAL 5nS/cm

BOTTOM TRACE (I_B)
VERTICAL 1mA/cm
HORIZONTAL 5nS/cm

Fig. 3.49
Storage time waveforms, I_B = 2mA.
Time Scale Expanded

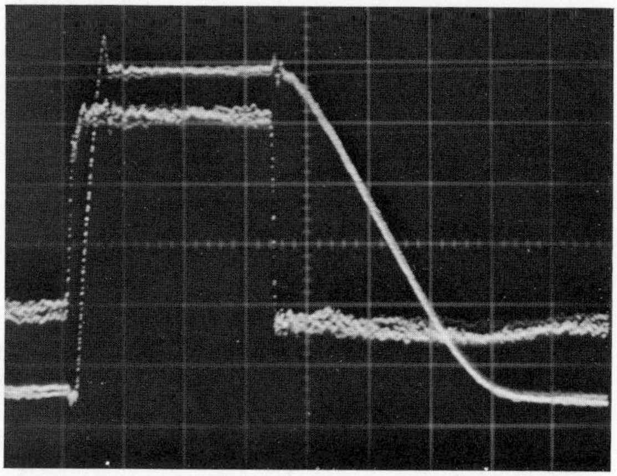

TOP TRACE (I_C)
VERTICAL 2mA/cm
HORIZONTAL 20nS/cm

BOTTOM TRACE (I_B)
VERTICAL 1ma/cm
HORIZONTAL 20nS/cm

Fig. 3.50
Storage time waveforms, I_B = 3mA.

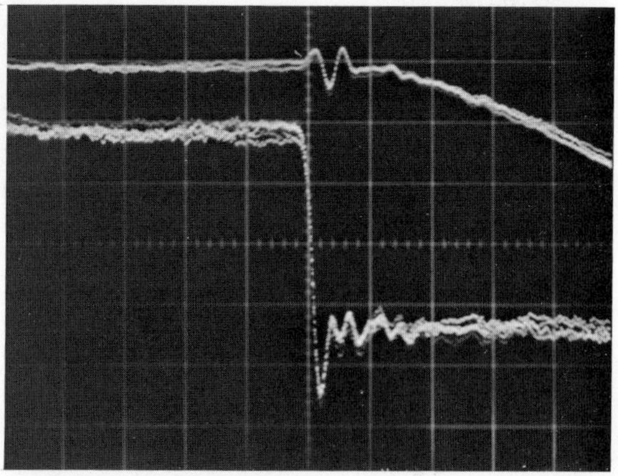

TOP TRACE (I_C)
VERTICAL 2mA/cm
HORIZONTAL 5nS/cm

BOTTOM TRACE (I_B)
VERTICAL 1mA/cm
HORIZONTAL 5nS/cm

Fig. 3.51
Storage time waveforms, I_B = 3mA.
Time Scale Expanded

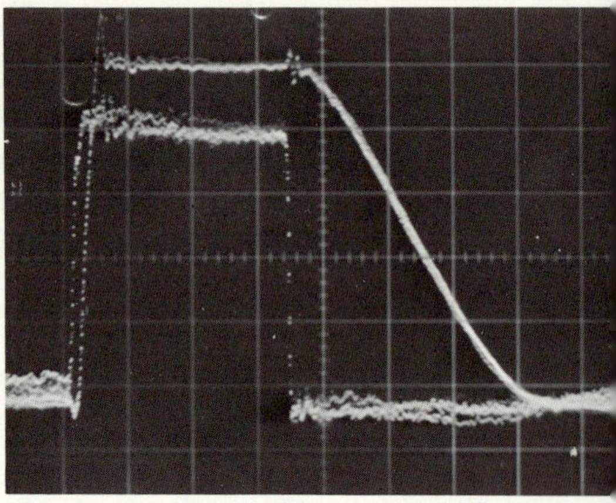

TOP TRACE (I_C)
VERTICAL 2mA/cm
HORIZONTAL 20nS/cm

BOTTOM TRACE (I_B)
VERTICAL 1mA/cm
HORIZONTAL 20nS/cm

Fig. 3.52
Storage time waveforms, I_B = 4mA.

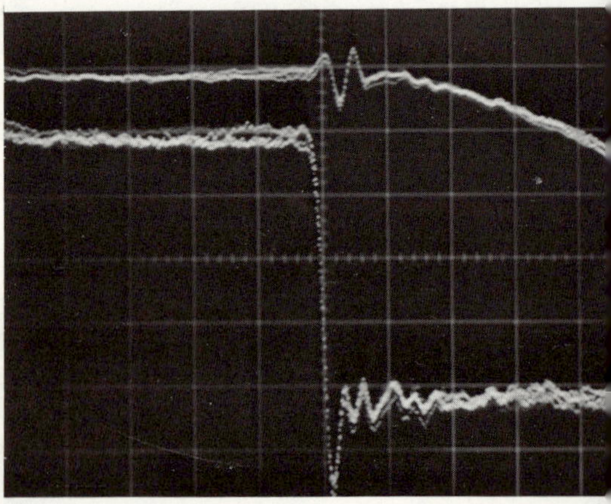

TOP TRACE (I_C)
VERTICAL 2mA/cm
HORIZONTAL 5nS/cm

BOTTOM TRACE (I_B)
VERTICAL 1mA/cm
HORIZONTAL 5nS/cm

Fig. 3.53
Storage time waveforms, I_B = 4mA.
Time Scale Expanded

TOP TRACE (I_C)
VERTICAL 2mA/cm
HORIZONTAL 20nS/cm

BOTTOM TRACE (I_B)
VERTICAL 1mA/cm
HORIZONTAL 20nS/cm

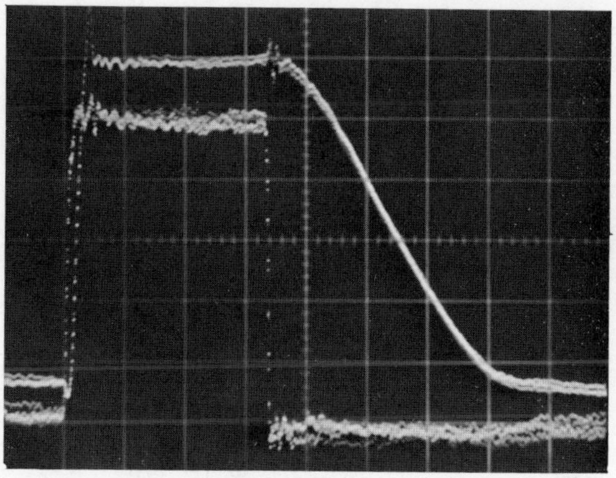

Fig. 3.54
Storage time waveforms, I_B = 5mA.

TOP TRACE (I_C)
VERTICAL 2mA/cm
HORIZONTAL 5nS/cm

BOTTOM TRACE (I_B)
VERTICAL 1mA/cm
HORIZONTAL 5nS/cm

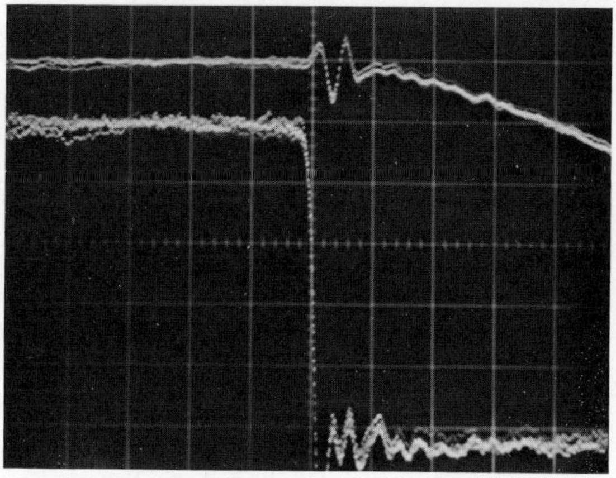

Fig. 3.55
Storage time waveforms, I_B = 5mA.
Time Scale Expanded

MEASURED PARAMETERS					PARAMETERS FROM SCEPTRE RUN				CALCULATED PARAMETERS		
I_B	I_{B_2}	I_{cSAT}	$t_{STORAGE}$	V_{input}	α_I	h_{FE_N}	h_{FE_I}	JC	t_S	T_{cI}	T_S
1 ma	−.4ma	10ma	8ns	2.7v	.02909	60.63	.02997	.7935ma	8.815ns	291.127ns	8.556ns
2	−.4	10	10	4.75	.05073	60.73	.03344	1.845	6.910	129.304	6.559
3	−.4	10	12	6.8	.06752	60.83	.07241	2.932	6.680	92.252	6.228
4	−.4	10	14	8.85	.07889	60.93	.08564	4.031	6.815	79.577	6.277
5	−.4	10	15	10.9	.09059	61.04	.09961	5.155	6.638	66.639	6.036

Fig. 3.56 Storage time parameters

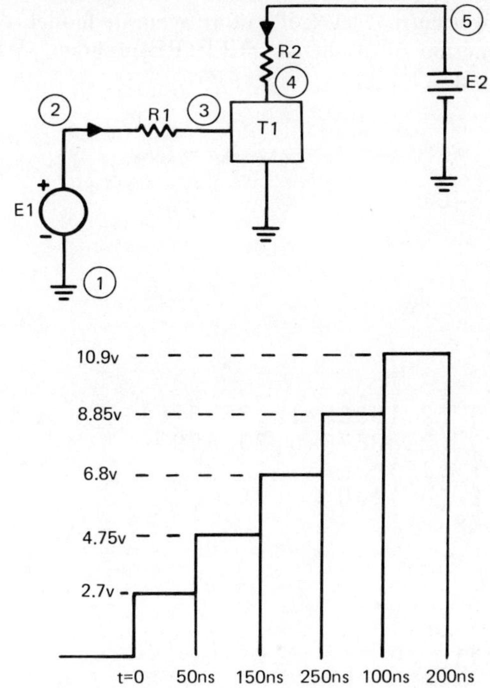

Fig. 3.57 Storage time test circuit (constant capacitance)

The preliminary model description with fixed capacitances is shown in Figure 3.58. The circuit description for the storage time test circuit is shown in Figure 3.59. The required parameters α_I, h_{FE_I}, h_{FE_N}, JC are obtained from the numerical outputs of Figures 3.60, 3.61, 3.62, 3.63, and 3.64, for each step of the input waveform after the transients have settled out. The outputs are tabularized in Figure 3.56.

Knowing these seven parameters, T_S is obtained from the following calculations:

1. Calculate τ_s, where:

$$\tau_s = \frac{t_{storage}}{\ell n \dfrac{I_B - I_{B2}}{\dfrac{I_{cSAT}}{h_{FE}} - I_{B2}}}$$

2. From τ_s, calculate the charge control parameter TCI using the relation:

$$TCI = \frac{\tau_s}{h_{FE_I}}$$

3. From TCI, calculate T_s using the relation:

$$T_s = \alpha_I \ TCI$$

The values calculated for τ_S, TCI, T_S are tabularized in Fig. 3.56. This technique may be extended to higher or lower current levels if a more accurate model is desired.

T_s is entered as a function of JC in the SCEPTRE program. Values for T_s and the corresponding JC are found in Fig. 3.56.

```
MODEL 2N2369 (B-E-C)
ELEMENTS
CE, 1-E = 10
CC, 1-2 = 1
RB, B-1 = .0498
RC, C-2 = .00274
R1, 1-E = 8.33E6
R2, 1-2 = 4.16E6
JE, 1-E = DIODE Q (2.1596E-12,38.376)
JC, 1-2 = DIODE Q (2.8367E-6,20.559)
JN, 2-1 = TABLE 1 (JE)*JE
JI, E-1 = TABLE 2 (JC)*JC
DEFINED PARAMETERS
P1 = T2(JC)
P2 = Q1(T1(JE))
P3 = Q1(T2(JC))
FUNCTIONS
Q1(A) = (A/(1.-A))
TABLE 1*
.001,.97276,.0514,.97276,.1026,.97466,.2047,.97704,.5103,
.97982, 1.0195,.98087,2.0353,.98265,5.0832,.98363,10.165,
.98376,20.325,.98400,30.510,.98328,40.697,.98287,50.92,
.98193, 61.111,.98182,71.386,.98058,81.65,.97979,91.963,
.97865, 102.28,.97771,500,.97771
TABLE 2
.1,.02045,.489,.02045,.727,.02751,1.25,.04,1.92,.05208,2.95,
.06779 3.88,.07731,5.38,.09293,9.03,.11074,15.0,.1333,25,.1333
```

WHERE:

$$P1 = \alpha_I$$

$$P2 = h_{FE_N} = \beta_N = \frac{\alpha N}{1 - \alpha N}$$

$$P3 = h_{FE_I} = \beta_I = \frac{\alpha I}{1 - \alpha I}$$

2N2369 TRANSISTOR WITH CONSTANT CAPACITANCE DESCRIPTION

Fig. 3.58

* SEE APPENDIX F regarding Tables such as TABLE 1 and TABLE 2.

```
STORAGE TIME TEST CIRCUIT (CONSTANT CAPACITANCE)
ELEMENTS
R1, 2-3 = 2
R2, 5-4 = 1
E1, 1-2 = TABLE 1 (TIME)
E2, 1-5 = 10
T1, 3-1-4 = MODEL 2N2369
OUTPUTS
E1,JCT1,P1T1(ALPHAI),P2T1(HFEN),P3T1(HFEI),IRCT1,IRBT1,PLOT
FUNCTIONS
TABLE 1
0,0,0,2.7,50,2.7,50,4.75,100.4,75,100,6.8,150,6.8,150,8.85,
200,8.85,200,10.9,250,10.9
RUN CONTROLS
RUN INITIAL CONDITIONS
MAXIMUM PRINT POINTS = 200
STOP TIME = 250
END

CONSTANT CAPACITANCE SIMULATION FOR CURRENT LEVELS
```

Fig. 3.59

Storage time test circuit data for constant capacitance run

TIME	4.00312+01	9.19219+01	1.42875+02
E1	2.70000+00	4.75000+00	6.80000+00
JCT1	7.93516−01	1.84523+00	2.93237+00
ALPHA I	2.90985−02	5.07319−02	6.75210−02
HFEN	6.06323+01	6.07318+01	6.08330+01
HFEI	2.99769−02	5.34432−02	7.24102−02
IRCT1	9.82119+00	9.85959+00	9.87971+00
IRBT1	9.45494−01	1.94435+00	2.94330+00
	Fig. 3.60	**Fig. 3.61**	**Fig. 3.62**

TIME	1.91922+02	2.41867+02
E1	8.85000+00	1.09000+01
JCT1	4.03191+00	5.15534+00
ALPHA I	7.88918−02	9.05905−02
HFEN	6.09356+01	6.10406+01
HFEI	8.56488−02	9.96147−02
IRCT1	9.89299+00	9.90287+00
IRBT1	3.94234+00	4.94144+00
	Fig. 3.63	**Fig. 3.64**

3.3.18 Transistor Data Summary

A summary of the required parameters for the bipolar transistor model of Fig. 3.16 is given in Table 3.2. The corresponding measured parameter values for the 2N2369 as developed in this section are also listed.

Table 3.2 Transistor data summary

REQUIRED TRANSISTOR PARAMETERS	2N2369 MEASURED VALUES
Θ_E	38.376 volts^{-1}
Θ_c	20.559 volts^{-1}
IES	2.1596×10^{-12} ma
ICS	2.8367×10^{-6} ma
αN	See Figs. 3.23, 3.24
αI	See Fig. 3.27
n_E	.3708
n_c	.2556
Φ_E	.9 volt
Φ_C	.9 volt
COE	4.289
COC	1.275
T_E	See Fig. 3.44
T_S	See Sec. 3.3.17
RB	.0498 Kilohms
RC	.00274 Kilohms
R1	8.33E6 Kilohms
R2	4.166E6 Kilohms

3.3.19 SCEPTRE Rise Time Test Circuit Example

In order to verify to some extent the validity of the transistor model, the test circuit used in Fig. 3.37 to determine collector current rise time is put into SCEPTRE form in Fig. 3.65.

The input wave form (E1) is an approximation of the actual voltage seen at point A in Fig. 3.37. The amplitude of the input waveform corresponds to the measured value which caused a peak collector current of 50 ma. The 2N2369 model and test circuit are shown coded in Figs. 3.66 and 3.67. The computer output I_{E2} in Fig. 3.68 corresponds to the collector current shown in Fig. 3.42. The 27% - 73% SCEPTRE rise time is 10.27 ns which compares very favorably to the measured value of 11 ns. It should be mentioned that the 10.27 ns is actually obtained from the SCEPTRE printed output and the plot of Fig. 3.68 is shown only for convenience.

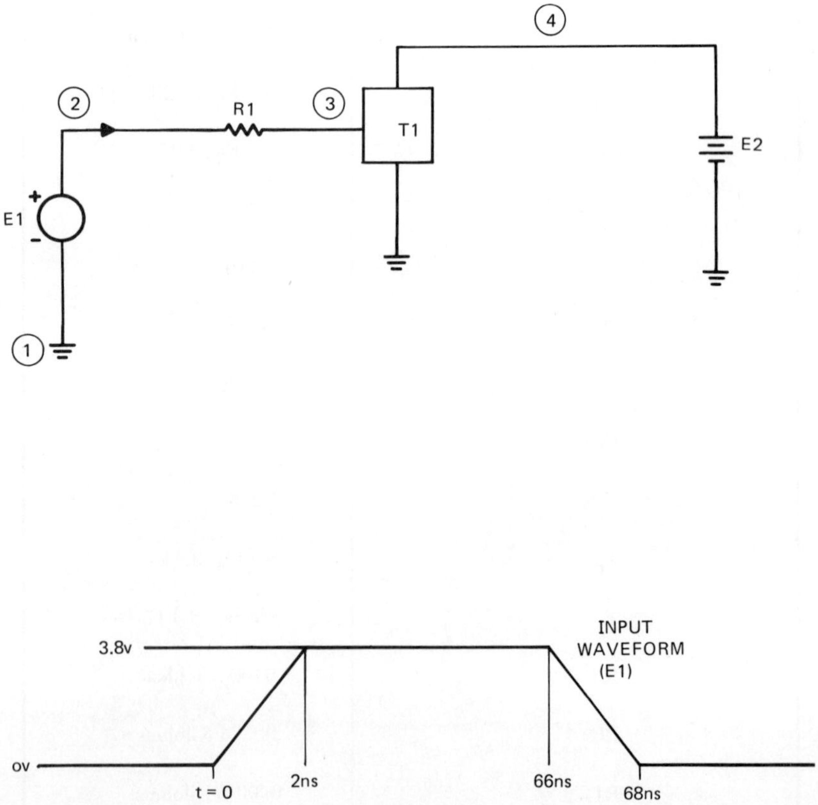

Fig. 3.65 Rise time test circuit (SCEPTRE form).

```
MODEL 2N2369 (B-E-C)
ELEMENTS
CE, 1-E = Q1(4.289,.9,VCE,.3708,38.376,T3(JE),JE,2.1596E-12)
CC, 1-2 = Q1(1.275,.9,VCC,.2556,20.559,T4(JC),JC,2.8367E-6)
RB, B-1 = .0498
RC, C-2 = .00274
R1, 1-E = 8.33E6
R2, 1-2 = 4.16E6
JE, 1-E = DIODE Q (2.1596E-12,38.376)
JC, 1-2 = DIODE Q (2.8367E-6,20.559)
JN, 2-1 = TABLE 1 (JE)*JE
J1, E-1 = TABLE 2 (JC)*JC
FUNCTIONS
Q1(A,B,C,D,E,F,G,H) = (A/(B-C)**D+E*F*(G+H))
TABLE 1
.001,.97276,.0514,.97276,.1026,.97466,.2047,.97704,.5103,
.97982,1.0195,.98087,2.0353,.98265,5.0832,.98363,10.165,
.98376,20.325,.98400,30.510,.98328,40.697,.98287,50.92,
.98193,61.111,.98182,71.396,.98058,81.65,97979,91.963,
.97865,102.28,97771,500,.97771
TABLE 2
.1,.02045,.489,.02045,.727,.02751,1.25,.04,1.92,.05208,2.95,
.06779,3.88,.07731,5.38,.09293,.11074,15.0,.1333,25,.1333
TABLE 3
.1,.4597,1.02,.4597,2.54,.3521,5.108,.2823,10.16,.2177,
25.41,.1803,50.92,.1441,100,.1441
TABLE 4
.1,8.556,.7935,8.556,1.845,6.559,2.932,6.228,4.031,6.277,
5.155,6.036,10,6.036
```

Fig. 3.66 Complete final 2N2369 model description

```
RISE TIME TEST CIRCUIT
ELEMENTS
R1, 2-3 = 3.3
C1, 4-3 = .67
C2, 3-1 = .07
C3, 4-1 = .67
E1, 1-2 = TABLE 1 (TIME)
E2, 1-4 = 2
T1, 3-1-4 = MODEL 2N2369
OUTPUTS
IE2,E1,IRBT1,IRCT1,CET1,JET1,PLOT
FUNCTIONS
TABLE 1
0,0,2,3.8,66,3.8,68,0,70,0
RUN CONTROLS
RUN INITIAL CONDITIONS
STOP TIME = 100
MAXIMUM PRINT POINTS = 200
END
```

Fig. 3.67 Rise time test circuit description

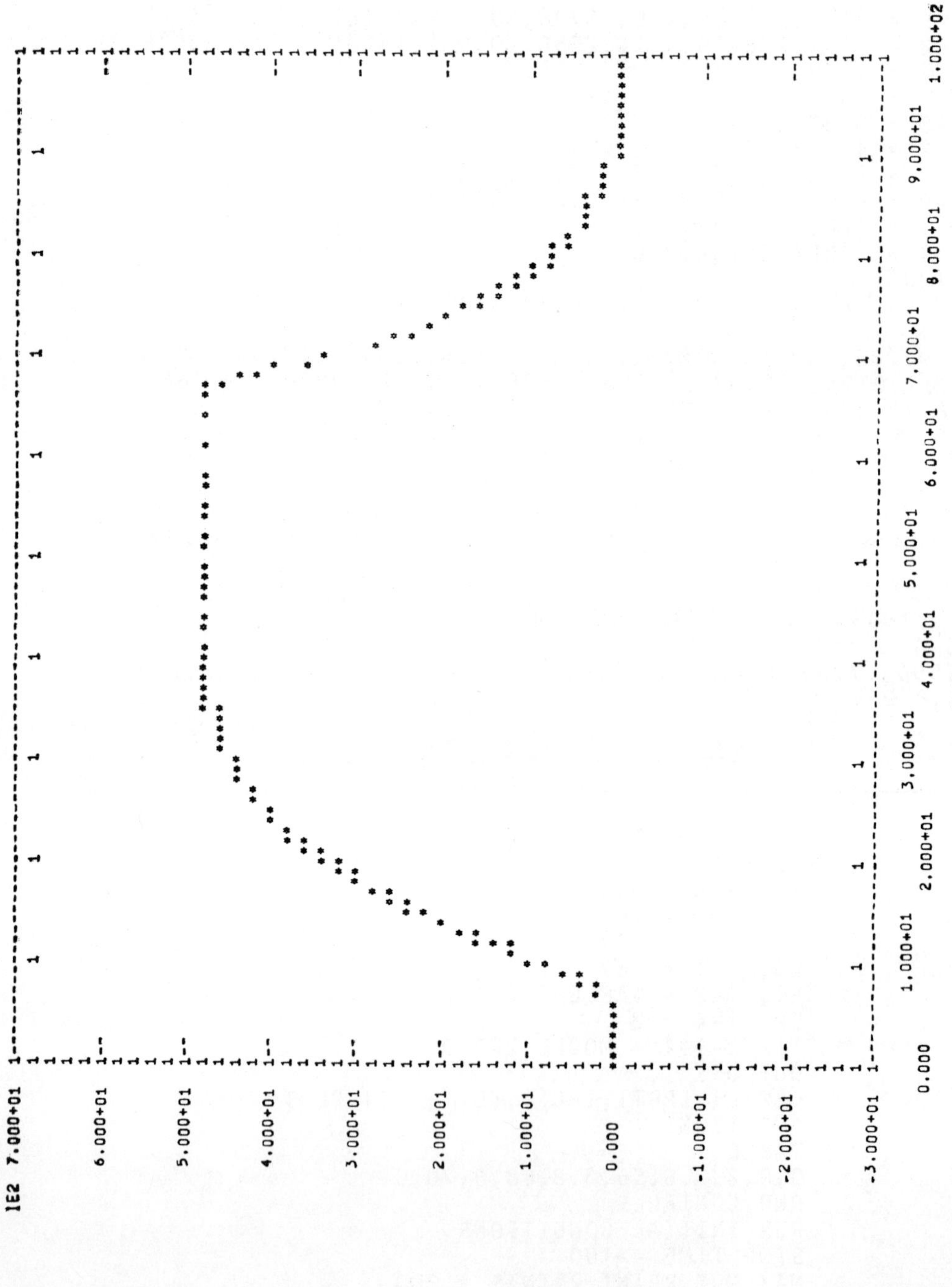

Fig. 3.68. Plot of IE2 vs. time.

3.3.20 SCEPTRE Storage Time Test Circuit Example

In order to test the validity of the model with respect to storage time, the storage time is set up in SCEPTRE form as shown in Fig. 3.69. This circuit includes all measured can capacitances and the test circuit stray capacitances (C_1 = .94pf, C_2 = 2.57pf, C_3 = 2.35pf).

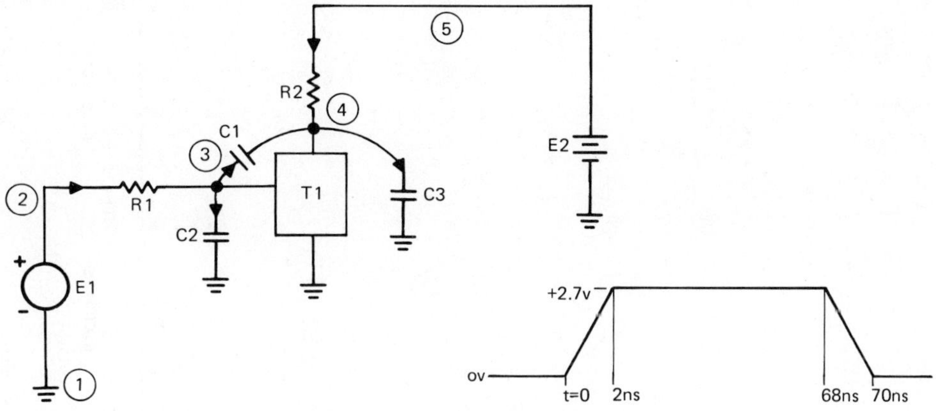

Fig. 3.69 Storage time test circuit (SCEPTRE form).

The input waveform corresponds to the measured input voltage for a base current of 1 ma.

The circuit description for the storage time test circuit of Fig. 3.69 is shown in Fig. 3.70. The computer outputs IRCT1 (Fig. 3.71), and IR1 (Fig. 3.72) correspond to the collector and base currents. The measured storage time from Fig. 3.47 is about 8 ns. The SCEPTRE storage time from the tabular data is also approximately 8 ns.

```
STORAGE TIME TEST CIRCUIT
ELEMENTS
R1, 2-3 = 2
R2, 5-4 = 1
C1, 3-4 = .94
C2, 3-1 = 2.57
C3, 4-1 = 2.35
E1, 1-2 = TABLE 1 (TIME)
E2, 1-5 = 10
T1, 3-1-4 = MODEL 2N2369
OUTPUTS
E1,IRBT1,IRCT1,IR1,IR2,VCET1,VCCT1,JCT1,JET1,JNT1,JIT1,PLOT
FUNCTIONS
TABLE 1
0,0,2,2.7,68,2.7,70,0,80,0
RUN CONTROLS
RUN INITIAL CONDITIONS
MAXIMUM PRINT POINTS = 200
STOP TIME = 100
END
```

Fig. 3.70 Storage time test circuit description

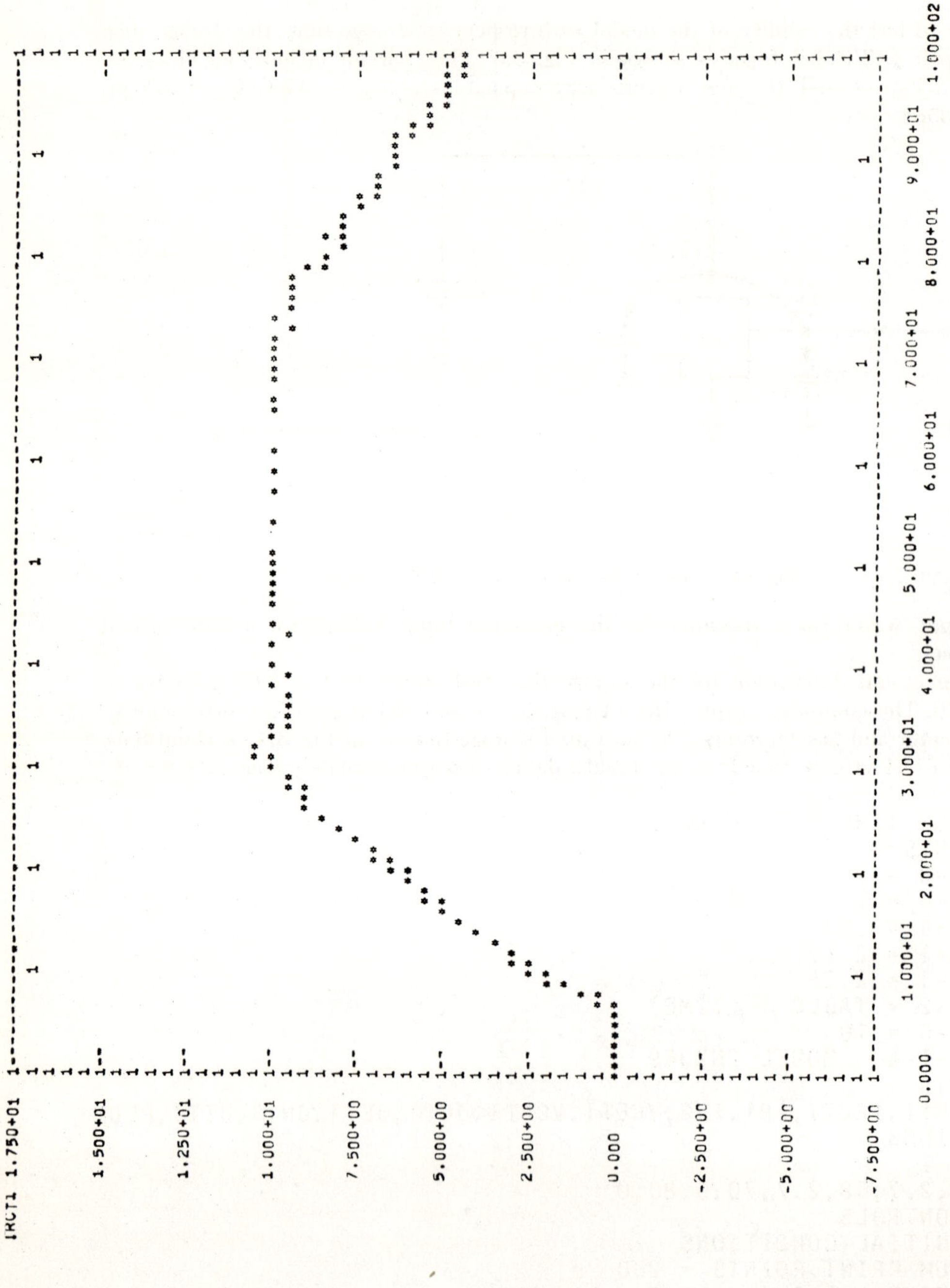

Fig. 3.71. Plot of 1RCT1 vs. time.

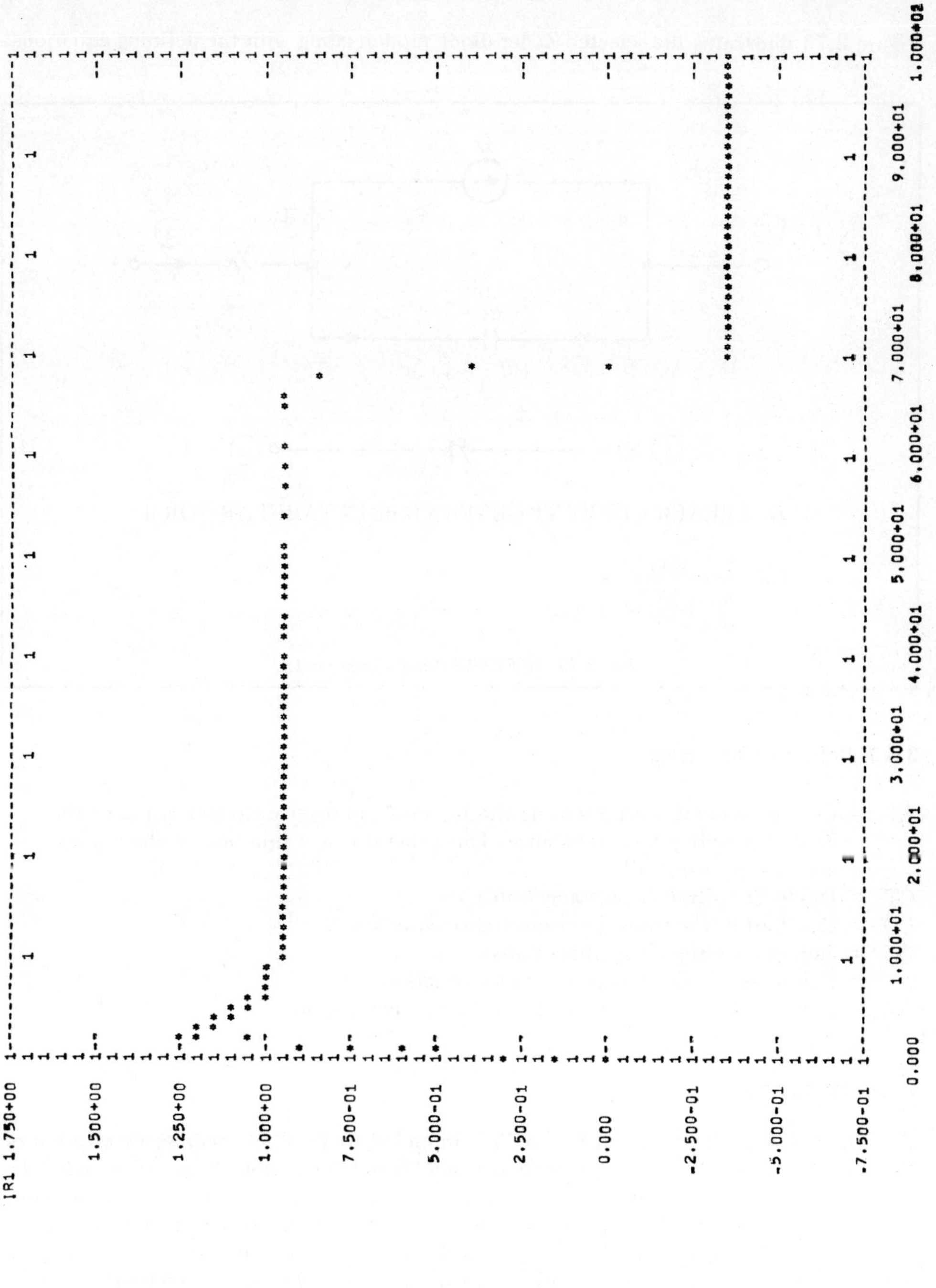

Fig. 3.72. Plot of IR1 vs. time.

3.4 SCEPTRE ZENER DIODE MODEL[5]

Figure 3.73 illustrates the selected Zener diode model, along with the defining equations.

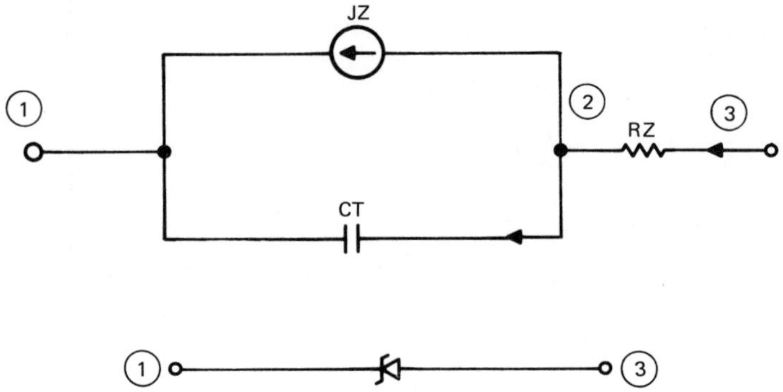

JZ = ZENER CURRENT GENERATOR IN TABULAR FORM

$$CT = \frac{CO}{(\phi - VCT)^{n}}$$

Fig. 3.73 SCEPTRE Zener diode model.

3.4.1 Definition of Terms

JZ = Current generator representing the forward and reverse characteristics of the Zener excluding bulk resistance. This generator is a function of the voltage across capacitor CT.
CT = Diode Transition Capacitance (pfd).
CO = Constant of the transition capacitance equation.
ϕ = Junction Contact Potential (Volts).
n = Exponent of transition capacitance equation.
RZ = Bulk resistance of the Zener in the breakdown region.

3.4.2 Introduction

The Zener model illustrated in Fig. 3.73 is intended to yield an accurate description of the Zener characteristics in the reverse biased and breakdown regions of operation only. The forward region is purposely simplified on the assumption that this is not a normal mode of operation for a Zener. The simplifications for the forward region include a deletion of the diffusion capacitance and the tabular entry of the forward region, without separating the bulk resistance from the forward biased junction characteristics. These effects are easy to include, if desired, utilizing the same techniques as in the junction diode.

3.4.3 Model Parameter Determination for the 1N754 Zener Diode

The measurement technique for each Zener diode parameter is discussed in the following sections.

3.4.4 JZ, Zener Current Generator and RZ, Bulk Resistance

The Zener current generator is represented by tabular entry of Zener current as a function of junction voltage. The junction voltage is obtained from the measured V-I characteristic by subtracting the drop due to bulk resistance at each tabular point. The model yields a piecewise linear approximation to the actual device characteristics. The accuracy of the model is determined by the number of points entered in the table describing JZ.

Tabular data for JZ is determined from the measured forward and reverse characteristics as shown in Figs. 3.74, 3.75, and 3.76. The test configurations used in obtaining this data are illustrated in Figs. 3.77 and 3.78. For the forward region measurements, the power supply is set at a fixed voltage (10v) and the variable resistor is adjusted to obtain the current level at which the forward voltage is to be read. For reverse region measurements, the power supply is adjusted to several values of reverse voltages which are less than the breakdown voltage and the corresponding currents are recorded. The power supply voltage is then carefully increased past the breakdown point. The supply is adjusted to obtain several values of current in the breakdown region and the corresponding Zener voltage is recorded. The resistance used in the reverse region test setup is necessary to provide current limiting, since the Zener current in the breakdown region increases rapidly with additional applied reverse voltage.

The measured values for voltage in the reverse region include the voltage drop due to reverse current flowing through the bulk resistance, RZ. This voltage must be subtracted from each of the measured values of voltage before entering reverse voltage values in the table describing JZ.

The bulk resistance is determined from the slope of the V-I curve in the breakdown region, as shown in Fig. 3.75. The slope is measured at high current levels, past the knee region.

RZ is calculated from the data in Fig. 3.75. For $V_2 = 6.815\text{v}$, $I_2 = 40$ ma and for $V_1 = 6.792\text{v}$, $I_1 = 35$ ma. Calculating RZ yields:

$$RZ = \frac{\Delta V}{\Delta I} = \frac{V_2 - V_1}{I_2 - I_1} = \frac{6.815\text{v} - 6.792\text{v}}{40\text{ ma} - 35\text{ ma}} = 4.6\Omega$$

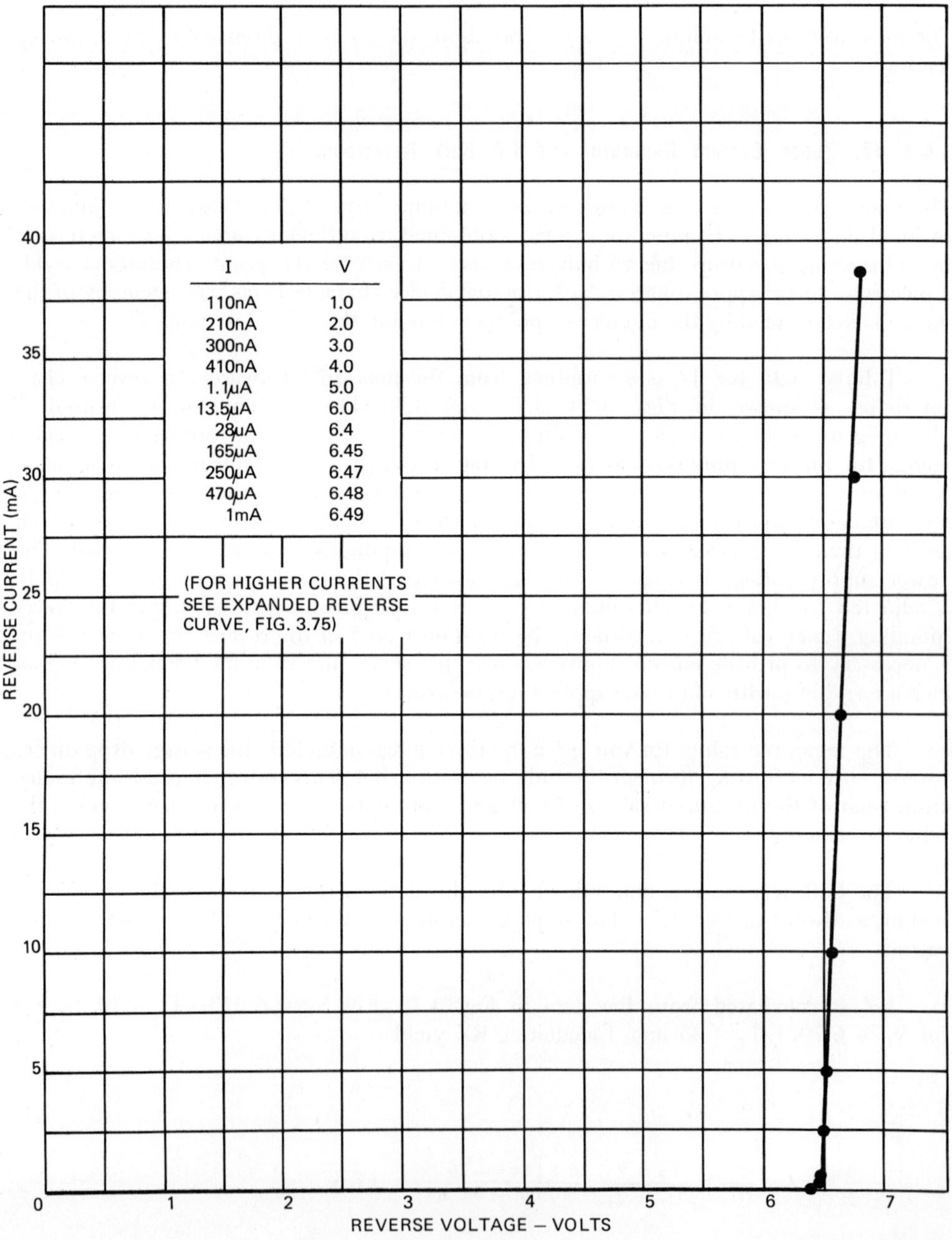

Fig. 3.74 1N754 reverse characteristic.

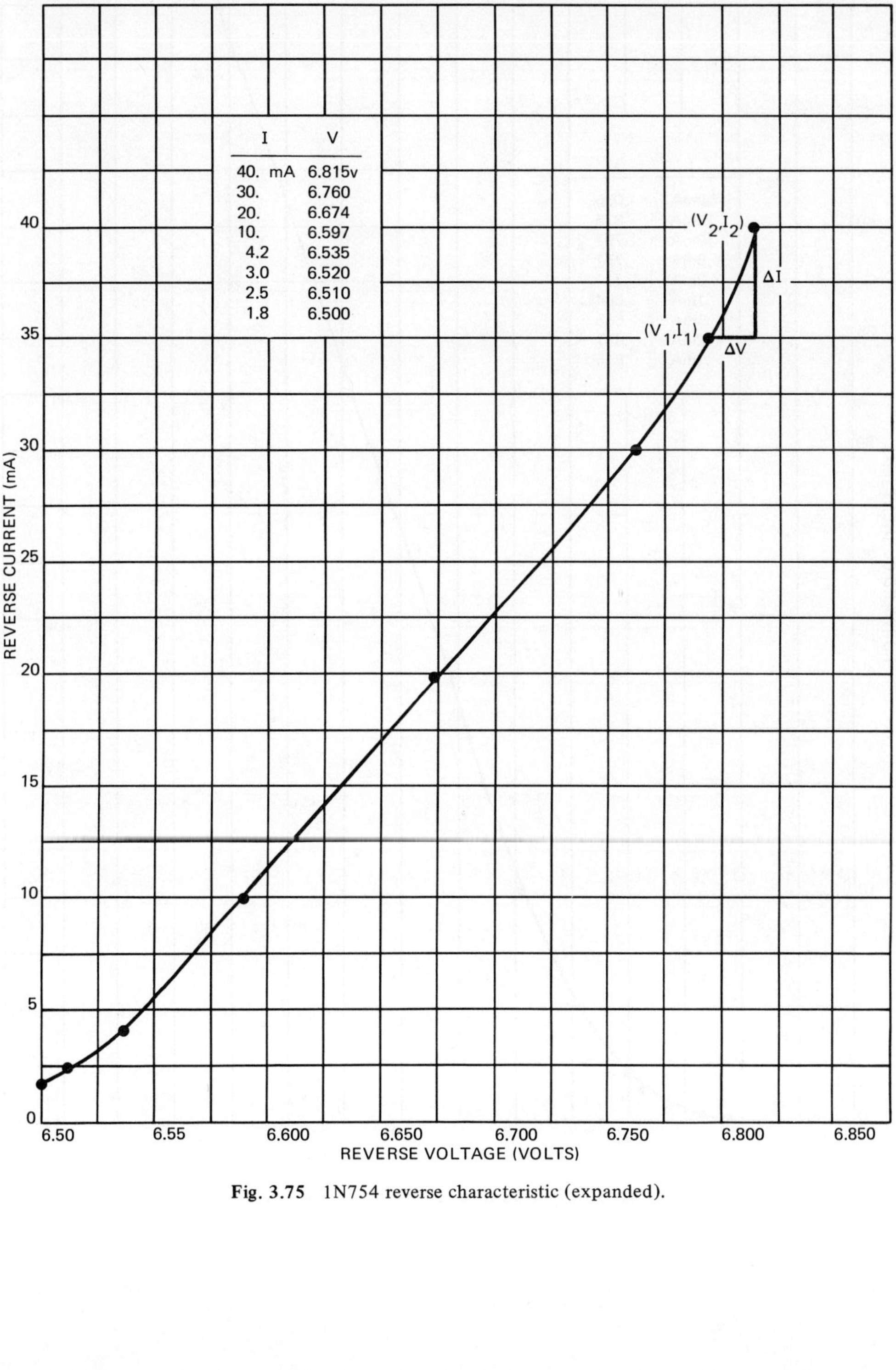

Fig. 3.75 1N754 reverse characteristic (expanded).

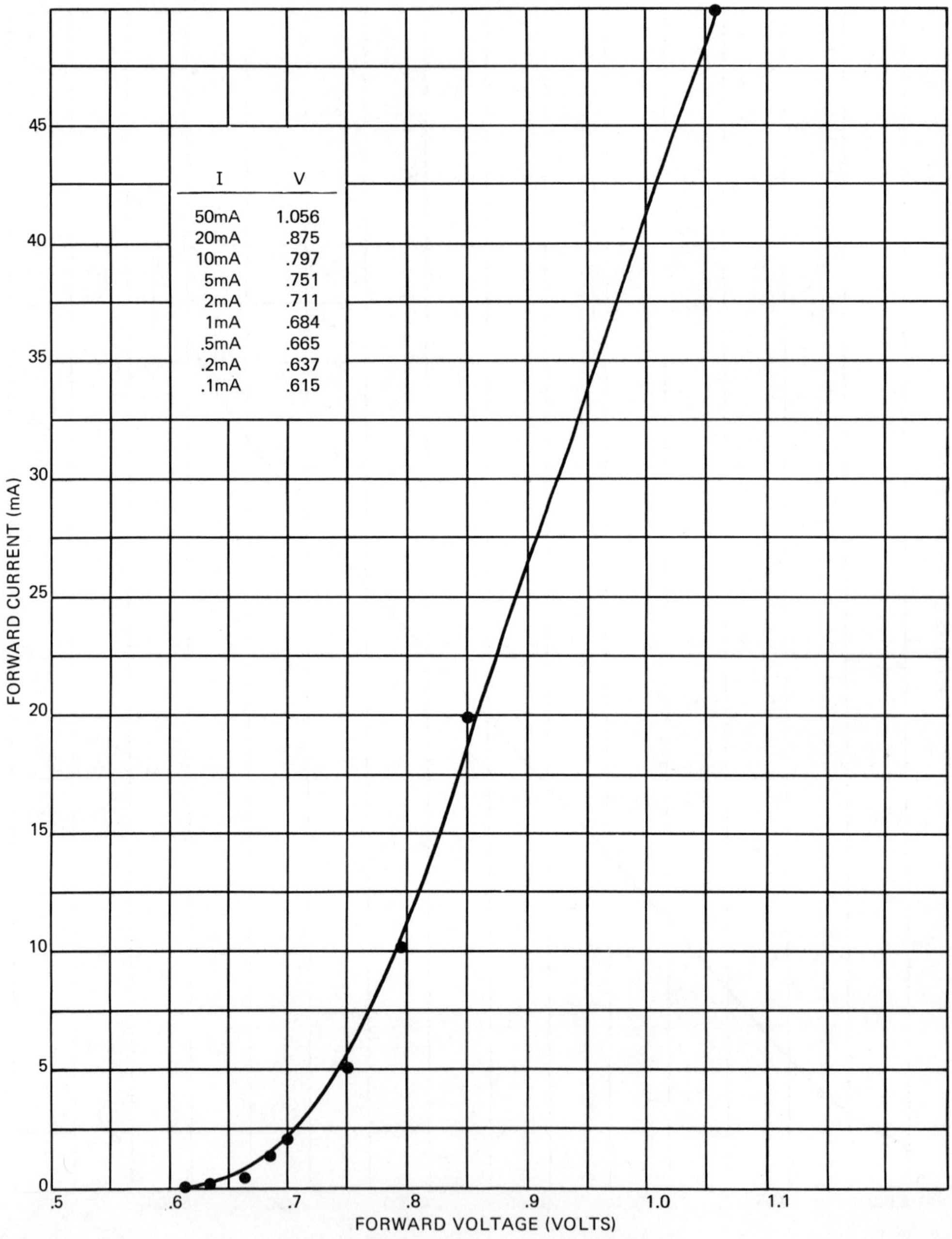

Fig. 3.76 Forward current vs. forward voltage.

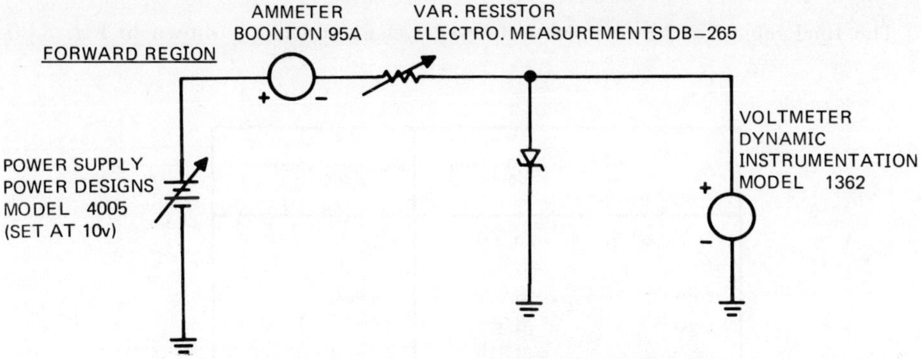

Fig. 3.77 Zener test circuit (forward region)

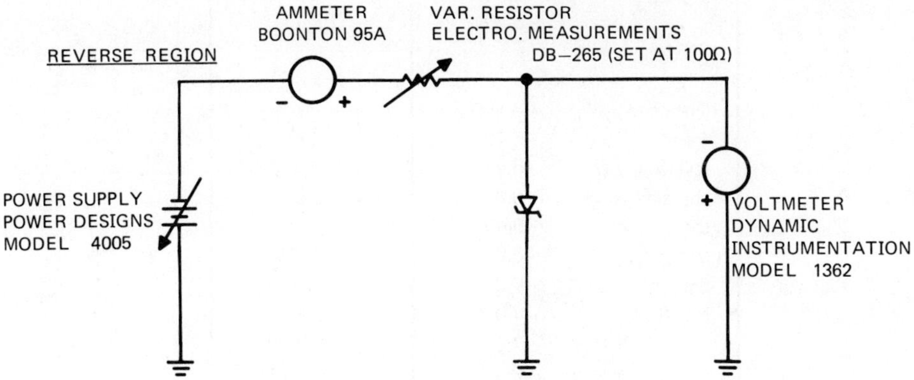

Fig. 3.78 Zener test circuit (reverse region)

Illustrated in Fig. 3.79 is a table with measured data points in the reverse region, showing ($I_{reverse}$ xR_{Zener}) drop at this current level, and with the corrected value of reverse junction voltage which is used as the independent table variable. It is seen from Fig. 3.79 that for currents less than 1.8 ma, the drop due to bulk resistance becomes negligible and the measured values of reverse voltage and current are entered directly in the table describing JZ.

REVERSE CURRENT (ma)	MEASURED REVERSE VOLTAGE (VOLTS)	$I_{Reverse} \times R_Z$ (VOLTS)	CORRECTED JUNCTION VOLTAGE $= V_{reverse} - I_{rev} R_Z$ (VOLTS)
40	6.815	.184	6.631
30	6.760	.138	6.622
20	6.674	.092	6.582
10	6.597	.046	6.551
4.2	6.535	.0193	6.515
3.0	6.520	.0143	6.505
2.5	6.510	.0115	6.498
1.8	6.500	.00828	6.491

Fig. 3.79 Corrected reverse region high current V–I characteristics

The final table for the forward and reverse characteristics is shown in Fig. 3.80.

	V_{Zener}(v)	I_{Zener}(ma)
Obtained	−6.631	−40
from	−6.622	−30
corrected	−6.582	−20
reverse	−6.551	−10
region	−6.515	− 4.2
table	−6.505	− 3.05
Fig. 3.79	−6.498	− 2.5
	−6.491	− 1.8
Obtained	−6.490	− 1
from	−6.48	− .470
reverse	−6.47	− .250
characteristics	−6.45	− .165
showing low	−6.40	− .028
current	−6.0	− .0135
region,	−5.0	− .0011
Fig. 3.74	−4.0	− .00041
	−3.0	− .00030
	−2.0	− .00021
	−1.0	− .00011
	0	0
Obtained	.615	.1
from	.637	.2
measured	.665	.5
forward	.684	1
characteristic,	.711	2
Fig. 3.76.	.751	5
	.797	10
	.875	20
	1.056	50

Fig. 3.80 Complete zener V–I table

3.4.5 CT, Transition Capacitance

The transition capacitance is obtained utilizing the same procedure as in the junction diode model. When measuring transition capacitance for the Zener, reverse voltage is applied only to the edge of breakdown. The capacitance remains relatively constant after breakdown since the junction voltage changes very little.

The measured values of transition capacitance versus reverse voltage are shown in Fig. 3.81.

For $V_1 = 0v$, $CT_1 = 376$ pf
$V_2 = 5v$, $CT_2 = 173$ pf

Assuming $\phi = 1.250$, then:

$$n = \frac{\ln CT_1 - \ln CT_2}{\ln(\phi - V_2) - \ln(\phi - V_1)} = \frac{\ln 376 - \ln 173}{\ln(1.25 + 5) - \ln(1.25 - 0)} = .4823$$

Using the value $CT_1 = 376$ pf at $V_1 = 0$:

$$CO = CT\,(\phi - V_1)^{.4823} = 376(1.25)^{.4823}$$

$$CO = 418.7$$

3.4.6 Zener Diode Data Summary

A summary of the required parameters of the Zener diode model of Fig. 3.73 is given in Table 3.3. The corresponding measured parameter values for the 1N754 Zener as developed in this section are also listed.

3.4.7 SCEPTRE Zener Diode Circuit Example

The Zener diode is used in a circuit which forces the device to operate over the full reverse and forward regions of operation. The test circuit is shown in Fig. 3.82. The SCEPTRE equivalent form of the test circuit is shown in Fig. 3.83. The circuit including the Zener model is shown coded in SCEPTRE form in Fig. 3.84 and 3.85. The output IRZD1(IZ), PLOT(P1(VZ)) is a plot of Zener current versus Zener voltage and is shown in Fig. 3.86. It is seen that the output illustrates the typical Zener diode curve. The curve is not an exact duplication of the tabular data because Fig. 3.86 represents a transient run. If the capacitive and other transient effects are to be nullified, a series of reruns can be made using the RUN INITIAL CONDITIONS ONLY cali. However, since the voltage input in this case is purposely very slowly varying with time, the transient curve very nearly duplicates the desired results.

Table 3.3 Summary of zener parameters

REQUIRED ZENER DIODE PARAMETERS	1N754 MEASURED VALUES
JZ	See Fig. 3.80
CO	418.7
ϕ	1.25 volts
n	.4823
RZ	4.6 kilohms

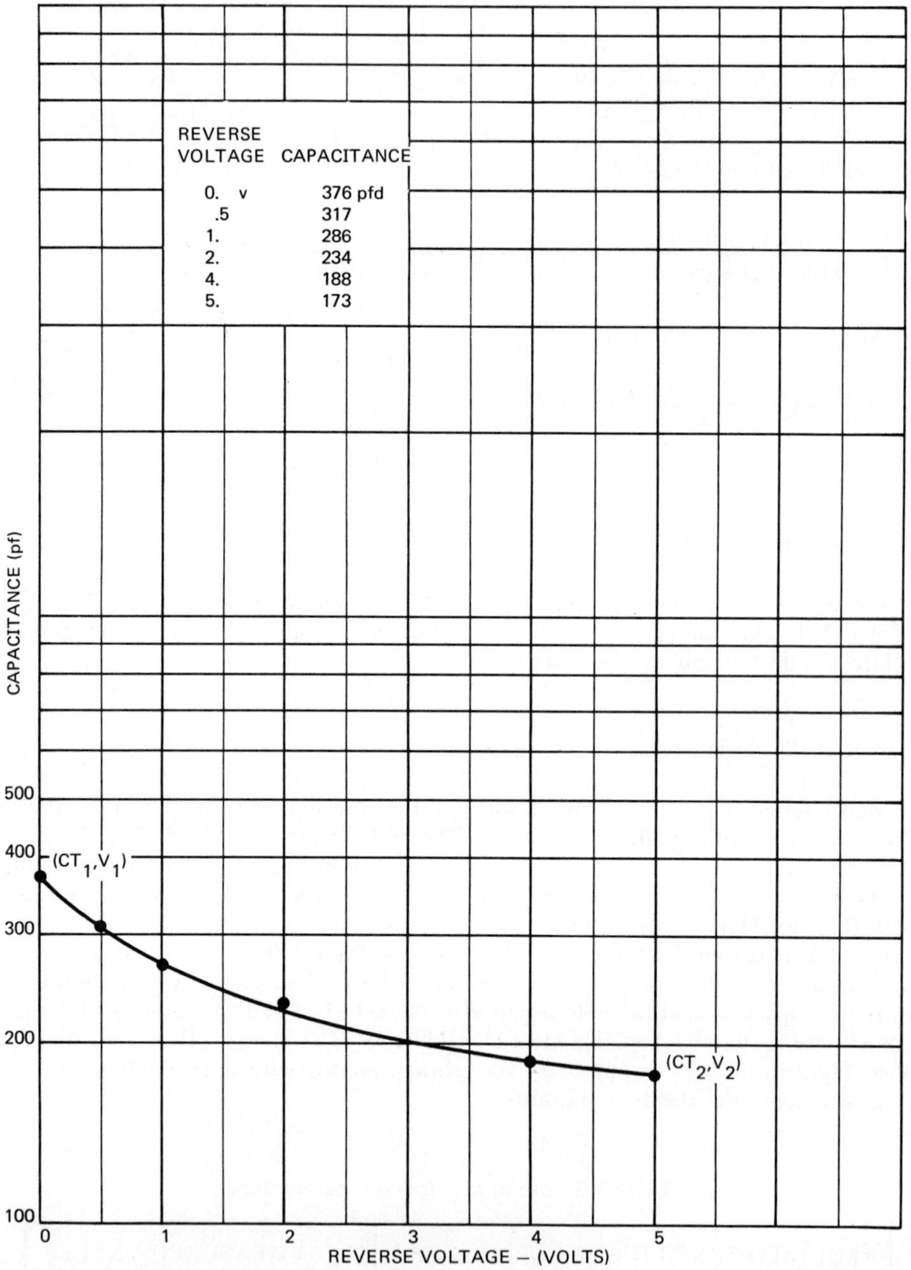

Fig. 3.81 1N754 Zener diode transition capacitance.

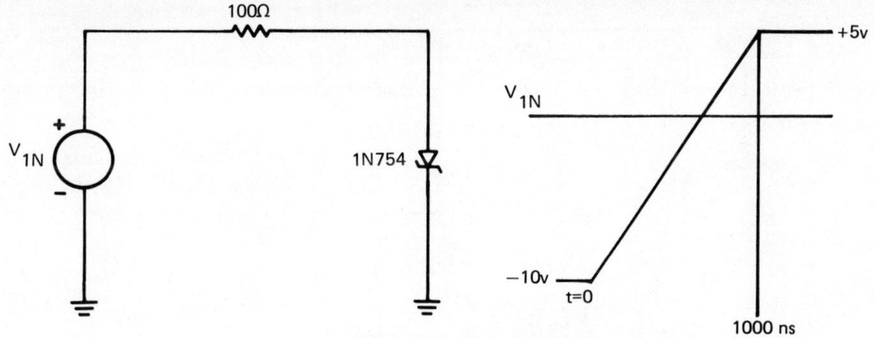

Fig. 3.82 Zener test circuit

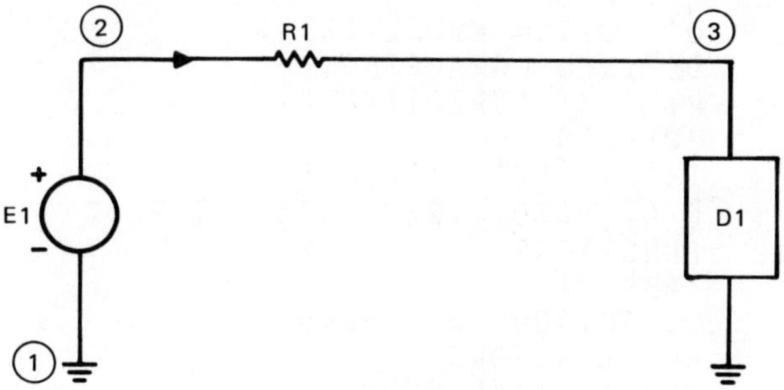

Fig. 3.83 Zener test circuit (sceptre form)

```
MODEL 1N754 (3-1)
ELEMENTS
RZ, 3-2 = .0046
CT, 2-1 = Q1(418.7,1.25,VCT,.4823)
JZ, 2-1 = DIODE TABLE 1
FUNCTIONS
Q1(A,B,C,D) = (A/(B-C)**D)
TABLE 1
-6.631,-40,-6.622,-30,-6.582,-20,-6.551,-10,-6.515,-4.2,
-6.47,-.25,-6.45,-.165,-6.4,-.028,-6.0,-.0135,-5.0,-.0011,
-4.0,-.00041,-3.0,-.00030,-2.0,-.00021,-1.0,-.00011,0,0,
.615,.1,.637,.2,.665,.5,.684,1,.711,2,.751,5,.797,10,.875,
20, 1.056,50
```

Fig. 3.84 In 754 Zener model description

```
ZENER TEST CIRCUIT
ELEMENTS
E1, 1-2 = TABLE 1 (TIME)
R1, 2-3 = .1
D1, 3-1 = MODEL 1N754
DEFINED PARAMETERS
P1 = X1 (VRZD1+VCTD1)
OUTPUTS
IRZD1 (IZ),PLOT(P1(VZ))
CTD1,VCTD1,IRZD1,JZD1,P1,PLOT
FUNCTIONS
TABLE 1
0,-10,1000,5,1050,5
RUN CONTROLS
RUN INITIAL CONDITIONS
STOP TIME = 1000
END
```

Fig. 3.85 Zener test circuit description

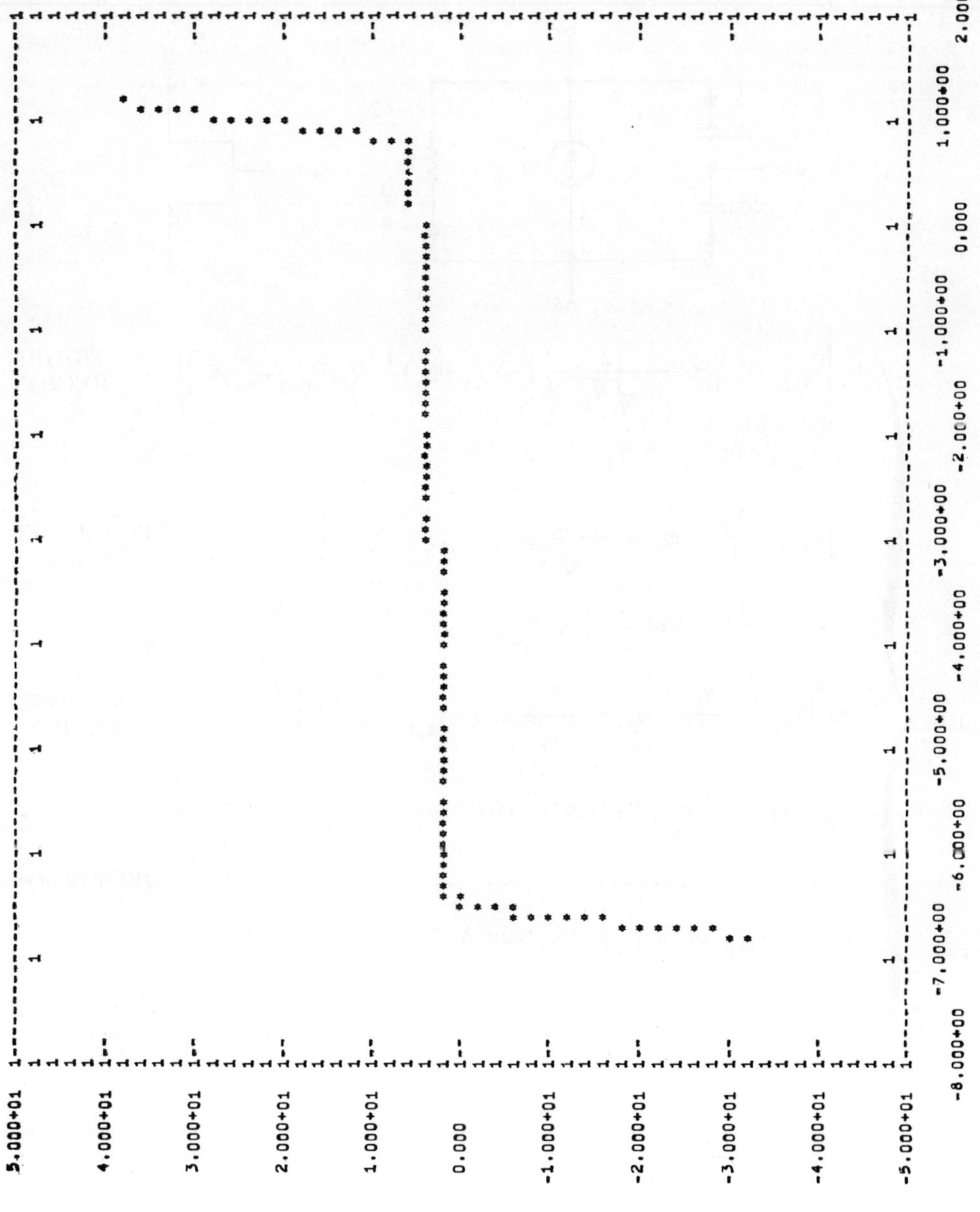

Fig. 3.86. Plot of IZ vs. VZ. (transient voltage sweep).

3.5 SCEPTRE JUNCTION FIELD EFFECT TRANSISTOR MODEL (P-CHANNEL)[6]

Figure 3.87 illustrates the selected JFET model, along with the defining equations.

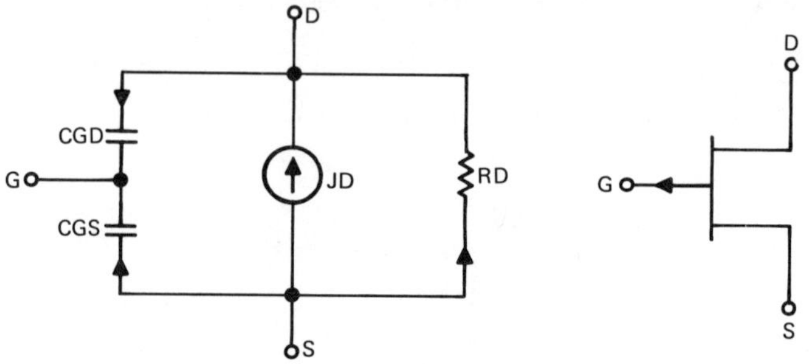

$$
JD=\begin{cases}
G_c\left\{V_{GS}-V_{GD}+\dfrac{2}{3\sqrt{V_p}}\left[(-V_{GS}+\Phi_B)^{\frac{3}{2}}-(-V_{GD}+\Phi_B)^{\frac{3}{2}}\right]\right\} & \text{TRIODE REGION} \\
V_{GD} \text{ and } V_{GS} \geq V_T & \\
G_c\left[V_{GS}+\dfrac{V_p}{3}-\Phi_B+\dfrac{2}{3\sqrt{V_p}}(-V_{GS}+\Phi_B)^{\frac{3}{2}}\right] & \text{PINCH-OFF REGION} \\
V_{GD} < V_T \text{ and } V_{GD} < V_{GS} & \\
-G_c\left[V_{GD}+\dfrac{V_p}{3}-\Phi_B+\dfrac{2}{3\sqrt{V_p}}(-V_{GD}+\Phi_B)^{\frac{3}{2}}\right] & \text{INVERTED REGION} \\
V_{GS} < V_T \text{ and } V_{GS} < V_{GD} & \\
0 & \text{CUT-OFF REGION} \\
V_{GD} = V_{GS} \text{ or } (V_{GD} < V_T \text{ and } V_{GS} < V_T) & \\
0 & \text{FORWARD BIASED REGION (JD is set equal to zero for this forbidden region)} \\
V_{GD} > \Phi_B \text{ or } V_{GS} > \Phi_B &
\end{cases}
$$

$$CGS = \frac{CG}{(\Phi_B - V_{GS})^{n_G}}$$

$$CGD = \frac{CD}{(\Phi_B - V_{GD})^{n_D}}$$

$$RD = \frac{I_T \cdot RO}{JD + K}$$

Fig. 3.87 Sceptre JFET model

3.5.1 Definition of Terms

JD = A current generator representing the drain to source current, excluding the variation of drain current with drain to source voltage in the PINCH–OFF region (ma).

CGS = The gate to source capacitance (pfd).

CGD = The gate to drain capacitance (pfd).

RD = The resistance which models the variation in drain current with drain to source voltage in the PINCH–OFF region. This resistance is a function of JD (kilohms).

P = The PINCH–OFF voltage (volts).

V_T = $\phi_B - V_p$

ϕ_B = The junction contact potential (volts)

V_{GS} = The gate to source voltage (volts).

Also,

V_{GS} = VCGS

V_{GD} = The gate to drain voltage (volts)

Also,

V_{GD} = VCGD

G_c – The channel conductance (ma/volt).

CG = Constant of the gate to source capacitance equation.

CD = Constant of the gate to drain capacitance equation.

n_G = Exponent of the gate to source capacitance equation.

n_D = Exponent of the gate to drain capacitance equation.

RO = The resistance which models the variation in drain current with drain to source voltage in the PINCH–OFF region with $V_{GS} = 0$(kilohms).

I_T = The drain to source current for $V_{GS} = 0$ and drain to source voltage $= -|V_T|$

K = Current chosen to reduce the drain to source voltage/RD ratio to a small leakage current when JD = 0 (ma).

$C_{iss} = CGS + C_{rss}$

$C_{rss} = CGD$

3.5.2 Introduction

The Junction Field Effect transistor model described in this section utilizes the FORTRAN subroutine feature to describe the drain current. The model covers all four of the device operating regions. Because of the type of dependencies involved in the JFET subroutine, the model will not converge for initial conditions. However, this limitation is no large problem and several methods to circumvent this deficiency are described. For example, if a particular circuit is to be used with a number of different inputs, with many reruns or further runs, the suggested procedure is to apply no input voltage and to run in the transient mode for an appropriate number of nanoseconds consistent with the estimated system time constants. For outputs, call out all capacitor voltages. These plots indicate if the voltages converged to steady state values and what these values are. These final voltages represent the initial condition values and are entered as such for all subsequent runs. In most cases the user can save additional computer time by computing the approximate I.C. values using short-cut hand calculations. Then the transient run does not have to run as long to obtain the accurate values, since it is starting from voltages very close to the actual final results.

Another procedure, which is quite simple and does not require two separate runs, is to once again calculate the approximate I.C. values. Enter these values into the program and allow an appropriate time on the transient run before the step, or other input is applied to the circuit. Therefore, instead of the input being applied at t = 0, which is the usual case, the input is applied at, say, t = 50 ns. Between time equal zero to fifty, the system is converging from the approximate values to the exact steady state values, and from that point on, the results are the same as any transient run, except "time" starts at t = 50ns. User experience makes the above manipulations quite routine and no problems result. This procedure is illustrated in Sec. 3.5.7 with an example. The subroutine used for the JFET model is shown at the end of this section in Fig. 3.108.

This model does not include the gate to source and gate to drain leakage resistances. These can easily be added if the application requires.

3.5.3 V_p, G_c, Drain to Source Current Generator Parameters

The parameters V_p and G_c are obtained from the JFET drain current versus drain to source voltage characteristic in the PINCH–OFF region. In this region, the drain current is relatively independent of drain to source voltage. The drain characteristic for the 2N2498 JFET is obtained using the test configuration in Fig. 3.88. The gate to source voltage is initially set to zero and the drain current recorded for drain to source voltages from −.5v to −15v. The process is then repeated for gate to source voltages of +.5, +1.5, and +2v. The data obtained is tabularized in Fig. 3.89 and plotted in Fig. 3.90.

The break point between the PINCH–OFF and TRIODE region for zero gate to source voltage is at $V_{DS} = -[\,|V_P| - |\Phi_B|\,]$ for a P channel device. The equations describing the P–Channel Model were originally derived for an N–Channel device. In order to adapt these equations for use with a P–Channel device and obtain the proper numerical values for the parameters, it is necessary to make frequent use of absolute value signs. The equation which describes the device operation in the PINCH–OFF region of $V_{GS} = 0$ is:

$$JD = I_{Drain} = G_c \left[\frac{|V_p|}{3} - |\Phi_B| + \frac{2}{3\sqrt{|V_p|}} \left(|\Phi_B|\right)^{\frac{3}{2}} \right]$$

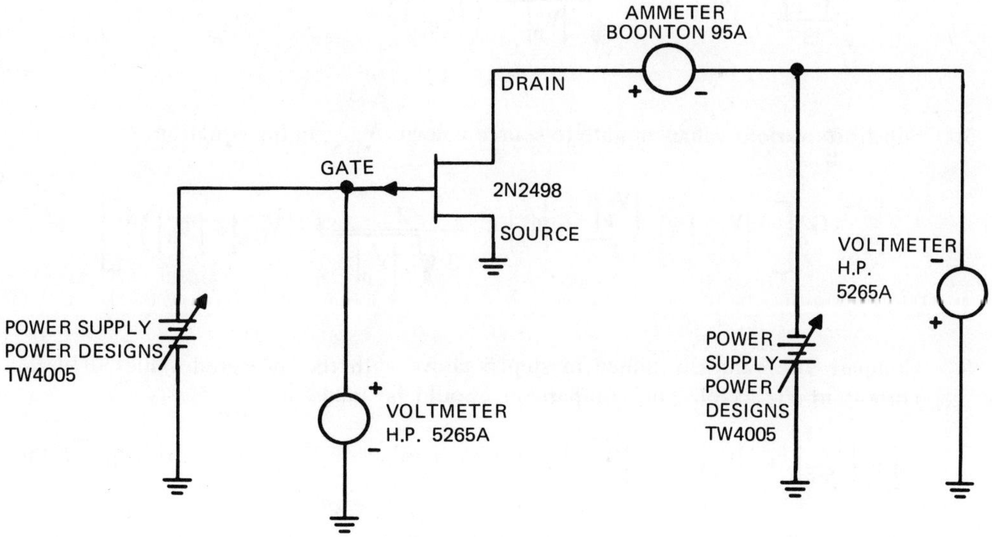

Fig. 3.88 JFET drain characteristic test

This equation states that in the PINCH–OFF region, the drain current is independent of the drain to source voltage. Utilizing this equation and the previously measured drain current characteristic, G_c and V_p are obtained by the following iterative process:

1. Assume a value for Φ_B.

   Φ_B should be in the range .6v to 1v.

2. From the measured drain characteristic, Fig. 3.90 with $V_{GS} = 0$, determine (approximately) the value of drain to source voltage where the break point between the PINCH–OFF and TRIODE region occurs.

3. Calculate $|V_p|$ from:

$$V_p = |\Phi_B| + |V_{DS}|$$

4. Calculate G_c from:

$$G_c = \frac{I_{Drain}\big|_{VDS = -[\,|V_p| - |\Phi_B|\,]}}{\frac{|V_p|}{3} - |\Phi_B| + \frac{2}{3\sqrt{|V_p|}}\left(|\Phi_B|\right)^{\frac{3}{2}}}$$

5. Substitute various values of gate to source voltage, V_{GS}, in the equation:

$$I_{Drain} = G_c\left[-|V_{GS}| + \frac{|V_p|}{3} - |\Phi_B| + \frac{2}{3\sqrt{|V_p|}}\left(-|V_{GS}| + |\Phi_B|\right)^{\frac{3}{2}}\right]$$

6. Compare the values obtained in step 5 above with the measured values of drain current in Fig. 3.89. The comparison should be made at $V_{DS} =$

   $-[-|V_{GS}| + |V_p| - |\Phi_B|]$ for each value of V_{GS}, since the edge of the

   PINCH–OFF region is a function of gate voltage.

7. If a good comparision results between the calculated drain current and the measured drain current, the procedure is finished. If the comparison is unsatisfactory, a new value of V_{DS} at which the break point occurs in Step 2 is assumed and the process is repeated. This procedure is readily implemented on a small digital computer. The break point between the PINCH–OFF and TRIODE region for the 2N2498 is determined from Fig. 3.90 with $V_{GS} = 0$. The value is seen to be between V_{DS}

V_{GS} =	0v		
V_{DS} =	− .5v	I_D =	−1.50 ma
	−1.0		−2.65
	−2.0		−4.00
	−2.5		−4.32
	−2.7		−4.40
	−3.0		−4.50
	−5.0		−4.75
	−10.0		−4.83
	−15.0		−4.85
V_{GS} =	+ .5v		
V_{DS} =	− .5v	I_D =	−1.18 ma
	−1.0		−2.05
	−2.0		−2.87
	−3.0		−3.10
	−5.0		−3.25
	−10.0		−3.33
	−15.0		−3.34
V_{GS} =	+ 1.0v		
V_{DS} =	− .5v	I_D =	− .850 ma
	− 1.0		−1.36
	− 2.0		−1.71
	− 3.0		−1.80
	− 5.0		−1.87
	−10.0		−1.92
V_{GS} =	+1.5v		
V_{DS} =	− .5v	I_D =	−.474 ma
	−1.0		−.650
	−2.0		−.745
	−3.0		−.785
	−5.0		−.820
	−10.0		−.858
	−15.0		−.870
V_{GS} =	2.0v		
V_{DS} =	− .5v	I_D =	−.071 ma
	−1.0		−.090
	−2.0		−.108
	−3.0		−.117
	−5.0		−.128
	−10.0		−.142
	−15.0		−.148

Fig. 3.89. JFET drain characteristic data.

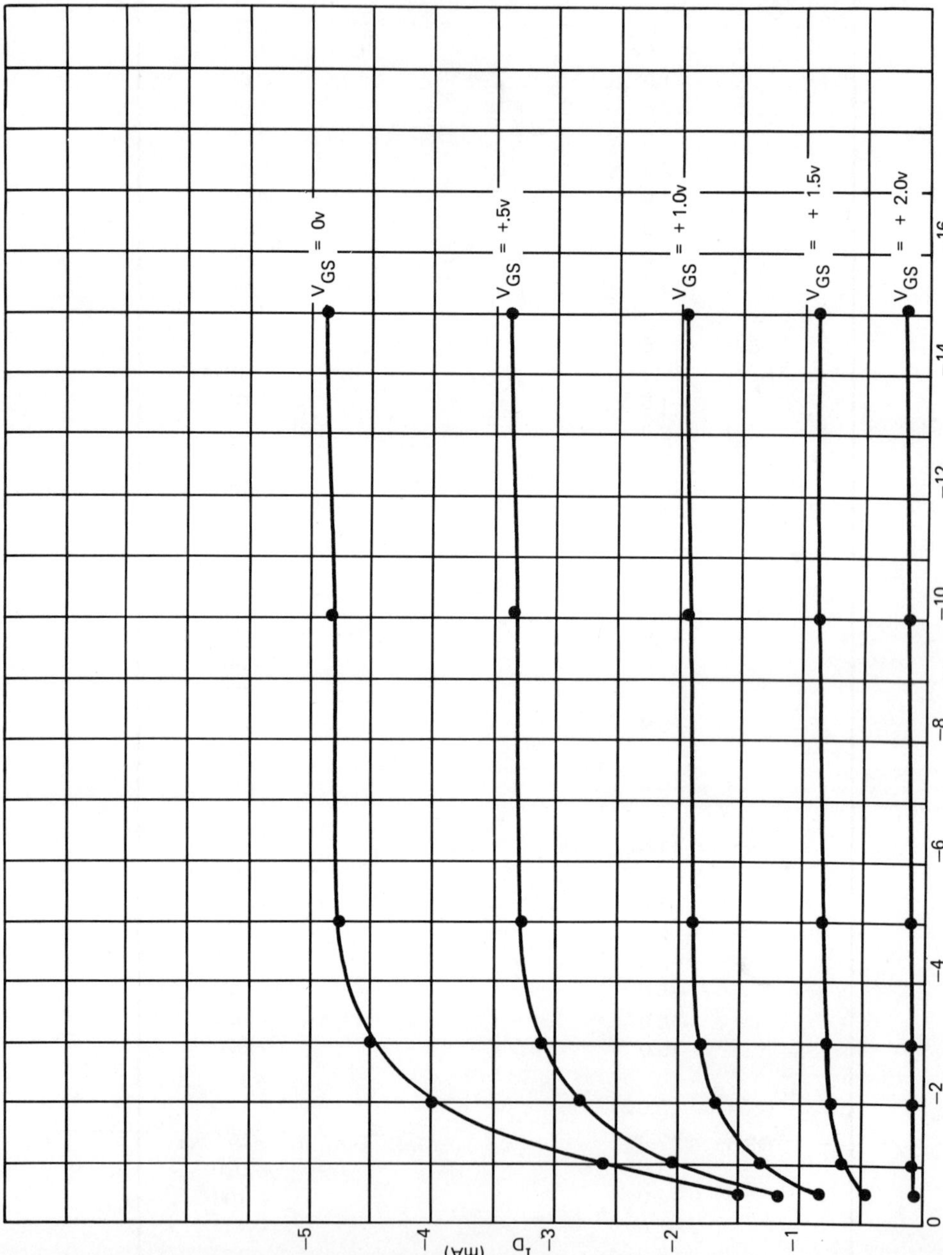

Fig. 3.90 2N2498 drain characteristic.

= −2v and V_{DS}= −4v. After several iterations*, the break point is determined to be V_{DS} = −2.7v, I_D = −4.4 ma. The corresponding values of $|V_P|$ and $|G_c|$ are $|V_P|$ = 3.7v and $|G_c|$ = 7.587 ma/v. The resulting drain currents for V = 0v, .5v, 1v, 1.5v, 2v are tabulated in Fig. 3.91. The measured value of drain current for each value of V_{GS} at $V_{DS} = -[-|V_{GS}| + |V_P| - |\Phi_B|]$ is also interpolated from Fig. 3.89.

V_{GS}	I_D Calculated	I_D Measured at $V_{DS} = -\left[-|V_{GS}| + |V_P| - |\Phi_B|\right]$
0	4.4ma	4.4ma
+ .5	2.808ma	2.91ma
+1.0	1.621ma	1.65ma
+1.5	.784ma	.680ma
+2.0	.260ma	.080ma

Fig. 3.91. Calculated vs. measured drain current.

These values represent the best fit to the device characteristics for this model. Inaccuracies at low current levels can be primarily attributed to the assumption of a step junction and rectangular geometry.

3.5.4 RO, I_T, K Parameters of the RD Equation

RO is determined from two points on the drain to source characteristic for V_{GS} = 0. The first point occurs at the break point between the saturation and triode region. For the 2N2498 this occurs at $I_D = I_T = -$ 4.4ma and V_{DS} = 2.7v. The second point is chosen on the V_{GS} = 0 curve to give the best simulation of the change in drain current with drain to source voltage in the saturation region. A value of $V_{DS} = -$ 4.85ma yields the closest fit to the actual curve. RO is then calculated from:

$$RO = \frac{\Delta V}{\Delta I} = \frac{15v \;-\; 2.7v}{4.85\text{ ma} \;-\; 4.4\text{ ma}} = 27.3 \text{ Kilohms.}$$

The value for K is determined from drain to source leakage current measurements in the cutoff region for critical applications. For most applications K is assigned an arbitrary value, less than 1μa. For the 2N2498 a value of K = .0001 ma is chosen.

*See Appendix J

3.5.5 CG, CD, n_F, n_D, Gate to Source and Gate to Drain Capacitance Parameters

The values for these parameters are obtained from measurements of C_{iss} as a function of gate to source voltage and C_{rss} as a function of gate to drain voltage. The test circuits for measuring C_{iss} and C_{rss} are shown in Figs. 3.92 and 3.93.

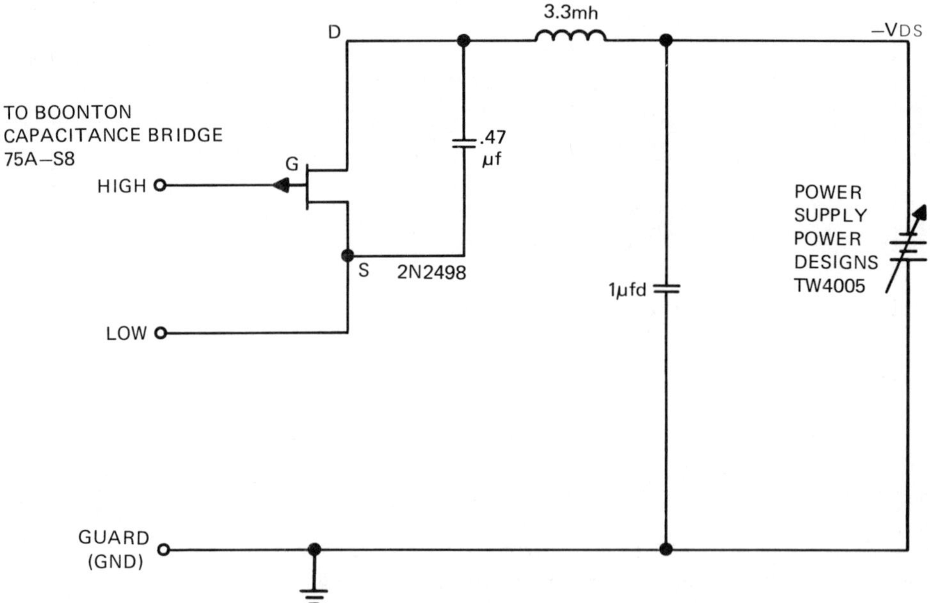

Fig. 3.92 C_{iss} measurement test circuit

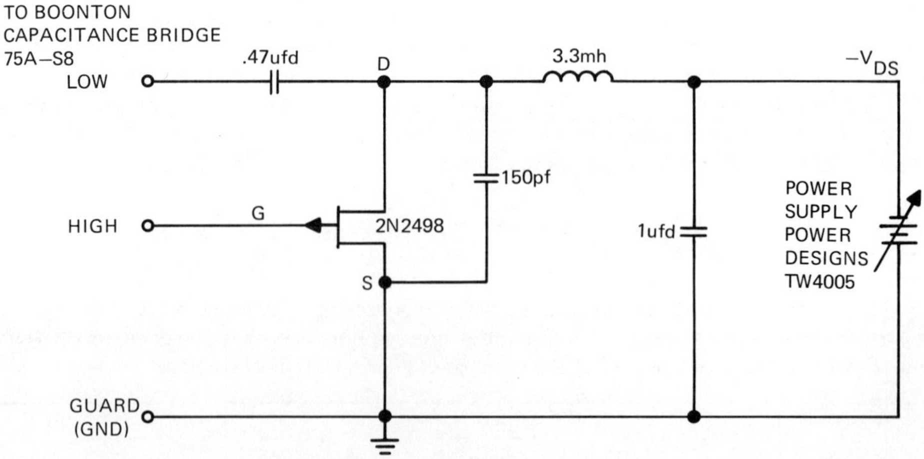

Fig. 3.93 C_{rss} measurement test circuit

In determining C_{iss}, the drain to source voltage is maintained at a constant value of −10v. The value of C_{iss} is then measured for gate to source voltages of 0v to + 9v, utilizing the internal supply of the BOONTON bridge. The values obtained for C_{iss} are tabularized in Fig. 3.94. In determining C_{rss}, the drain to source voltage and gate to source voltage are varied to yield gate to drain voltages from 0v to −19v. The values for V_{DS}, V_{GS}, V_{GD}, and C_{rss} are tabularized in Fig. 3.95. The gate to source capacitance is given by CGS = C_{iss} − C_{rss}. The values for CGS are calculated and tabularized in Fig. 3.94, for corresponding gate to source and drain to source biases. The gate to drain capacitance is given by CGD = C_{rss}. CGS and CGD are plotted as a function of reverse bias voltage in Figs. 3.96 and 3.97.

Knowing CGS and CGD as a function of voltage, the parameters CG, CD, n_G, n_D are obtained utilizing the procedure used in determining the coefficients of the transition capacitances for the bipolar transistor. The value of ϕ_B has previously been determined as 1v.

Considering two points (V1,CGS1), (V2,CGS2) on the curve of CGS versus voltage, the value for n_G is obtained from:

$$n_G = \frac{\ell n\ CGS_1 - \ell n\ CGS_2}{\ell n(\phi_B - V_2) - \ell n(\phi_B - V_1)} = \frac{\ell n 4.38 - \ell n 3.27}{\ell n(2) - \ell n(1)} = .4216$$

CG is then calculated from the point V = −.5v, CGS = 3.63pf using:

$$CG = CGS(\phi_B - V)^n = 3.63(1 + .5)^{.4216}$$

or, $CG = 4.31$

Considering two points (V1,CGD1), (V2,CGD2) on the curve of CGD versus voltage, the value of n_D is obtained from:

$$n_D = \frac{\ell n\ CGD_1 - \ell n\ CGD_2}{\ell n(\phi_B - V_2) - \ell n(\phi_B - V_1)} = \frac{\ell n 3.43 - \ell n 2.56}{\ell n 3 - \ell n 1.5}$$

or, $n_D = .4221$

CD is then calculated from the point V = − .5, CGD = 3.43 using,

$$CD = CGD(\phi_B - V)^n = 3.43(1 + .5)^{.4221}$$

$$CD = 4.07$$

V_{DS}	V_{GS}	C_{iss}	$C_{GS} = C_{iss}\big\vert_{V_{DS}=-10v} - Crss\big\vert_{V_{DS}=-10v}$	
−10v	0 v	5.9 pf	5.9−1.52	= 4.38 pf
−10v	.5	5.13	5.13−1.50	= 3.63 pf
−10v	1.0	4.74	4.74−1.47	= 3.27 pf
−10v	2.0	4.28	4.28−1.43	= 2.85 pf
−10v	5.0	3.45	3.45−1.36	= 2.09 pf
−10v	10.0	3.02	3.02−1.29	= 1.73 pf

Fig. 3.94 Gate to source capacitance data

V_{DS}	V_{GS}	$V_{GD} = V_{DS} - V_{GS}$	$C_{rss} = C_{GD}$
− .5v	0 v	− .5v	3.43 pf
− 1.0	0	− 1.0	3.02
− 2.0	0	− 2.0	2.56
− 5.0	0	− 5.0	1.89
−10.0	0	−10.0	1.52
−10.0	.5	−10.5	1.50
−10.0	1.0	−11.0	1.47
−10.0	2.0	−12.0	1.43
−10.0	5.0	−15.0	1.36
−10.0	9.0	−19.0	1.29

Fig. 3.95 Gate to drain capacitance data

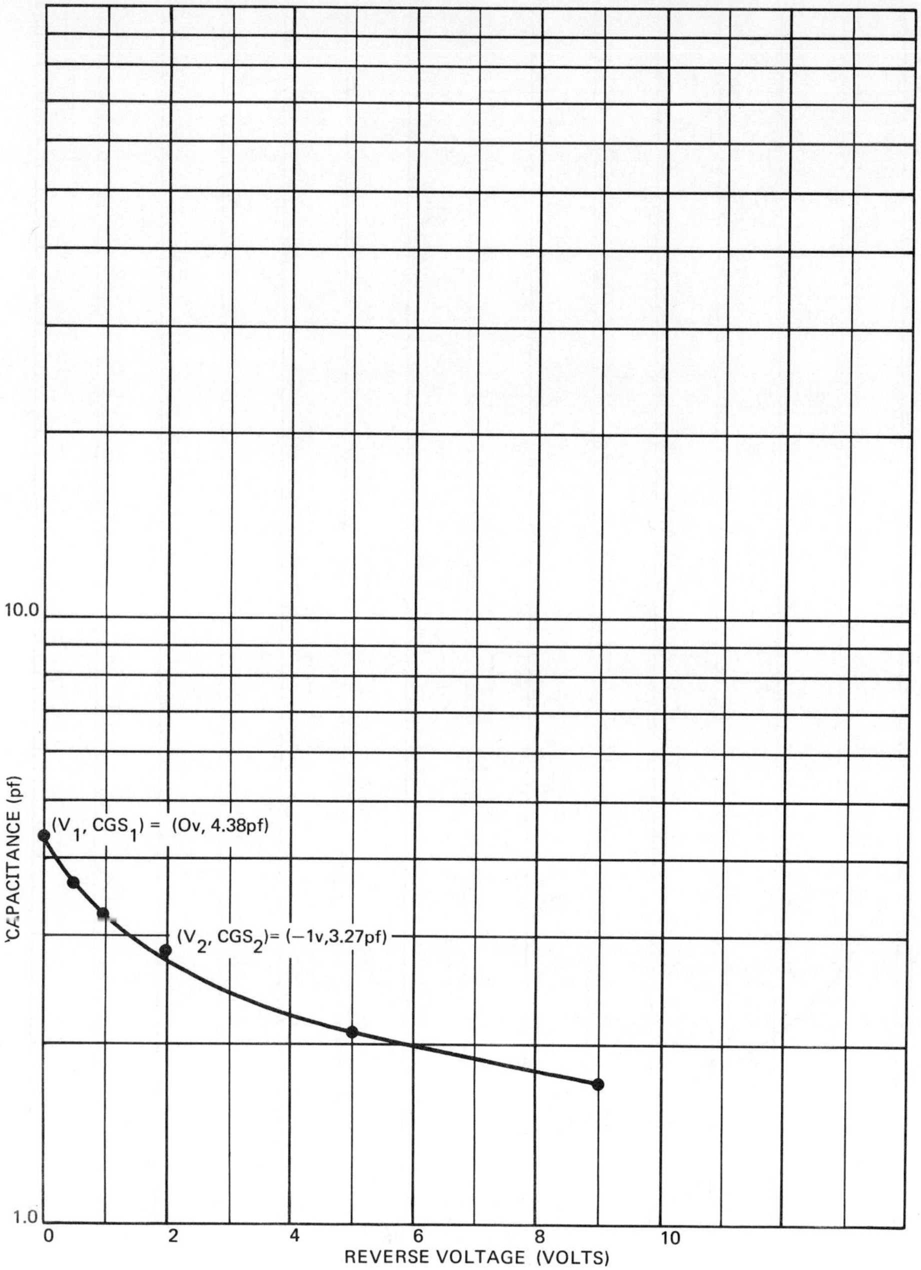

Fig. 3.96 CGS curve data

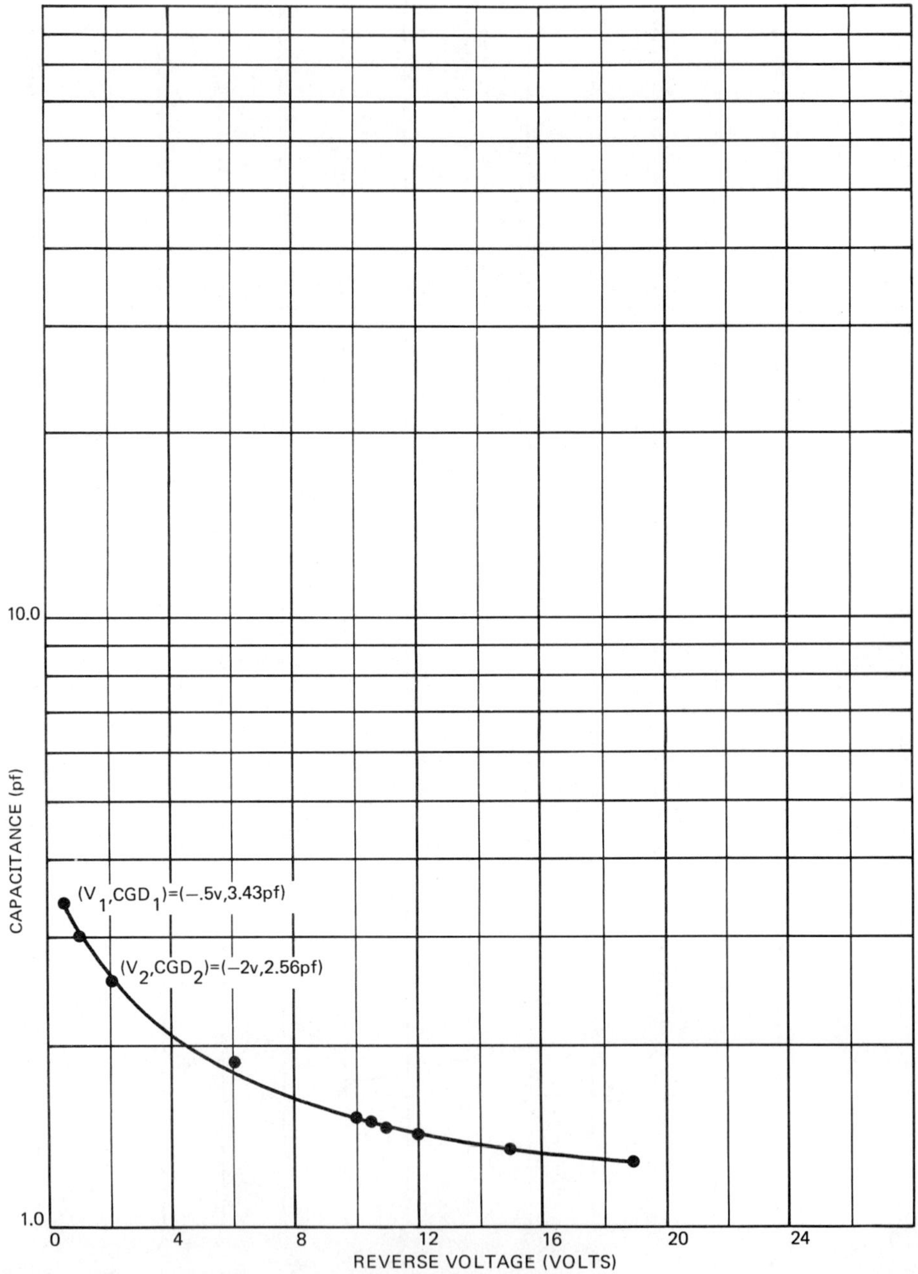

Fig. 3.97 CGD curve data

3.5.6 Summary of Model Parameters

A summary of the required parameters for the JFET transistor model of Fig. 3.87 is given in Table 3.4. The corresponding measured parameter values for the 2N2498 as developed in this section are also listed.

Table 3.4 Summary of JFET parameters

REQUIRED JFET PARAMETERS	2N2498 MEASURED VALUES
V_p	3.7 volts
G_C	7.587 ma/volt
ϕ_B	1 volt
CG	4.31
CD	4.07
n_G	.4216
n_D	.4221
I_T	4.4 ma
RO	27.3 Kilohms
K	.0001 ma

3.5.7 SCEPTRE Circuit Example-JFET Switch

The 2N2498 transistor from which the model parameters are obtained is set up in the test circuit shown in Fig. 3.98. The circuit input voltage and the output drain to ground voltage are shown in Fig. 3.100. The input voltage and the output gate voltage are shown in Fig. 3.101.

The test circuit is shown in SCEPTRE form in Fig. 3.99. In addition to the capacitances in the transistor model, stray capacitances measured from drain to ground (C1) and gate to ground (C2) are included in the SCEPTRE circuit. These capacitances were measured to be 7.5 pf and 2.6 pf respectively. The measurements are taken with the device removed from the test socket. These stray capacitances are due to test circuit wiring, BNC connector capacitance, scope probe capacitance, and test socket capacitance.

As discussed in the introduction, there are two approaches to obtaining initial conditions. One of them is shown here for illustration and also to verify the measurement techniques. The approach where the I.C. calculations and a transient input are combined into one run is shown. First, the approximate initial capacitor voltages are calculated. Second, a conservative estimate of 50 ns is used for the input settling time to allow the approximate values to converge to the actual I.C. values.

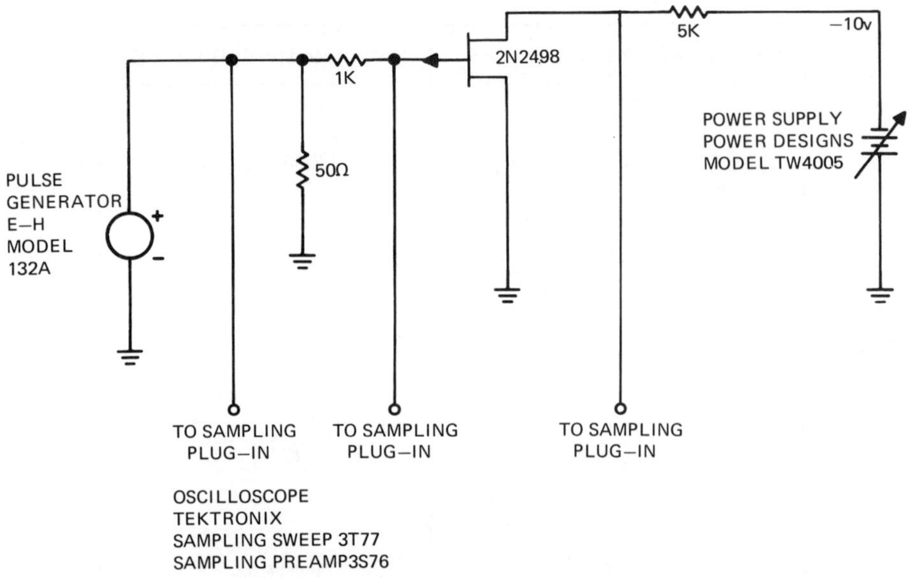

Fig. 3.98 JFET switch test circuit.

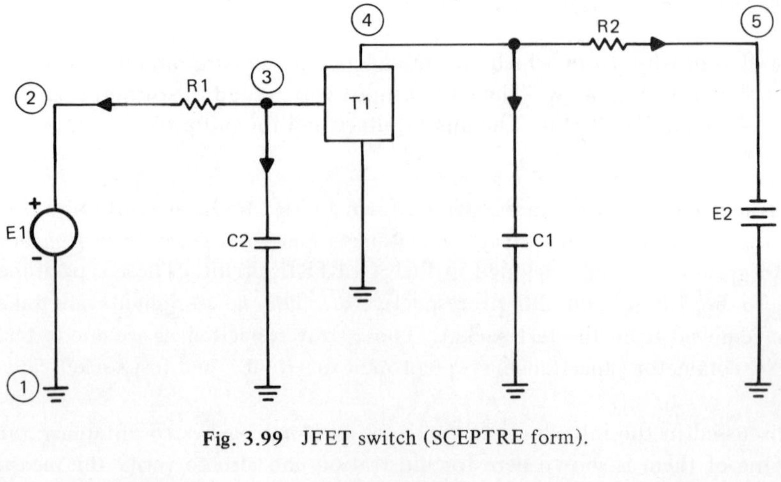

Fig. 3.99 JFET switch (SCEPTRE form).

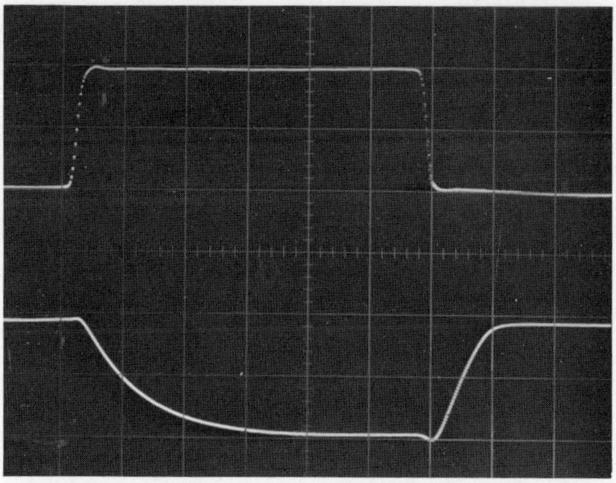

JFET switch waveforms.
Fig. 3.100

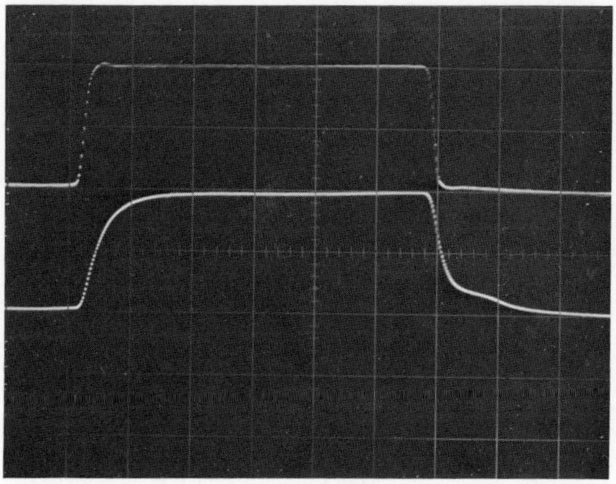

JFET switch waveforms.
Fig. 3.101

The listing of the JFET model and the circuit description are shown in Figures 3.102 and 3.103. Figure 3.104 represents the input voltage to the JFET SWITCH. The 50 ns time delay is obvious from this Figure. Figure 3.105 represents one of the capacitor voltages and indicates that the approximate values are very close to the actual values. In fact, on the graphical scale it is difficult to see that the values changed at all. The tabular data, however, reveals about a 7% change over the 50 ns period.

Figures 3.106 and 3.107 illustrate the output voltage (drain to ground) and the gate voltage. A comparison of Figures 3.106 and 3.107 with Figures 3.100 (bottom trace) and 3.101 (bottom trace) shows close correlation between the test results and the SCEPTRE simulation.

```
MODEL 2N2498 (G-S-D)
ELEMENTS
CGS, S-G = Q1(4.31,1.,VCGS,.4216)
CGD, D-G = Q1(4.07,1.,VCGD,.4221)
JD, S-D = FJFET(VCGS,VCGD,3.7,7.587,1.0)
RD, S-D = Q2(27.3,4.4,JD,.0001)
FUNCTIONS
Q1(A,B,C,D) = (A/(B-C)**D)
Q2(A,B,C,D) = ((A*B)/(C+D))
```

Fig. 3.102 2N2498 JFET Model description

```
JFET SWITCH
ELEMENTS
R1, 3-2 = 1
R2, 4-5 = 5
C1, 4-1 = 7.5
C2, 3-1 = 2.6
E2, 5-1 = 10
E1, 1-2 = TABLE 1 (TIME)
T1, 3-1-4 = MODEL 2N2498
DEFINED PARAMETERS
P1 = (VR2-E2)
OUTPUTS
E1,P1(VOUT),VCGST1,VCGDT1,VC2,JDT1,PLOT
INITIAL CONDITIONS
VC1 = -.65
VC2 = 0
VCGST1 = 0
VCGDT1 = -.65
FUNCTIONS
TABLE 1
0,0,50,0,60,4.85,330,4.85,340,0,350,0
RUN CONTROLS
STOP TIME = 400
END
```

Fig. 3.103 JFET Test circuit description

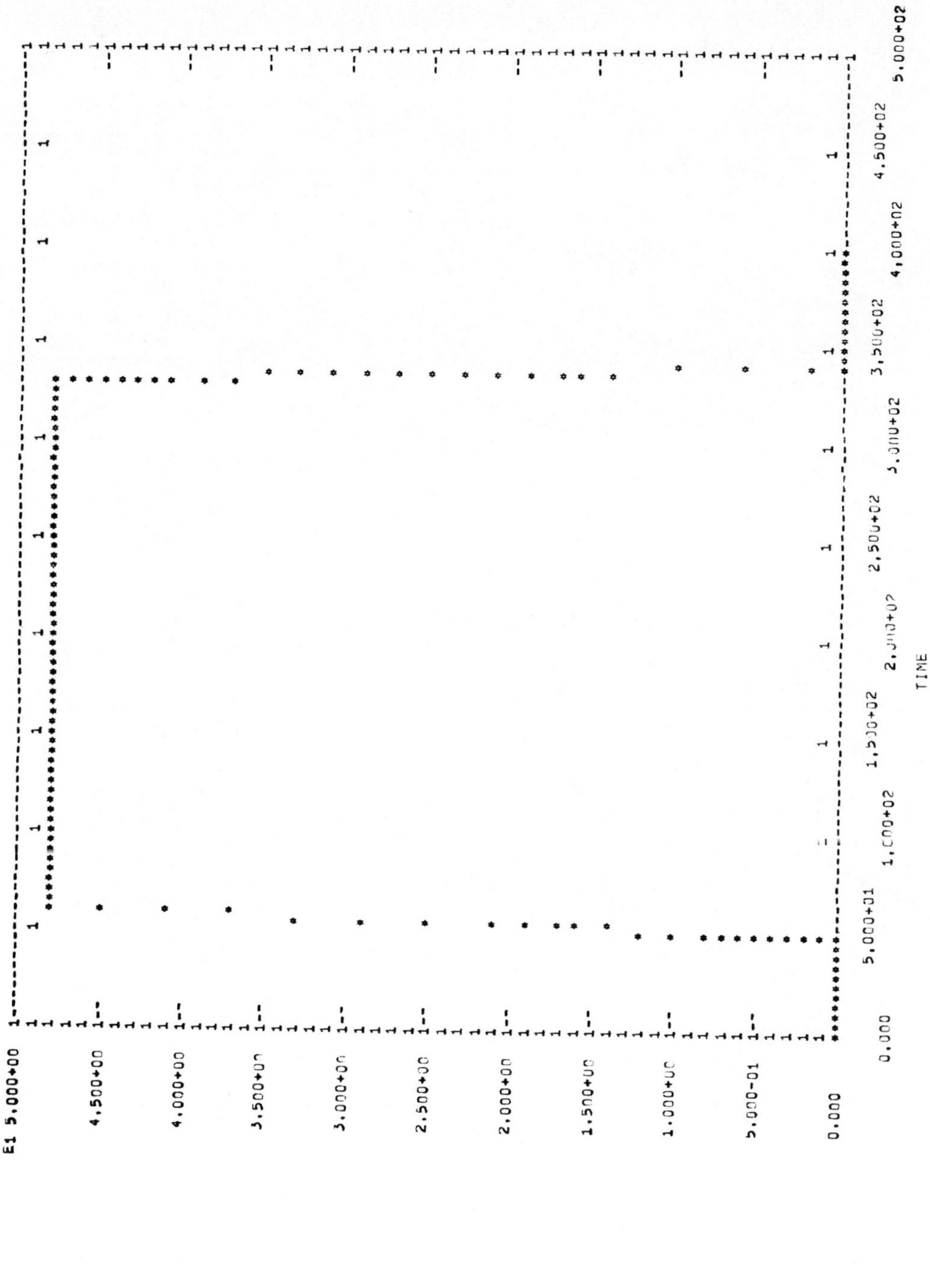

Fig. 3.104. Plot of E1 vs. time.

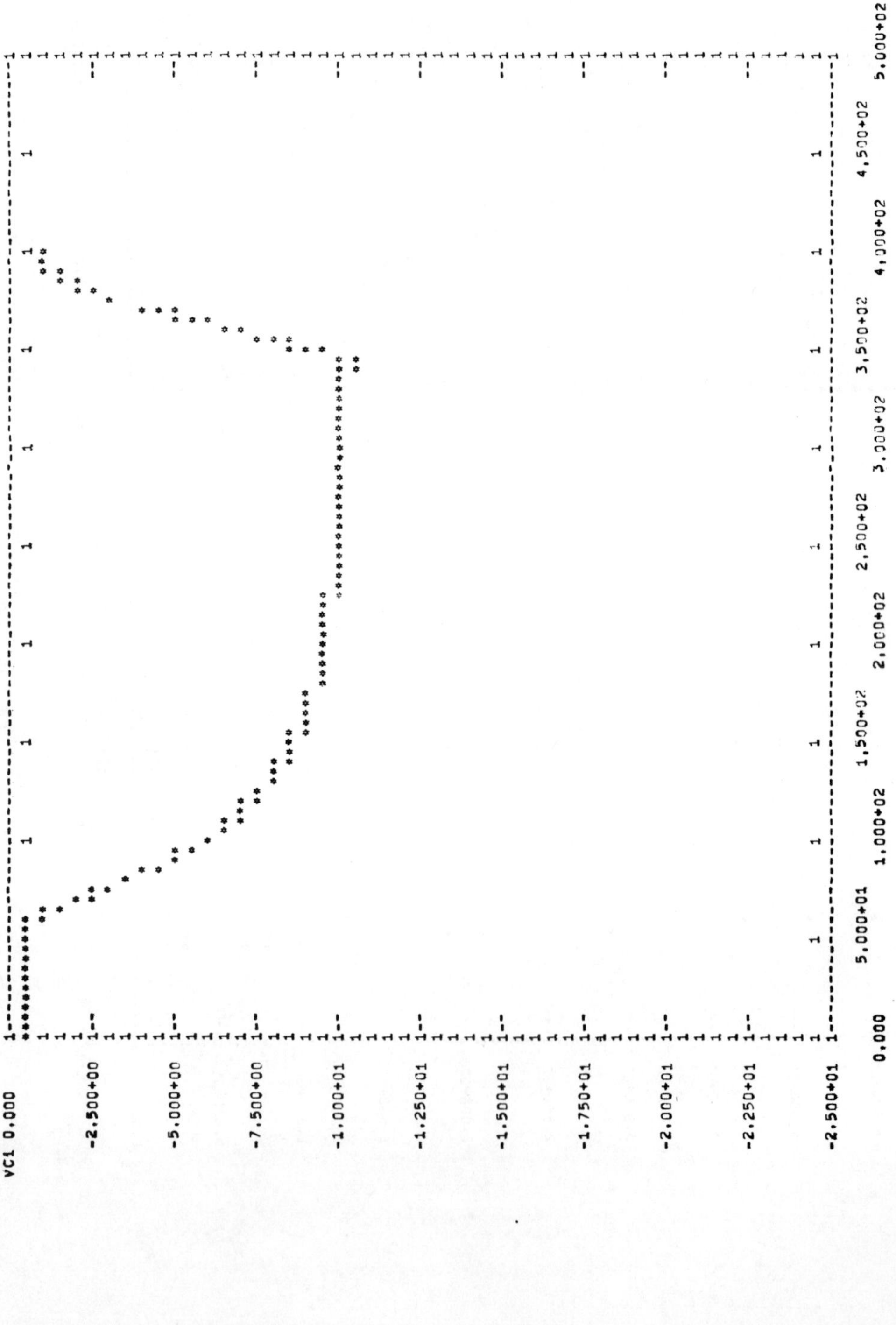

Fig. 3.105. Plot of VC1 vs. time.

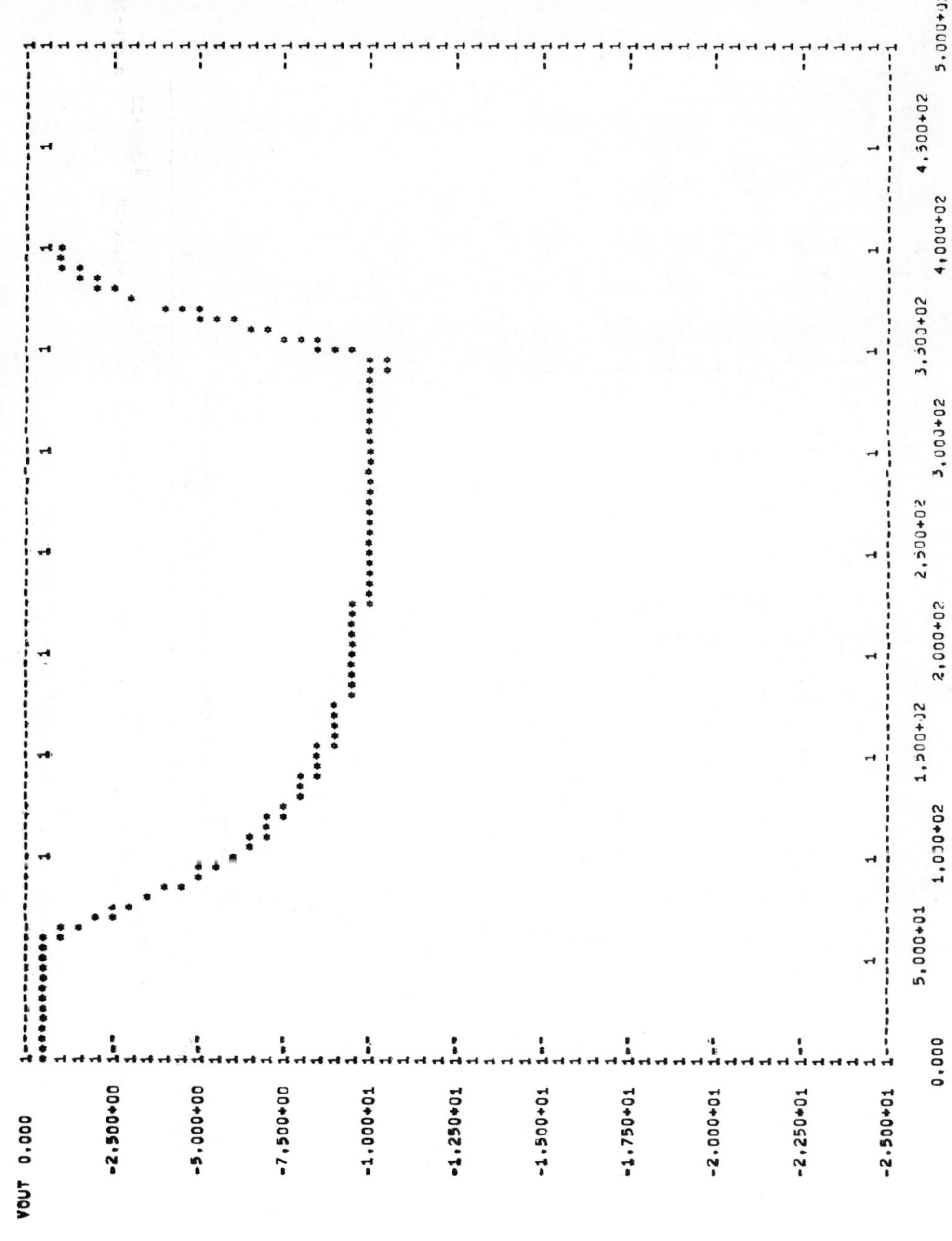

Fig. 3.106. Plot of VOUT vs. time.

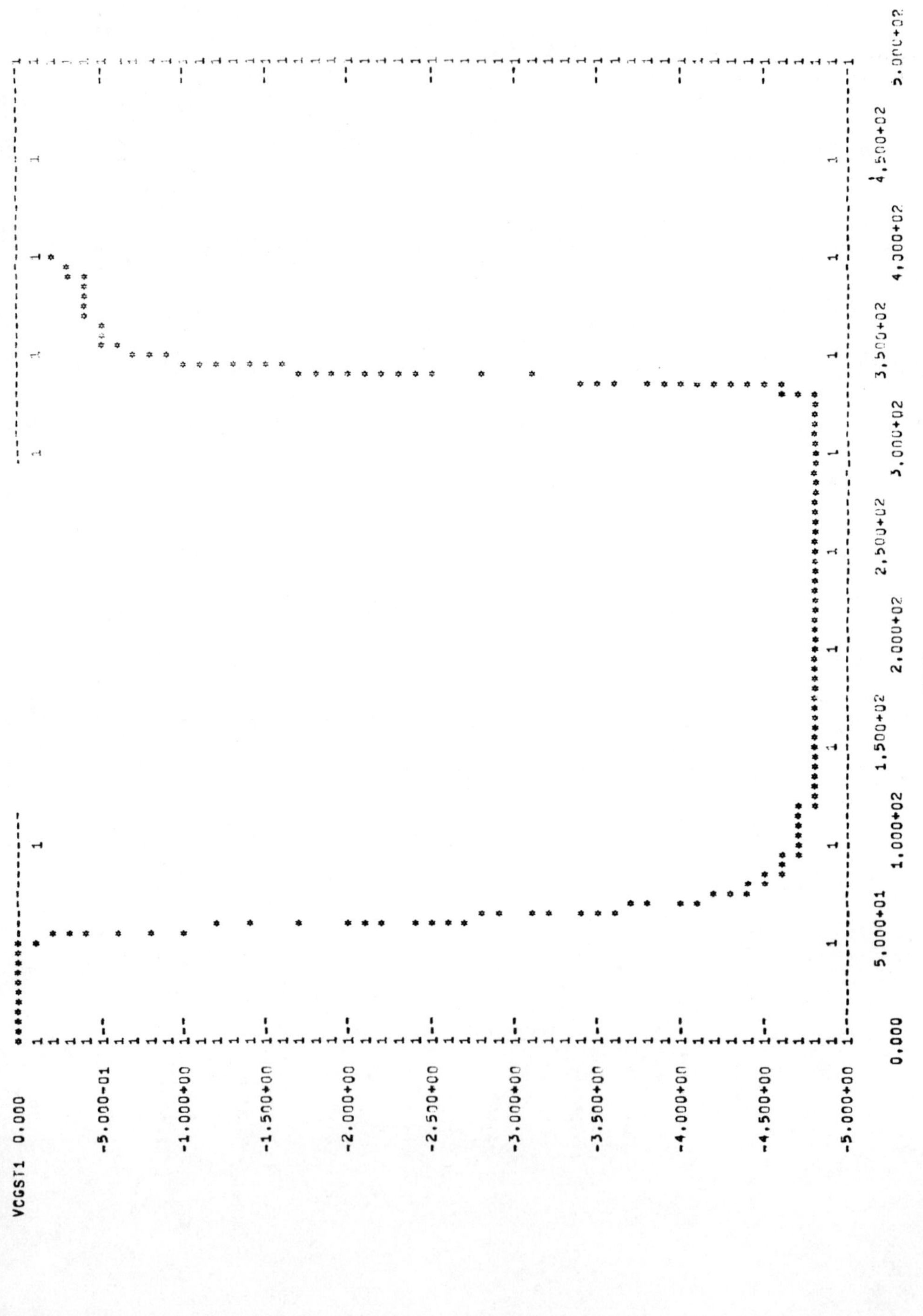

Fig. 3.107. Plot of VCGST1 vs. time.

```
      FUNCTION FJFET (VGS, VGD, VP, GC, PHI)
C     JFET SUBROUTINE
C     STEP JUNCTION AT GATE TO CHANNEL INTERFACE
C     CHANNEL CURRENT IS SET TO ZERO UPON FORWARD BIAS OF THE GATE TO
C     SOURCE OR GATE TO DRAIN JUNCTIONS GREATER THAN THE BUILT IN
C     POTENTIAL --PHI.
      VT = PHI-VP
      XK = 2./(3.*SQRT(VP))
C     CHECK FOR FORWARD BIAS GREATER THAN PHI.
      IF (VGD .GT. PHI .OR. VGS .GT. PHI) GO TO 10
C     CHECK FOR CUTOFF
      IF (VGD .EQ. VGS .OR. (VGD .LT. VT .AND. VGS .LT. VT)) GO TO 10
C     CHECK FOR CHANNEL PINCHOFF.
      IF (VGD .GT. VT .AND. VGS .GT. VT) GO TO 20
C     CHANNEL IS PINCHED-OFF
C     DETERMINE IF THE JFET IS IN THE NORMAL OR INVERTED MODE
      IF (VGS .LT. VT) GO TO 30
C     JFET IS IN THE NORMAL MODE.
      FJFET = GC*(VGS+VP/3. -PHI+XK*(-VGS+PHI)**1.5)
      RETURN
C     GATE TO SOURCE OR GATE TO DRAIN JUNCTION FORWARD BIASED OR CUTOFF.
   10 FJFET = 0
      RETURN
C     CHANNEL IS NOT PINCHED-OFF
   20 FJFET = GC*(VGS-VGD+XK*((-VGS+PHI)**1.5-(-VGD+PHI)**1.5))
      RETURN
C     JFET IS IN THE INVERTED MODE.
   30 FJFET = -GC*(VGD+VP/3. -PHI+XK*(-VGD+PHI)**1.5)
      RETURN
      END
```

JFET subprogram.

Fig. 3.108

3.6 SCEPTRE MOS FIELD EFFECT TRANSISTOR MODEL (P-CHANNEL)[6]

Figure 3.109 illustrates the selected MOSFET model, along with the defining equations.

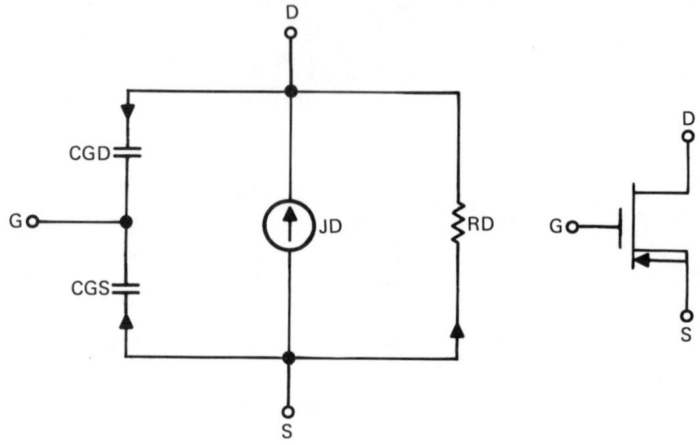

$$JD = \begin{cases} \frac{B}{2}\left[V_{GS}^2 - V_{GD}^2 - 2V_T\,(V_{GS} - V_{GD})\right] \text{------------ TRIODE REGION} \\ \quad V_{GS} \geq V_T \quad \text{and} \quad V_{GD} \geq V_T \\ \frac{B}{2}\left[V_{GS} - V_T\right]^2 \text{-------------------------- PINCH–OFF REGION} \\ \quad V_{GS} \geq V_T \quad \text{and} \quad V_{GD} \leq V_T \\ -\frac{B}{2}\left[V_{GD} - V_T\right]^2 \text{----------------------------- INVERTED REGION} \\ \quad V_{GS} \leq V_T \quad \text{and} \quad V_{GD} \geq V_T \\ 0 \text{------------------------------------- CUT–OFF REGION} \\ \quad V_{GS} \leq V_T \quad \text{and} \quad V_{GD} \leq V_T \end{cases}$$

CGS = Constant

CGD = Constant

$$RD = \frac{I_T \cdot R0}{JD + K}$$

Fig. 3.109. SCEPTRE MOSFET model.

3.6.1 Definition of Terms

JD = A current generator representing the drain to source current, excluding the variation of drain current with drain source voltage in the PINCH–OFF region (ma).

CGS = The gate to source capacitance (pfd).

CGD = The gate to drain capacitance (pfd).

RD = The resistance which models the variation in drain current with drain to source voltage in the PINCH–OFF region. This resistance is a function of JD (Kilohms).

V_T = The threshold voltage (volts).

B = The device conductance parameter (ma/v^2)

V_{GS} = The gate to source voltage (volts).

Also, V_{GS} = VCGS

V_{GD} = The gate to drain voltage (volts).

Also, V_{GD} = VCGD

RO = The resistance which models the variation in drain current with drain to source voltage in the PINCH–OFF region for a specified value of V_{GS}.

I_T = The drain to source current with:

V_{GS} = The value specified in determining RO.

$V_{GD} = V_T$

K = Current chosen to reduce the drain to source voltage/RD ratio to a small leakage current when JD = 0.

3.6.2 Introduction

The MOS field effect transistor model described in this section utilizes the FORTRAN subroutine feature to describe the drain current as a function of gate to drain voltage. The model covers all four regions of device operation.

Due to the dependencies involved, no initial conditions run may be accomplished. Initial conditions are easily obtained however, in the same manner as previously discussed in Sec. 3.5.2.

The gate to drain and gate to source capacitances were measured for several devices as a function of voltage and found to be essentially constant; therefore, CGS and CGD are assumed constant in this model.

The model assumes that the substrate and source are at the same potential and that the drain to substrate capacitance is negligible. The stray capacitance from drain to ground is usually much more significant that the back biased drain to substrate junction capacitance.

The subroutine utilized with the MOSFET model is illustrated at the end of this section in Fig. 3.123. This model does not include the gate to source, gate to drain and the drain to substrate leakage resistance. These can easily be included if required.

3.6.3 V_T, B, Drain to Source Current Generator Parameters

The parameters V_T and B are obtained from the MOSFET drain current versus drain to source voltage characteristic in the saturation region. In this region, the drain current is relatively independent of drain to source voltage. The drain characteristic for the 2N3608 MOSFET is obtained using the test configuration in Fig. 3.110.

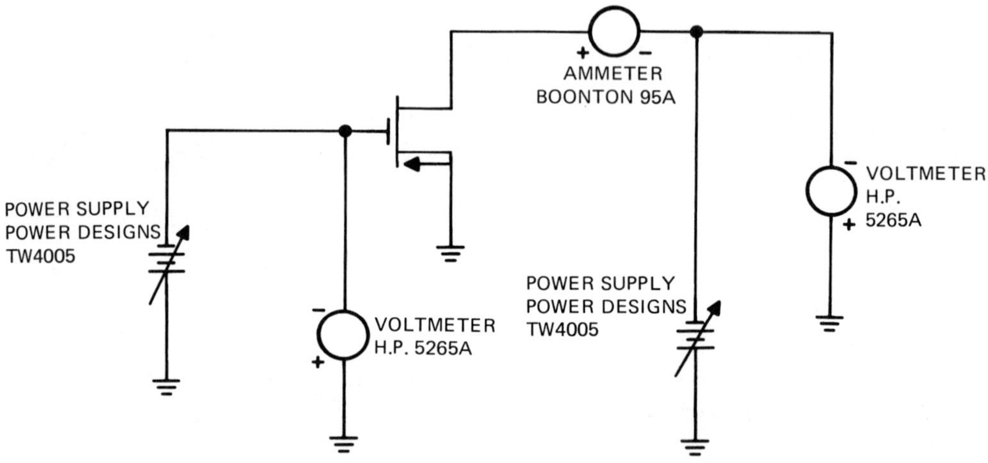

Fig. 3.110 MOSFET drain characteristic test.

The gate to source voltage is adjusted to a convenient value past the threshold voltage and the drain current recorded for values of drain to source voltage from .5v to 12v. This process is repeated for V_{GS} = –6v, –7v, –8v, –9v, –10v for the 2N3608. The data obtained is tabularized in Fig. 3.111 and plotted in Fig. 3.112.

The break point between the PINCH-OFF and TRIODE region occurs at the value of drain to source voltage where $V_{DS} = -[|V_{GS}| - |V_T|]$, for a P-Channel device. The equation which describes the device operation in the saturation region is:

$$JD = I_{Drain} = \frac{B}{2}\left(|V_{GS}| - |V_T|\right)^2$$

This equation states that in the PINCH-OFF region, the drain current is independent of drain to source voltage. Utilizing this equation and the previously measured drain current characteristic, V_T and B are obtained by the following iterative process:

1. For a fixed value of V_{GS} estimate V_{DS} at the break point between the PINCH-OFF and TRIODE region.

2. Calculate $|V_T| = |V_{GS}| - |V_{DS}|$ At the break point.

3. Using the equation:

$$B = \frac{2\, I_{Drain}}{\left(|V_{GS}| - |V_T|\right)^2}$$

   calculate a value for B using a measured value of drain current on or near the break point between the saturation and triode regions for the value of V_{GS} used in Step 2.

4. Calculate the drain current for several values of V_{GS} using:

$$I_{Drain} = \frac{B}{2}\left(|V_{GS}| - |V_T|\right)^2$$

5. Compare the calculated values for I_{Drain} with the measured values. Measured values should be taken at:

$$V_{DS} = -\left(|V_{GS}| - |V_T|\right)$$

6. If the comparison is favorable, the procedure is finished. If the comparison is unfavorable, a new value for V_{DS} , at the break point, is assumed and the procedure is repeated.

After several iterations, the values $|V_T| = 4.45$v, and B = .3570 are found to fit the measured device characteristic quite well. These parameters are determined for an initial $V_{GS} = -10$v and a test value of $V_{DS} = -[|V_{GS}| - |V_T|] = -5.5$v. The value of I_D at $V_{DS} = -7.5$v is used in the initial B calculation, since no data is available at $V_{DS} = -5.5$v. This approximation is found to introduce little error.

V_{GS} = −6 v	
V_{DS} = − .5v	I_D = − .218 ma
−1.0	− .290
−2.0	− .330
−3.0	− .350
−5.0	− .380
−7.5	− .405
−12.0	− .445
V_{GS} = −7 v	
V_{DS} = − .5v	I_D = − .460 ma
−1.0	− .740
−2.0	− .945
−3.0	−1.05
−5.0	−1.12
−7.5	−1.21
−12.0	−1.30
V_{GS} = −8 v	
V_{DS} = − .5v	I_D = − .680 ma
−1.0	−1.23
−2.0	−1.82
−3.0	−2.03
−5.0	−2.21
−7.5	−2.38
−12.0	−2.55
V_{GS} = −9 v	
V_{DS} = − .5	I_D = − .890
−1.0	−1.68
−2.0	−2.68
−3.0	−3.20
−5.0	−3.55
−7.5	−3.80
−12.0	−4.05
V_{GS} = −10v	
V_{DS} = − .5v	I_D = −1.12 ma
− 1.0	−2.08
− 2.0	−3.42
− 3.0	−4.40
− 3.5	−4.70
− 4.0	−4.90
− 4.5	−5.10
− 5.0	−5.18
− 7.5	−5.50
− 12.0	−5.82

Fig. 3.111 Drain characteristic data.

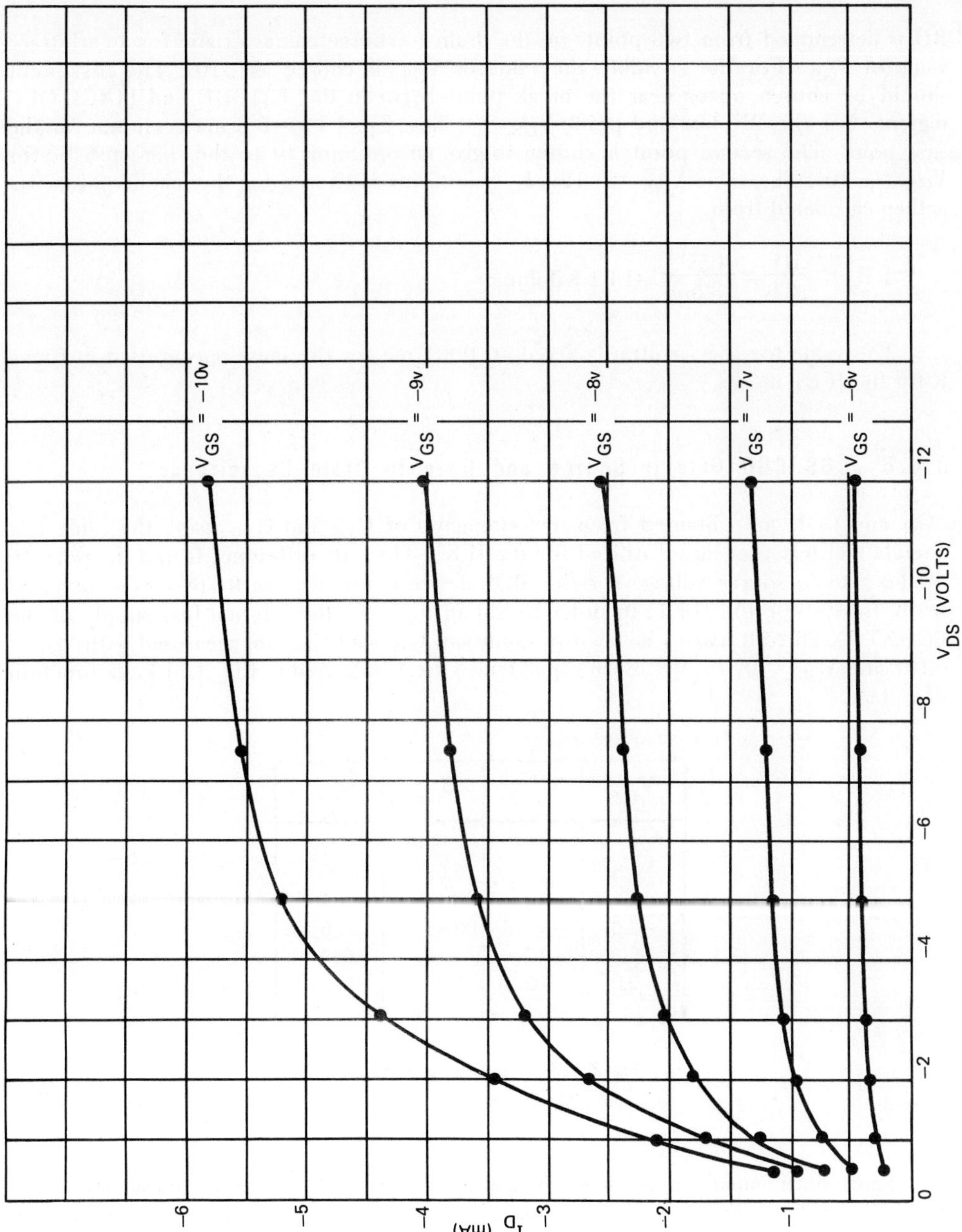

Fig. 3.112. 2N3608 drain characteristic.

3.6.4 RO, IT, K Parameters of the RD Equation

RO is determined from two points on the drain to source characteristic for an arbitrary value of V_{GS}. For the 2N3608, the value of V_{GS} is chosen as −10v. The first point should be chosen on or near the break point between the TRIODE and PINCH-OFF regions. For the 2N3608 the point $V_{DS} = -7.5$v, $I_D = I_T = 5.5$ ma is chosen for the first point. The second point is chosen to give an optimum fit to the characteristic for $V_{GS} = -10$v. The value $V_{DS} = -12$v, $I_D = 5.82$ ma is chosen for the second point. RO is then calculated from:

$$RO = \frac{\Delta V}{\Delta I} = \frac{4.5\text{v}}{.32\text{ma}} = 14.1 \text{ Kilohms.}$$

The value for K is arbitrarily fixed at .0001 ma for the same reason used in fixing K for the JFET model.

3.6.5 CGS, CGD Gate to Source and Gate to Drain Capacitance

CGS and CGD are obtained from measurements of C_{iss} and C_{rss} using the same test circuits and test procedure utilized for the JFET. The only difference is that the polarity of the gate to source voltage for the MOS device is varied from 0v to −9v as opposed to 0v to +9v for the JFET. In order to obtain the −9v, the internal bias supply of the BOONTON 75A-58 had to be slightly modified. C_{iss} and C_{rss} are measured with $V_{DS} = -10$v and $V_{GS} = 0$v to −9v. Both capacitances are tabularized in Fig. 3.113 as a function of voltage.

V_{GS}	V_{DS}	C_{iss}	C_{rss}
0. v	−10v	3.55pf	.595pf
− .5	−10	3.53	.600
−1.0	−10	3.49	.606
−2.0	−10	3.45	.615
−5.0	−10	3.34	.650
−9.0	−10	3.40	.700

Fig. 3.113 Capacitance parameters.

Since both capacitances only change by a few tenths of a picofarad, they are assumed constant. The value of C_{iss} and C_{rss} at the average gate to source voltage are used in calculating CGS and CGD:

$$C_{iss} \ @ \ V_{GS} = -5\text{v} = 3.34\text{pf}$$

$$C_{rss} \ @ \ V_{GS} = -5\text{v} = .650\text{pf}$$

Calculating CGS and CGD:

$$CGS = C_{iss} - C_{rss} = 3.34 - .65 = 2.69\text{pf}$$

$$CGS = C_{rss} = .65\text{pf}$$

Sometimes realistic values for C_{iss} and C_{rss} may be obtained directly from manufacturer data sheets. Since these values are constants, they may be used.

3.6.6 Summary of Model Parameters

A summary of the required parameters for the MOSFET of Fig. 3.109 is given in Table 3.5. The corresponding measured parameter values for the 2N3608 as developed in this section are also listed.

TABLE 3.5 Summary of MOSFET parameters.

REQUIRED MOSFET PARAMETERS	2N3608 MEASURED VALUES
V_T	4.45 volts
B	.3570 ma/volt^2
CGS	2.69pf
CGD	.65pf
I_T	5.55 ma
RO	14.1 Kilohms
K	.0001 ma

3.6.7 SCEPTRE Circuit Example-MOSFET Switch

The 2N3608 transistor from which the model parameters are obtained is set up in the test circuit shown in Fig. 3.114. The circuit input voltage and drain to ground output voltages are shown in Fig. 3.116. The input voltage and gate-ground voltage are shown in Fig. 3.117.

The test circuit is shown in SCEPTRE form in Fig. 3.115. In addition to the capacitances in the transistor model, stray capacitances measured from drain to ground (C1) and gate to ground (C2) are included in the SCEPTRE circuit. The drain to ground capacitance and gate to ground capacitances are measured to be 7.5 pf and 2.6 pf respectively. The measurements are taken with the device removed from the test socket. These

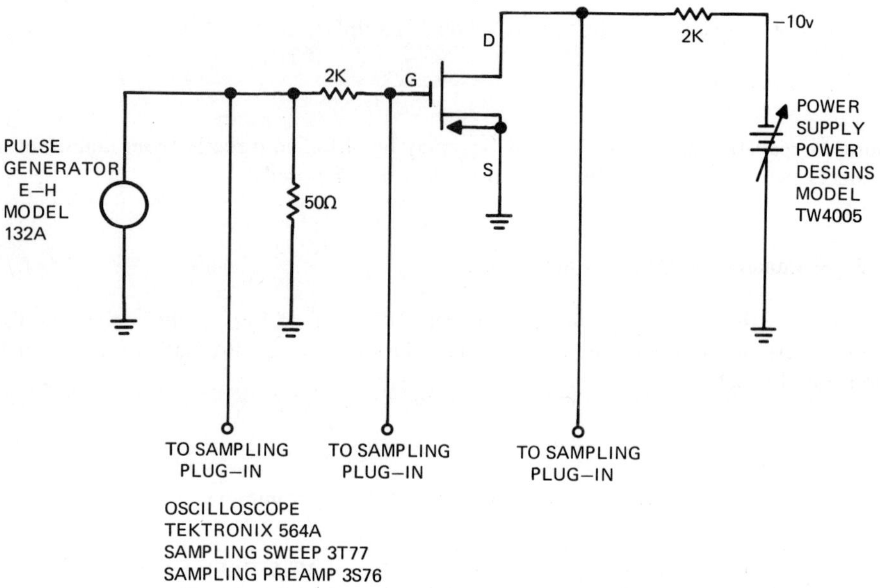

Fig. 3.114 MOSFET switch test circuit.

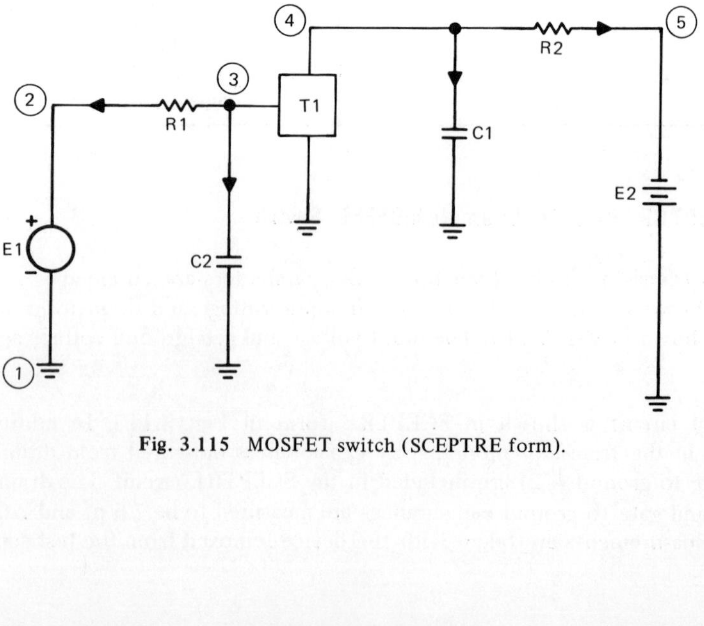

Fig. 3.115 MOSFET switch (SCEPTRE form).

TOP TRACE
INPUT PULSE
VERTICAL 5V/cm
HORIZONTAL 20nS/cm

BOTTOM TRACE
OUTPUT PULSE (DRAIN VOLTAGE)
VERTICAL 5V/cm
HORIZONTAL 20nS/cm

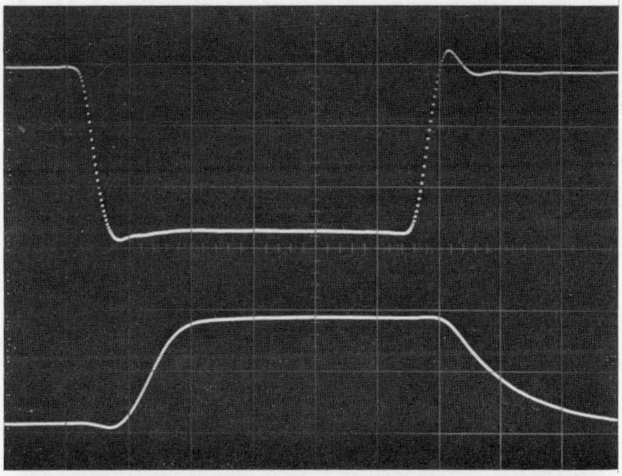

MOSFET switch waveforms.
Fig. 3.116

TOP TRACE
INPUT PULSE
VERTICAL 5V/cm
HORIZONTAL 20nS/cm

BOTTOM TRACE
GATE PULSE
VERTICAL 5V/cm
HORIZONTAL 20nS/cm

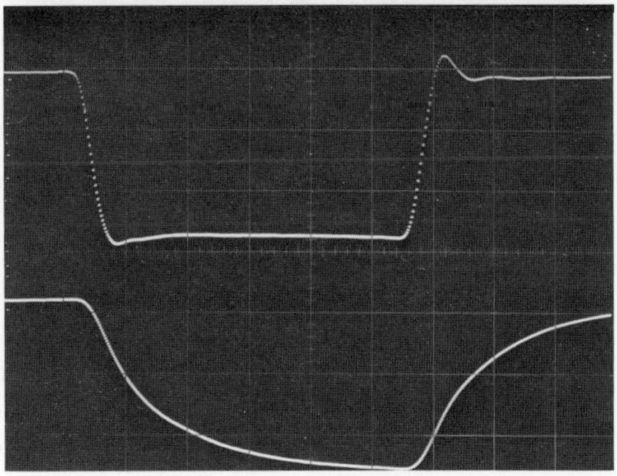

MOSFET switch waveforms.
Fig. 3.117

capacitances are due to wiring, BNC connector capacitance, scope probe capacitance, and test socket capacitance. The SCEPTRE listing of the MOSFET model and the Circuit Description are shown in Figs. 3.118 and 3.119.

The initial conditions and the transient analysis could be handled in either of the two ways discussed in the JFET section. The only technique illustrated here is the approach which combines the I.C. convergence into a transient run. To repeat, the steps are, first to calculate the approximate initial condition. Secondly, estimate the time from these approximate values to accurate convergence. Finally, allow the input to remain zero until this estimated time is reached, at which time the appropriate input voltage is applied. The transient run for the circuit of Fig. 3.115 uses an input, E1, approximately equal to the waveform of Fig. 3.116 (top trace). The output plots of the transient run are shown in Fig. 3.120, 3.121 and 3.122. These outputs correspond to the input voltage, the output voltage (drain to ground) and the gate voltage. A comparison of Figs. 3.121 and 3.122 with Figs. 3.111 (bottom trace) and 3.117 (bottom trace) indicates the close correlation between the test results and the SCEPTRE simulation.

```
MODEL 2N3608 (G-S-D)
ELEMENTS
CGS, S-G = 2.69
CGD, D-G = .65
RD, S-D = Q1(5.50,14.1,JD,.0001)
JD, S-D = FMOS(VCGS,VCGD,4.45,.3570).
FUNCTIONS
Q1(A,B,C,D) = (A*B)/(C+D))
```

Fig. 3.118 2N3608 MOSFET model description.

```
MOSFET SWITCH
ELEMENTS
R1, 3-2 = 2
R2, 4-5 = 2
C1, 4-1 = 7.5
C2, 3-1 = 2.6
E2, 5-1 = 10
E1, 1-2 = TABLE 1 (TIME)
T1, 3-1-4 = MODEL 2N3608
DEFINED PARAMETERS
P1 = X1 (VR2-E2)
OUTPUTS
E1,P1(VOUT),VCGST1,VCGDT1,VC1,VC2,JDT1,PLOT
INITIAL CONDITIONS
VC1 = -10
VC2 = 0
VCGST1 = 0
VCGDT1 = -10
FUNCTIONS
TABLE 1
0,0,50,0,60,-13.5,130,-13.5,140,0,150,0
RUN CONTROLS
STOP TIME = 200
END
```

Fig. 3.119 MOSFET test circuit description.

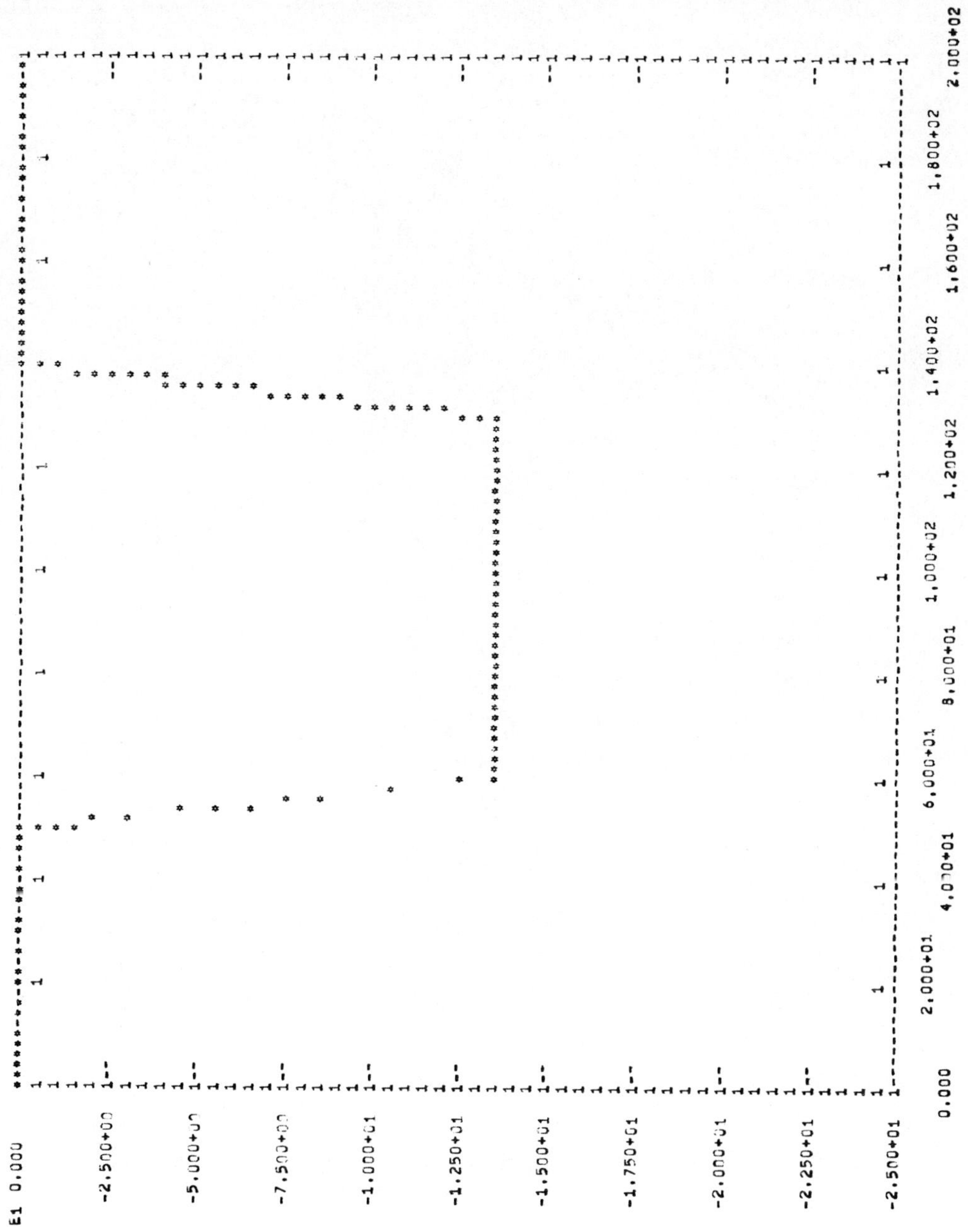

Fig. 3.120. Plot of E1 vs. time.

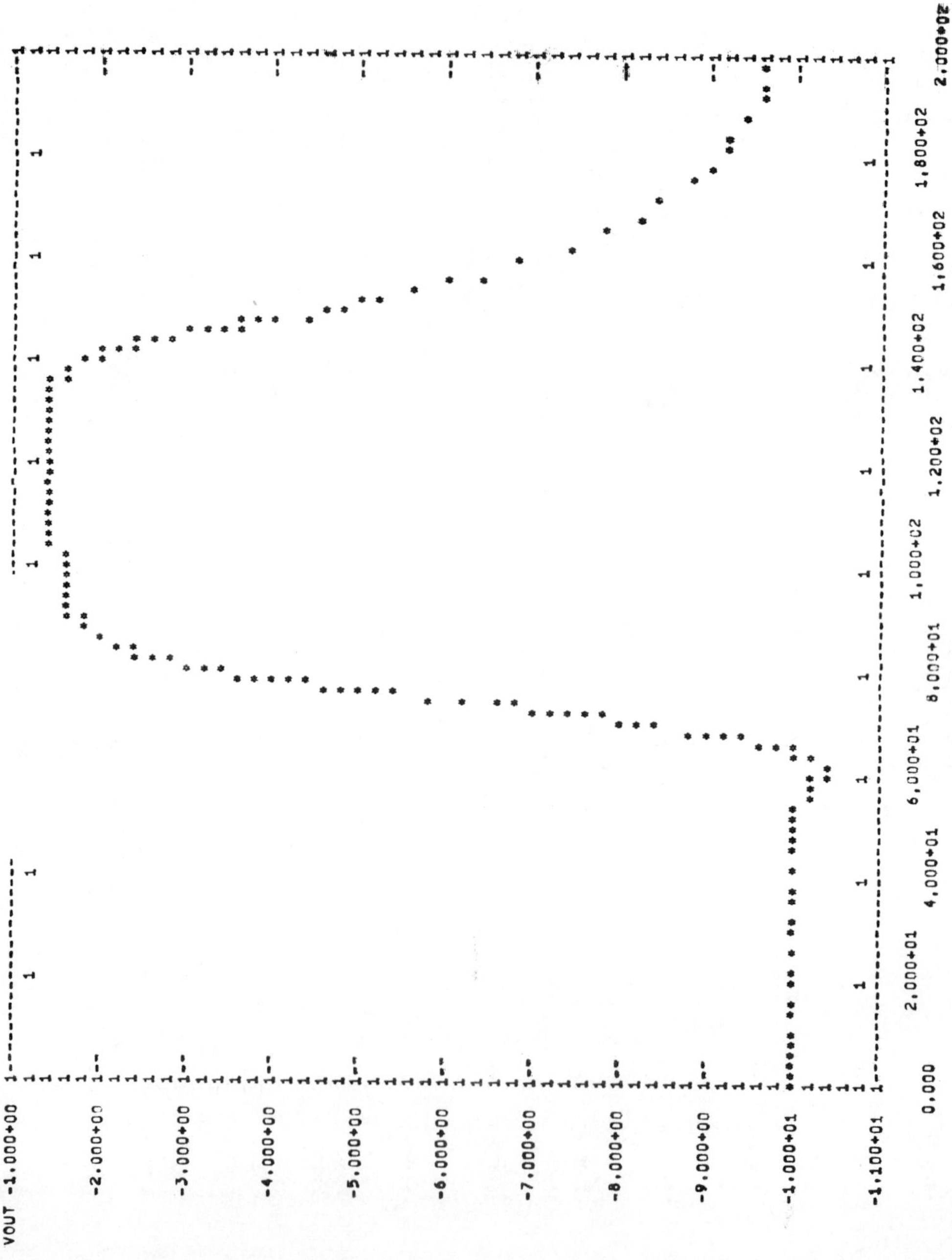

Fig. 3.121. Plot of VOUT vs. time.

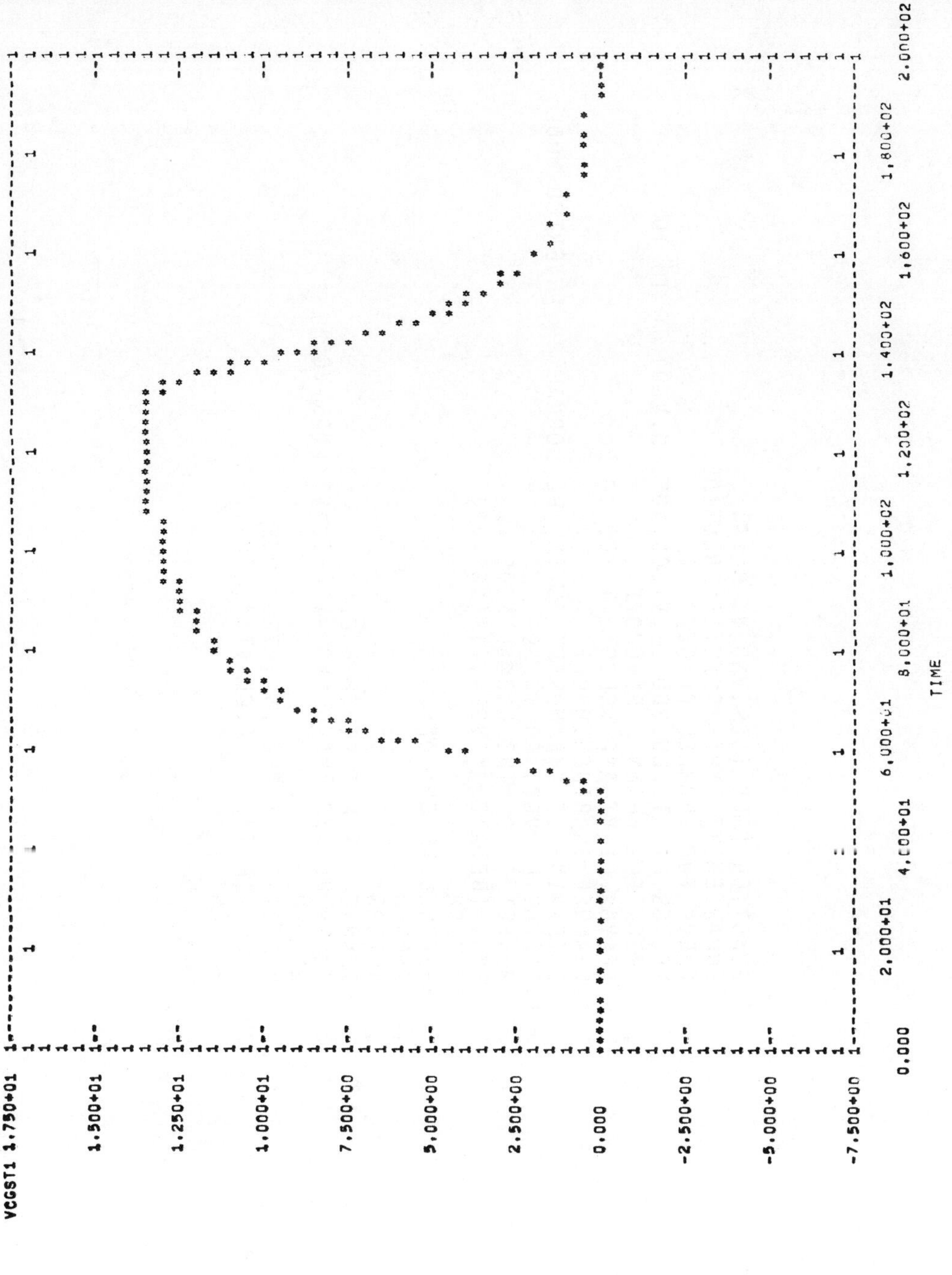

Fig. 3.122. Plot of VCGST1 vs. time.

```
      FUNCTION FMOS (VGS,VGD,VT,BETA)
C     ENHANCEMENT TYPE MOSFET SUBROUTINE
C     CHECK FOR CHANNEL CUTOFF
      IF(VGS.LT.VT.AND.VGD,LT.VT.OR.VGS.EQ.VGD) TO TO 10
C     CHECK FOR CHANNEL PINCHOFF
      IF(VGS.GT.VT.AND.VGD.GT.VT) GO TO 20
C     CHANNEL IS PINCHED-OFF
C     DETERMINE IF THE MOSFET IS IN THE NORMAL OR INVERTED MODE
      IF(VGS.LT.VGD) GO TO 30
C     MOSFET IS IN THE NORMAL MODE
      FMOS=(BETA/2.)*(VGS-VT)*(VGS-VT)
      RETURN
C     CHANNEL IS CUT OFF.
   10 FMOS=0
      RETURN
C     CHANNEL IS NOT PINCHED-OFF
   20 FMOS=(BETA/2.)*(VGS*VGS-VGD-2.*VT*(VGS-VGD))
      RETURN
C     MOSFET IS IN THE INVERTED MODE
   30 FMOS=-(BETA/2.)*(VGD-VT)*(VGD-VT)
      RETURN
      END
```

Fig. 3.123 MOSFET subroutine.

3.7 SCEPTRE UNIJUNCTION TRANSISTOR MODEL [7, 8]

Figure 3.124 illustrates the selected UNIJUNCTION model, along with the defining equations.

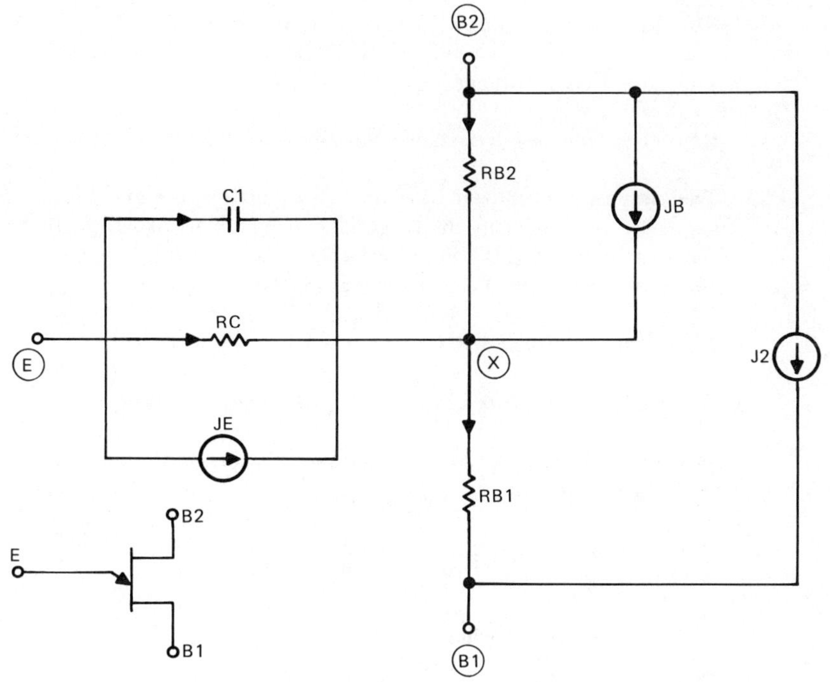

$$JB = \alpha I_E$$

$$JE = I_S(e^{\Theta \cdot VC1} - 1)$$

$$RB1 = \frac{[\eta_{VK} - a(VBB - VK)][RBB_{VK} + b(VBB - VK)]}{\left(\dfrac{I_E}{I_{E_1}}\right)}$$

$$RB2 = \{1 - [\eta_{VK} - a(VBB - VK)]\}[RBB_{VK} + b(VBB - VK)]$$

$$C1 = KD\ (JE + IS) + CX$$

Fig. 3.124 SCEPTRE UJT model.

3.7.1 Definition of Terms

η_{VK} = η the intrinsic standoff ratio measured at VBB = VK

VBB = The voltage between base two and base one (B2 and B1) (Volts).

VK = The test value of VBB at which RBB_{VK} and η_{VK} are specified (Volts).

RB1 = The base one resistance, a function of emitter current, I_E (Kilohms).

RB2 = The base two resistance, a constant (Kilohms).

RBB_{VK} = RBB, the interbase resistance measured at VBB = VK (Kilohms).

C1 = The sum of the emitter-base one diode diffusion capacitance and CX, a capacitor necessary in the SCEPTRE program to provide a finite capacitance across the E-B1 junction when the diffusion capacitance is zero. The diffusion capacitance is given by KD(JE + IS)(pfd).

α = A constant relating current generator JB to IE.

JE = A current source representing the emitter-base one diode.

JB = A current source representing the modulation of the base two region by the emitter current.

J2 = A zero value current source which provides an easy method of obtaining the interbase voltage, VBB, i.e., VBB = VJ2.

IE = The emitter current where,
$IE = JE^*$ (ma).

IE_1 = A constant determined from measurements (ma).

IS = DIODE SATURATION CURRENT (ma).

a = A constant which relates n to the interbase voltage, VBB.
$\eta = \eta_{VK} - a(VBB - VK)$ (Volts^{-1}).

b = A constant which relates RBB to interbase voltage, VBB.
$RBB = RBB_{VK} + b(VBB - VK)$ $\frac{(\text{Kilohms})}{\text{Volt}}$

N = The exponent of the equation which relates RB1 to the emitter current.

F = Frequency parameter (GHZ).

*IE is actually composed of JE plus the current through C1 and RC. Unfortunately, if IE is represented as the sum of these three currents, a computation delay occurs due to certain program algorithms. To obtain a working model, IE is considered equal to JE. Errors from this simplification occurs at very low current levels or very high frequencies, conditions which the UJT does not normally encounter in operation.

Θ = Constant of the emitter-base one diode equation (Volts^{-1}).

KD = Diffusion capacitance constant (pfd/ma)

$$KD = \frac{F}{2\pi\Theta}$$

3.7.2 Model Parameter Determination for the 2N2646 Unijunction Transistor

The measurement technique for each UJT parameter is discussed in the following sections.

3.7.3 Θ, Emitter-Base One Diode Equation Constant

The value for Θ is determined utilizing the same technique described for the junction diode. Base two is left open for this measurement. It must be emphasized that due to the high bulk resistances involved, the current levels chosen for determining Θ should be less than 1μa.

The test circuit used in measuring the emitter base-one forward characteristic is shown in Fig. 3.125. The resulting test data is shown in Fig. 3.126. The voltage across the emitter-base one junction is measured for I = .05μa to I = 1μa.

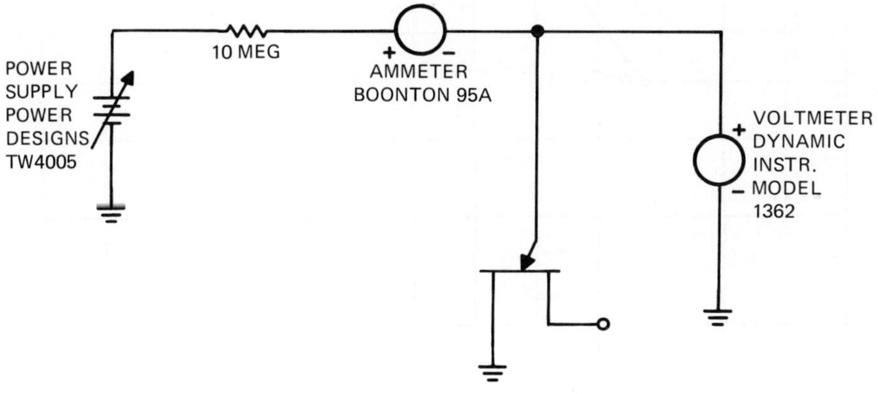

Fig. 3.125 UJT diode test circuit.

Choosing two points (V_1, I_1), (V_2, I_2) from the diode characteristic, an expression for Θ is:

$$\Theta = \frac{\ln\left(\frac{I_2}{I_1}\right)}{V_2 - V_1} = \frac{\ln\left(\frac{1 \times 10^{-6}}{.05 \times 10^{-6}}\right)}{.365 - .243} = 24.55 \text{ VOLTS}^{-1}$$

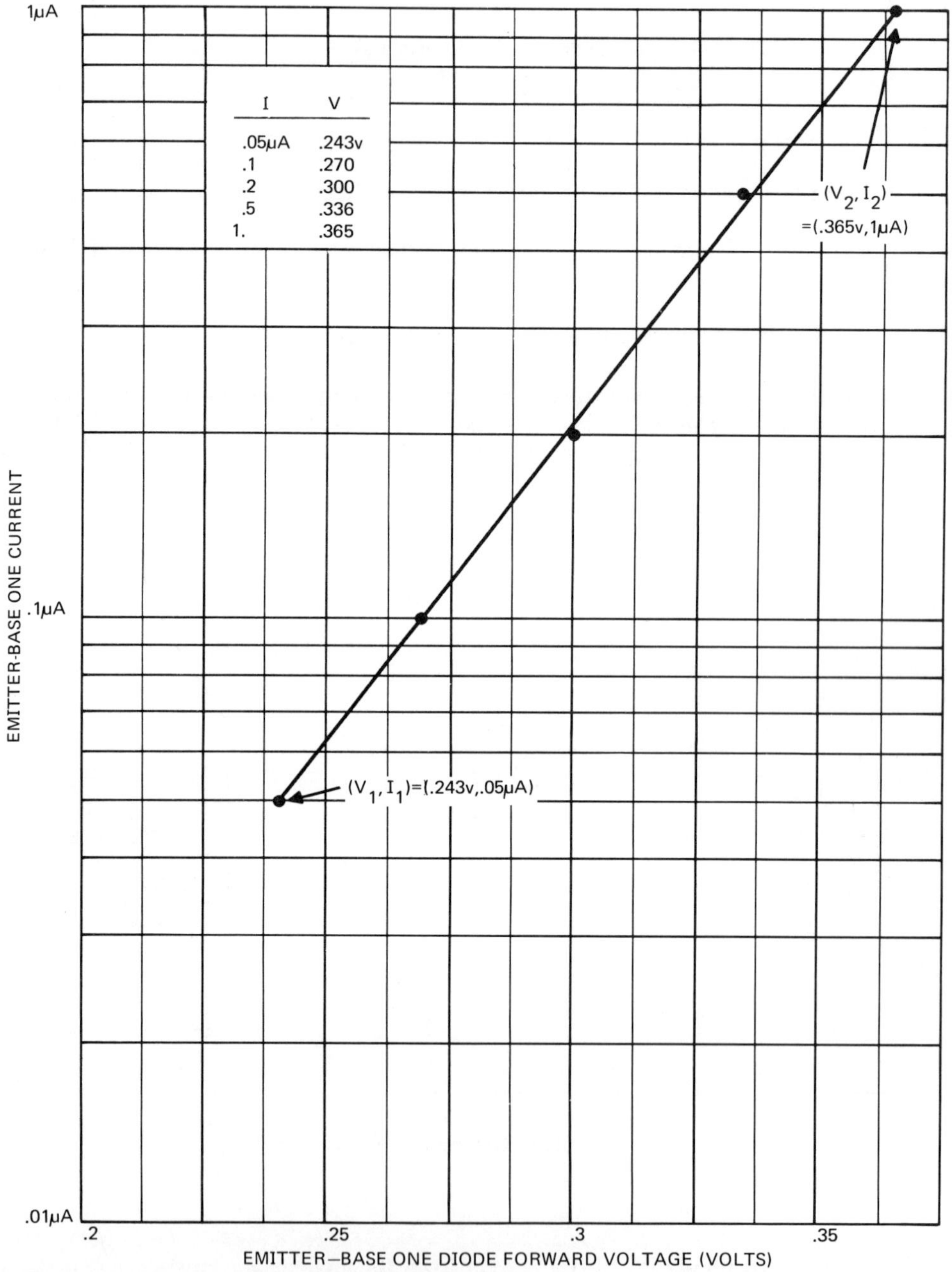

Fig. 3.126. Emitter-base one voltage vs. current.

3.7.4 IS, Diode Saturation Current

The diode saturation current is determined by choosing a single point (V, I) from the linear portion of the diode curve and substituting in the diode equation.

Picking I = .05μa, V = .243v yields:

$$IS = \frac{I}{e^{\theta V} - 1} = \frac{.05 \times 10^{-6}}{e^{(24.55)(.243)} - 1} = 1.281 \times 10^{-10} \text{amp}$$

or, $IS = 1.281 \times 10^{-7}$ ma.

3.7.5 η_{VK}, Intrinsic Standoff Ratio at VBB−VK and a, the Constant Which Relates η to VBB

Values for η are obtained at VBB = 5, 10, 15, 20, and 25 volts utilizing the test circuit shown in Fig. 3.127.

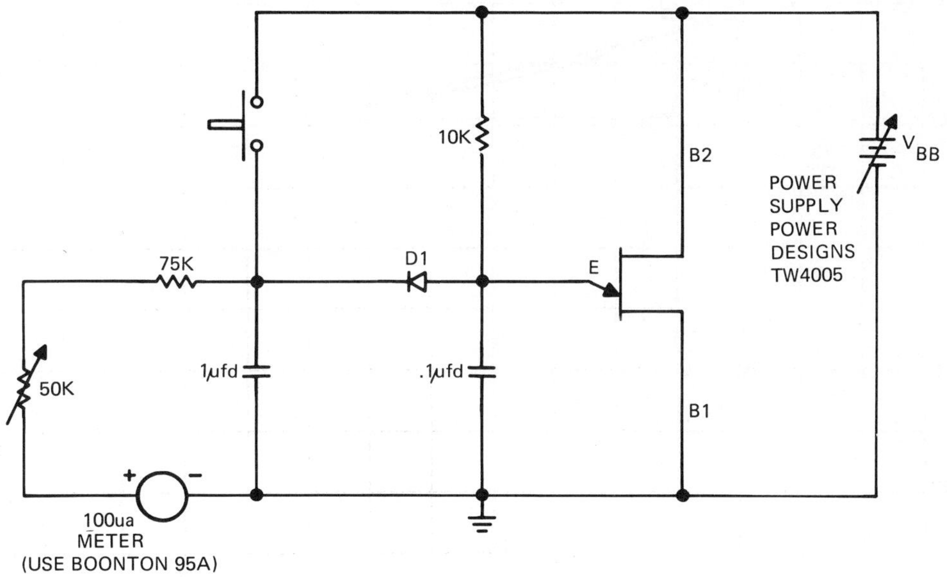

Fig. 3.127. η test circuit.

To obtain η, the "cal" button is pushed and the 50K pot adjusted to make the meter read full scale. The "cal" button is then released and the value of η is read directly from the meter (1.0 = FULL SCALE). This procedure is repeated for each value of VBB at which η is to be measured. The meter must be recalibrated for each value of VBB. Diode D1 is hand-picked to have a characteristic similar to the emitter base one diode which has already been measured.

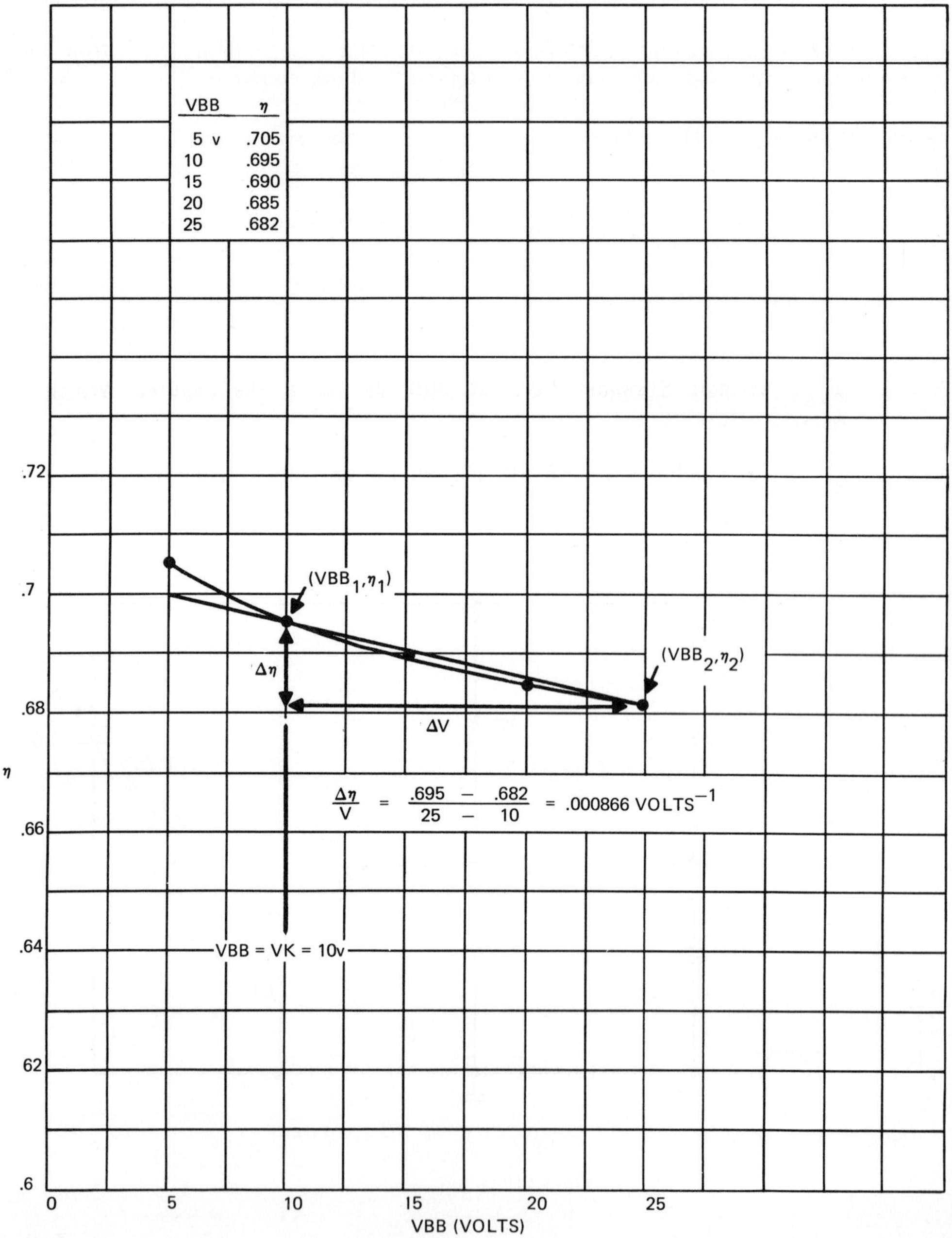

Fig. 3.128 *η* vs. VBB.

The values of η versus VBB are plotted in Fig. 3.128. The value of VK is chosen (arbitrarily) near the center of the range of VBB over which η is measured. VK is chosen to be 10v. At this value $\eta = \eta_{VK} = .695$.

The value for the constant "a" is obtained from the negative of the slope of the η versus VBB curve in Fig. 3.128. A straight line approximation to the curve is made to obtain the slope. From Fig. 3.128, choosing two points on the straight line approximation (V_{BB1}, η_1), (V_{BB2}, η_2) yields:

$$a = \frac{(\eta_2 - \eta_1)}{(V_2 - V_1)} = \frac{-(.682 - .695)}{25 - 10} = .000866 \text{ volts}^{-1}$$

3.7.6 RBB_{VK}, Interbase Resistance at VBB=VK, and b, the Constant Which Relates RBB to VBB

Values for RBB are obtained at VBB = 5, 10, 15, 20, 25 volts utilizing the test circuit shown in Fig. 3.129. For each value of VBB applied the base two-base one current is recorded. The values of RBB are obtained by calculating:

$$RBB = \frac{VBB}{I_{B2}}$$

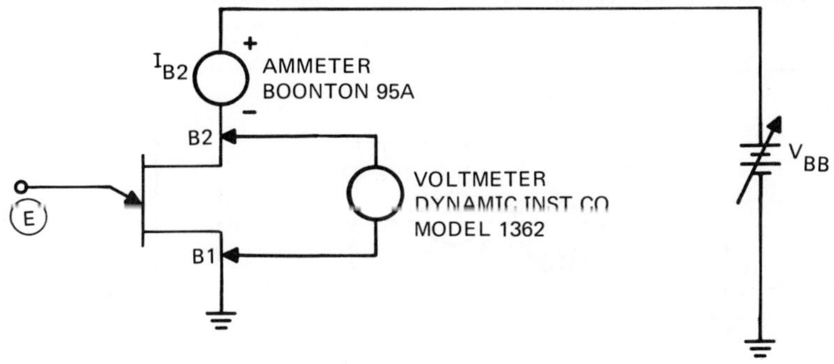

Fig. 3.129 RBB test circuit.

A table with VBB, I_{B2}, and calculated values for RBB is given in Fig. 3.130. RBB is plotted as a function of VBB in Fig. 3.131. Since VK has already been chosen as 10V:

$$RBB_{VK} = RBB\Big|_{VBB = VK = 10v} = 6.75K$$

The value for the constant "b" is obtained from the slope of the RBB versus VBB curve in Fig. 3.131. A straight line approximation to the curve is made in order to obtain

VBB	RBB	I_{B2}
5v	.760ma	6.58K
10v	1.48ma	6.75K
15v	2.10ma	7.14K
20v	2.70ma	7.4K
25v	3.30ma	7.6K

Fig. 3.130 RBB test data.

the slope. From Fig. 3.131, choosing two points on the straight line approximation (V_{BB1}, R_{BB1}), (V_{BB2}, R_{BB2}) yields:

$$b = \frac{R_{BB2} - R_{BB1}}{V_{BB2} - V_{BB1}} = \frac{7.6K - 6.58K}{25v - 5v} = .051 \frac{K\Omega}{VOLT}.$$

3.7.7 N, The Exponent of the Equation Relating RB1 to IE

N is obtained from the emitter-base one V–I characteristic in the saturation (high emitter current) region. At two points in the saturation region I_E and V_{E-B1} are measured using the test configuration shown in Fig. 3.132. The value of VBB applied is chosen equal to the previously determined value of VK for convenience.

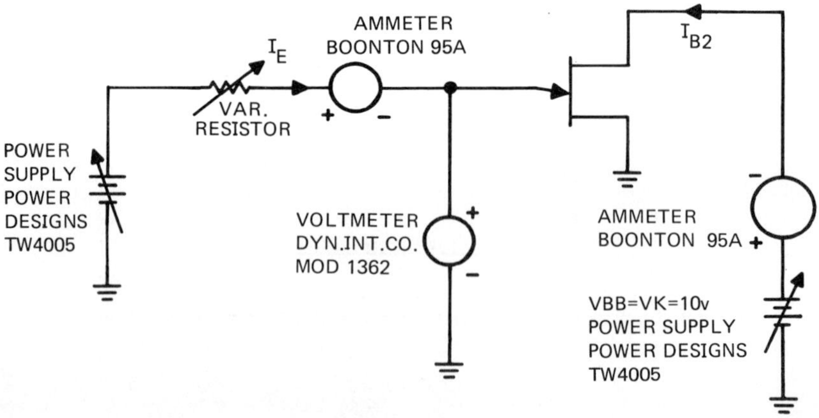

Fig. 3.132 Saturation region test circuit.

The variable resistor in Fig. 3.132 is adjusted to obtain two arbitrary values of I_E in the saturation region. The values chosen are $I_E = 20$ ma and $I_E = 50$ ma. The correspond-

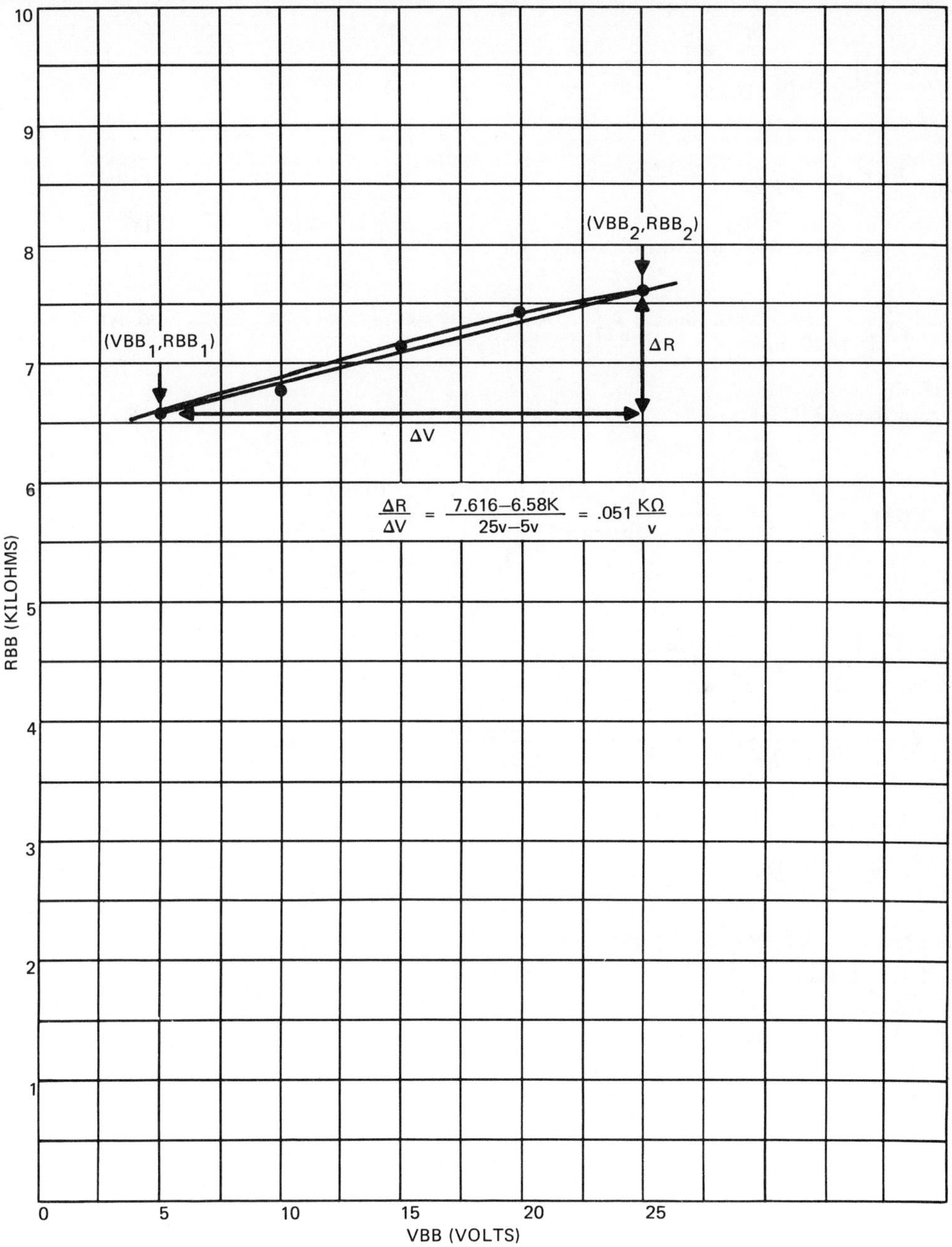

Fig. 3.131 RBB vs. VBB.

ing values of emitter-base one voltage are measured:

$$V_{EB1}\Big|_{I_E = 20\text{ ma}} = 1.918\text{v}$$

$$V_{EB1}\Big|_{I_E = 50\text{ ma}} = 2.327\text{v}$$

I_{B2} is also measured and found to be,

$$I_{B2}\Big|_{I_E = 20\text{ ma}} = 6.40\text{ma},\ I_{B2}\Big|_{I_E = 50\text{ ma}} = 7.25\text{ ma}$$

From the SCEPTRE model in Fig. 3.124, it is readily seen that:

$$RB1\Big|_{I_E = 20\text{ ma}} = \frac{V_{EB1}\Big|_{I_E = 20\text{ma}} - V_{JE}\Big|_{I_E = 20\text{ ma}}}{20\text{ma} + I_{B2}\Big|_{I_E = 20\text{ma}}}$$

also, $V_{JE}\Big|_{I_E = 20\text{ ma}}$ from,

$$V_{JE} = \frac{1}{\Theta}\,\ell n\left(\frac{I}{IS}\right) = \frac{1}{24.55}\,\ell n\left(\frac{20 \times 10^{-3}}{1.281 \times 10^{-10}}\right) = .7684\text{v}$$

Substituting:

$V_{E-B1} = 1.918\text{v}$

$V_{JE} = .7684\text{v}$

$I_E = 20\text{ ma}$

and $I_{B2} = 6.4\text{ ma}$

yields:

$$R_{B1}\Big|_{I_E = 20\text{ ma}} = \frac{1.918 - .7684}{20 \times 10^{-3} + 6.4 \times 10^{-3}} = 43.5\Omega$$

Similarly:

$$R_{B1}\Big|_{I_E = 50\text{ ma}} = \frac{V_{E-B1}\Big|_{I_E = 50\text{ ma}} - V_{JE}\Big|_{I_E = 50\text{ ma}}}{I_E + I_{B2}\Big|_{I_E = 50\text{ ma}}}$$

obtaining:

$$V_{JE}\Big|_{I_E = 50\text{ ma}} = .8058$$

from:

$$V_{JE} = \frac{1}{24.55}\ \ell n\left(\frac{50 \times 10^{-3}}{1.281 \times 10^{-10}}\right) = .8058\ .$$

Substituting:

$$V_{E-B1} = 2.327\text{v}$$

$$V_{JE} = .8058\text{v}$$

$$I_E = 50\text{ ma}$$

$$I_{B2} = 7.25\text{ ma}$$

yields:

$$R_{B1}\Big|_{I_E = 50\text{ ma}} = \frac{2.327 - .8058}{50 \times 10^{-3} + 7.25 \times 10^{-3}} = 27.32$$

With VBB = VK, the general expression for RB1 from the SCEPTRE model reduces to:

$$RB1 = \frac{RB10}{\left(1 + \dfrac{I_E}{I_{E1}}\right)^N} \quad \text{where, } RB10 = \eta_{VK}RBB_{VK}\ .$$

From the above equation, an expression for N is obtained in terms of RB1 at $I_E = 20$ ma and $I_E = 50$ ma. This equation is valid for $I_E >> I_{E1}$. This is the case in the saturation region. (This can readily be shown by examining what happens to RB1 as the emitter current increases. It is known that RB1 decreases rapidly and $I_E >> I_{E1}$, for this to occur):

$$N = \frac{\ell n\left(\dfrac{RB1|_{I_E = 20\text{ ma}}}{RB1|_{I_E = 50\text{ ma}}}\right)}{\ell n\left(\dfrac{I_E = 50\text{ ma}}{I_E = 20\text{ ma}}\right)} = \frac{\ell n\left(\dfrac{43.5}{27.32}\right)}{\ell n\left(\dfrac{50 \times 10^{-3}}{20 \times 10^{-3}}\right)}$$

$$N = .5076$$

3.7.8 IE1, A Constant Determined from Measurements

This parameter is obtained from the equation for RB1 with VBB = VK:

$$RB1 = \frac{RB10}{\left(1 + \frac{I_E}{I_{E1}}\right)^N}$$

For $I_E >> I_{E1}$, it is seen that $RB1 = \frac{RB10}{\left(\frac{I_E}{I_{E1}}\right)^N}$

Rearranging, $I_{E1} = \frac{I_E}{[RB10/RB1]^{1/N}}$

Calculating $RB10 = \eta_{VK} R_{BB_{VK}} = .695\ (6.75) = 4.69125K$.

Substituting the value of $RB1 = 43.5\Omega$ at $I_E = 20$ ma, N = .5076, and RB10 = 4.69125K yields,

$$I_{E1} = \frac{20 \times 10^{-3}}{\left(\frac{4.6912 \times 10^3}{43.50}\right)^{\left(\frac{1}{.5076}\right)}} = 1.9788 \times 10^{-6} \text{ amps}$$

or,

$$I_{E1} = 1.9788 \times 10^{-3} \text{ ma.}$$

3.7.9 α, The Constant Relating Current Generator JB to I_E

The value for α is determined from I_E, I_{B2}, RB1, RB2 at a point in the saturation region. It is seen from the SCEPTRE model that the following equation is valid for the saturation region.

$$VBB = VRB1 + VRB2 = (V_{E-B1} - V_{JE}) + (I_{B2} - JB)\ RB2 =$$

$$(I_E + I_{B2})\ RB1 + (I_{B2} - JB) RB2$$

Solving for JB:

$$JB = \alpha I_E = I_{B2}\ \frac{VBB}{RB2} - \frac{\left(I_E + I_{B2}\right)\ RB1}{RB2}$$

Solving for α:

$$\alpha = \frac{1}{I_E}\left[I_{B2} - \frac{VBB - \left(I_E + I_{B2}\right) RB1}{RB2}\right]$$

Since certain data is already available, α is calculated at $I_E = 20$ ma. Corresponding to this point, $I_{B2} = 6.40$ ma, RB1 = 43.5 and VBB = VK = 10v. The value of RB2 must be calculated. Since this is a special case where VBB = VK, RB2 is found from:

$$RB2 = (1 - \eta_{VK})RBB_{VK} = (1 - .695)\,(6.75K) = 2.0587K.$$

Substituting these values in the equation for α yields:

$$\alpha = \frac{1}{20 \times 10^{-3}} \left[6.4 \times 10^{-3} - \frac{10 - (20 \times 10^{-3} + 6.4 \times 10^{-3})43.5}{2.05875 \times 10^{3}} \right]$$

$$\alpha = .105024.$$

3.7.10 KD, The Diffusion Capacitance Constant

KD is obtained from storage time measurements for the emitter-base one diode, Base two is open circuited for this measurement. The measurement technique and test setup are identical to that used for determining junction diode storage time.

A photo of the emitter-base one diode reverse recovery characteristic is shown in Fig. 3.133. The top trace in the photo is the emitter-base one diode current with the vertical axis corresponding to 10 ma/cm. The bottom trace is the negative input pulse applied to the diode. The vertical axis corresponds to 2v/cm. The time scale (horizontal axis) is 50ns/cm. From the storage time is seen to be approximately 80ns. The forward and reverse diode currents are $I_F = I_R = 10$ ma.

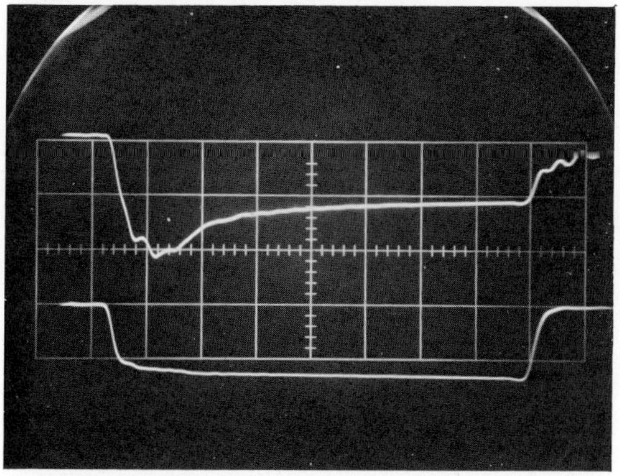

TOP TRACE

VERTICAL 10mA/cm
HORIZONTAL 50nS/cm

BOTTOM TRACE

VERTICAL 2V/cm
HORIZONTAL 50nS/cm

Emitter-base recovery characteristics.
Fig. 3.133

Calculating:

$$F = \frac{1}{2\pi} \, \frac{\ln\left(1 + \frac{I_F}{I_R}\right)}{T_s} = \frac{1}{2\pi} \, \frac{\ln\left(1 + \frac{10 \times 10^{-3}}{10 \times 10^{-3}}\right)}{80 \times 10^{-9}} = 1.379 \times 10^{6} \text{ HZ}.$$

or,

$$F = 1.379 \times 10^{-3} \text{ GHZ}.$$

Calculating KD:

$$KD = \frac{\Theta}{2\pi F} = \frac{24.55}{2\pi(1.379 \times 10^{-3})} = 2.834 \times 10^3 \text{ pfd/ma.}$$

3.7.11 CX, The Capacitance Required to Provide a State Variable

CX is required to produce a state variable across the emitter base-one junction when the diffusion capacitance approaches zero. Without this capacitance, extremely long computation times result, if the diode is initially reverse biased and the KD(JE + IS) term is very small. CX is selected to be equal to .01 times the smallest capacitor in the external circuit. In the example to follow, the UJT is used in a relaxation oscillator with the only capacitor equal to .2 μfd. Therefore, CX is chosen as .01x.2μfd or, CX = .002 x 10^6 pfd. This does not affect the circuit operation but does save considerable computer time.

3.7.12 RC, Emitter-Base One Diode Leakage Resistance

RC is obtained by applying a reverse bias to the emitter-base one diode with base two open and measuring the resulting current. This is a simplified procedure in relation to that used for leakage resistance in the bipolar device, but this technique has been sufficient for most applications.

For a reverse voltage of 15v, a current of approximately 1μa is measured. The leakage resistance is now calculated from:

$$RC = \frac{V_{Reverse}}{I_{Reverse}} = \frac{15v}{1\mu a} = 15 \times 10^6 \Omega$$

or,

$$RC = 1.5 \times 10^4 \text{ KILOHMS.}$$

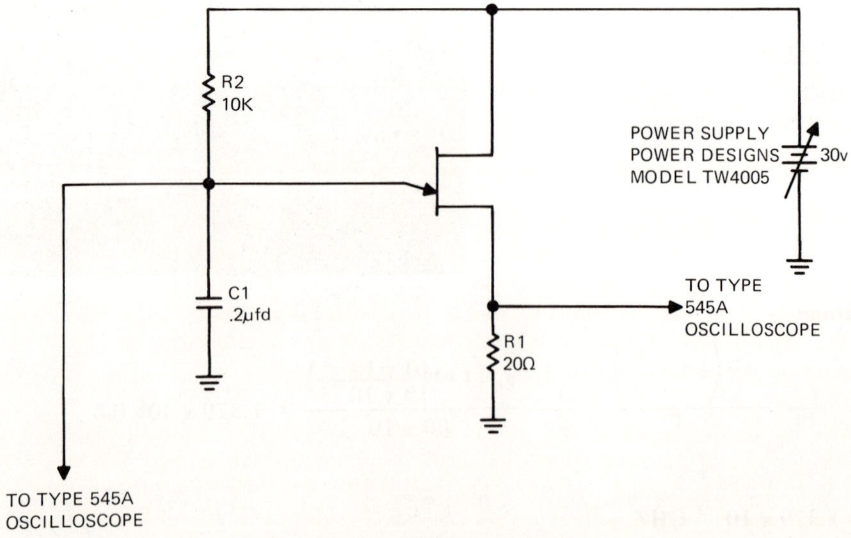

Fig. 3.134 UJT oscillator test circuit.

3.7.13 Unijunction Data Summary

A summary of the required parameters for the UNIJUNCTION model of Fig. 3.124 is given in Table 3.6. The corresponding measured parameter values for the 2N2646 as developed in this section are also listed.

Table 3.6 Summary of UJT parameters.

REQUIRED UJT PARAMETERS	2N2646 MEASURED VALUES
VK	10 volts
Θ	24.55 volts^{-1}
I_s	1.281 x 10^{-7} ma
η_{VK}	.695
a	.000866 volts^{-1}
RBB_{VK}	6.75 Kilohms
b	.051 $\frac{\text{kilohm}}{\text{volt}}$
N	.5076
I_{E1}	1.9788 x 10^{-3} ma
α	.105024
KD	2.834 x 10^3 pfd/ma
CX (Application Dependent)	002 x 10^6 pfd
RC	1.5 x 10^4 kilohms

3.7.14 SCEPTRE Unijunction Relaxation Oscillator Example

The 2N2646 UJT from which the model parameters were extracted is incorporated into the relaxation oscillator shown in Fig. 3.134. Laboratory measurement of the voltage across C1 yielded the period and amplitude of the oscillation to be 2.26 ms and 18.5 volts respectively, from the top trace in Fig. 3.135. The top trace is the voltage across C1, with the vertical axis = 10v/cm and the horizontal axis = 1ms/cm. The bottom trace shows the voltage across R1 with the vertical axis equal to 5v/cm.

The oscillator circuit of Fig. 3.134 is put into SCEPTRE form in Fig. 3.136. The 2N2646 model and the relaxation oscillator circuit incorporating the 2N2646 are shown coded in SCEPTRE format in Figs. 3.137 and 3.138. The SCEPTRE run output in Fig. 3.139 corresponds to the voltage across C1 and the output in Fig. 3.140 corresponds to

TOP TRACE

VERTICAL 10V/cm
HORIZONTAL 1mS/cm

BOTTOM TRACE

VERTICAL 5V/cm
HORIZONTAL 1mS/cm

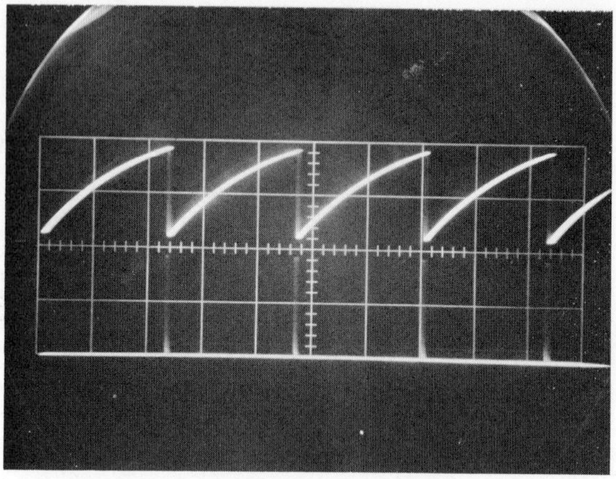

VC1 and VR1 vs. time.
Fig. 3.135

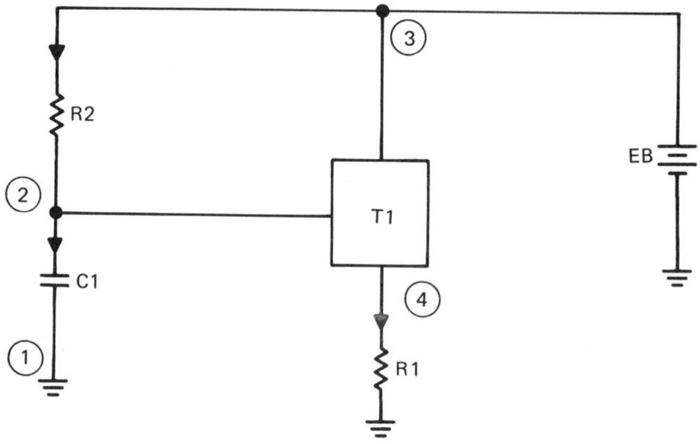

Fig. 3.136 UJT oscillator (SCEPTRE form).

```
MODEL 2N2646 (B1-E-B2)
ELEMENTS
JB, B2-X = .1054*JE
JE, E-X = DIODE Q(1.281E-7,24.55)
J2, B2-B1 = 0
C1, E-X = Q1(.002E6,2.834E3,JE,1.281E-7)
RB1, X-B1 = Q2(.695,.000866,VJ2,10.,6.75,.051,
1.,JE,PIE1,.5076)
RB2, B2-X = Q3(1.,.695,.000866,VJ2,10.,6.75,.051)
RC, E-X = 1.5E4
DEFINED PARAMETERS
PIE1 = 1.9788E-3
FUNCTIONS
Q1(A,B,C,D) = (A+B*(C+D))
Q2(A,B,C,D,E,F,G,O,P,Q)=
(((A-B*(C-D))*(E+F*(C-D)))/((G+(O/P))**Q))
Q3(A,B,C,D,E,F,G) = ((A-B+C*(D-E))*(F+G*(D-E)))
```

Fig. 3.137 2N2646 UJT model description.

the voltage across R1. The period is determined to be 2.164 ms after the initial transient. The peak voltage across C1 is seen to be 20.6 and the peak voltage across R1 is shown to be 13v. The difference between the computed and measured values may be attributed partially to the measurement technique used in obtaining η. By iterating several times between the measurements and the SCEPTRE results, the user can converge to a more accurate value for η.

```
UNIJUNCTION OSCILLATOR
ELEMENTS
C1, 2-1 =.2E6
R1, 4-1 = .02
R2, 3-2 = 10
T1, 4-2-3 = MODEL 2N2646
EB, 1-3 = 30
OUTPUTS
VC1,VR1,JET1,PLOT
RUN CONTROLS
STOP TIME = 5E6
MINIMUM STEP SIZE = 1.E-10
END
```

Fig. 3.138 UJT test circuit.

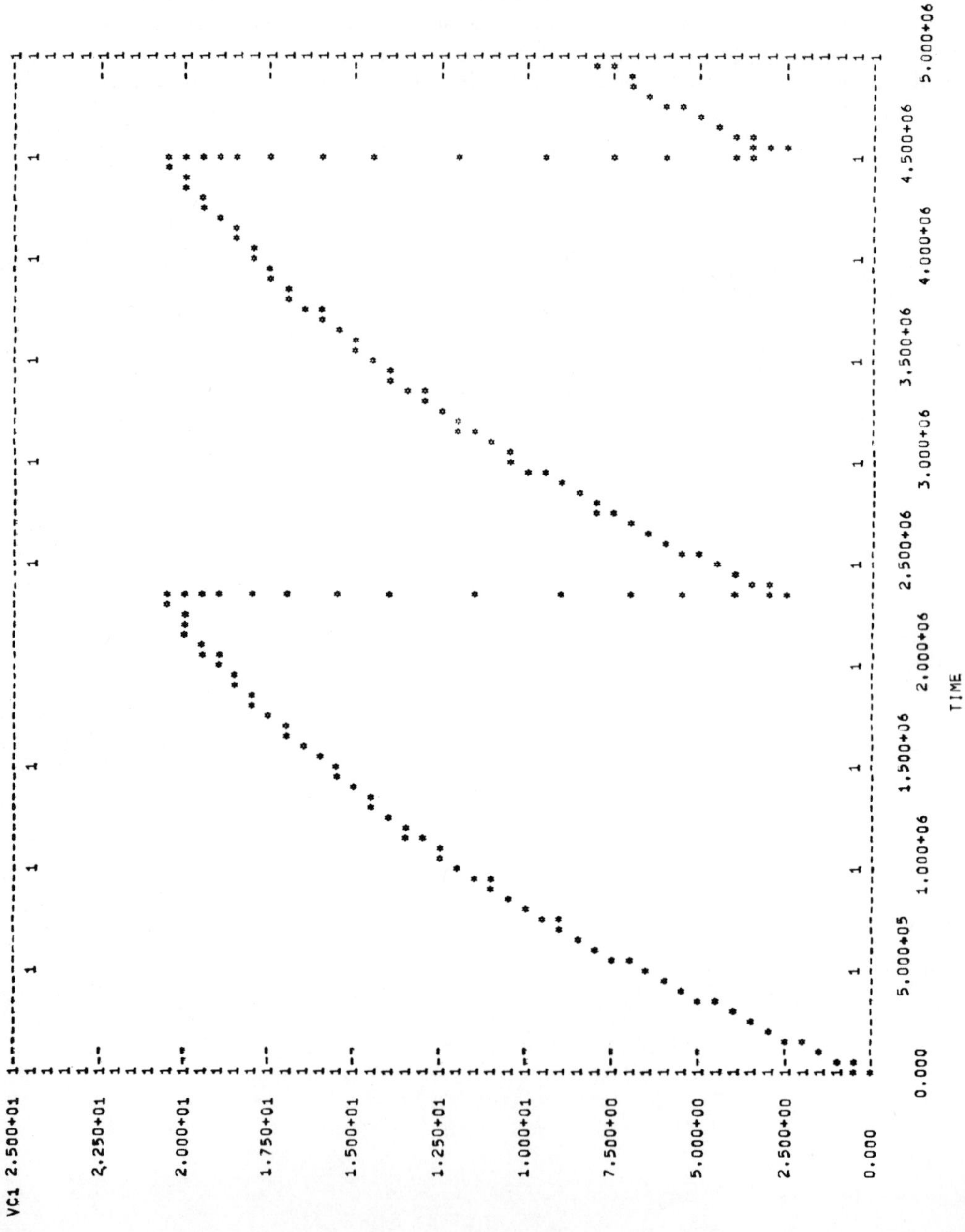

Fig. 3.139. Plot of VC1 vs. time.

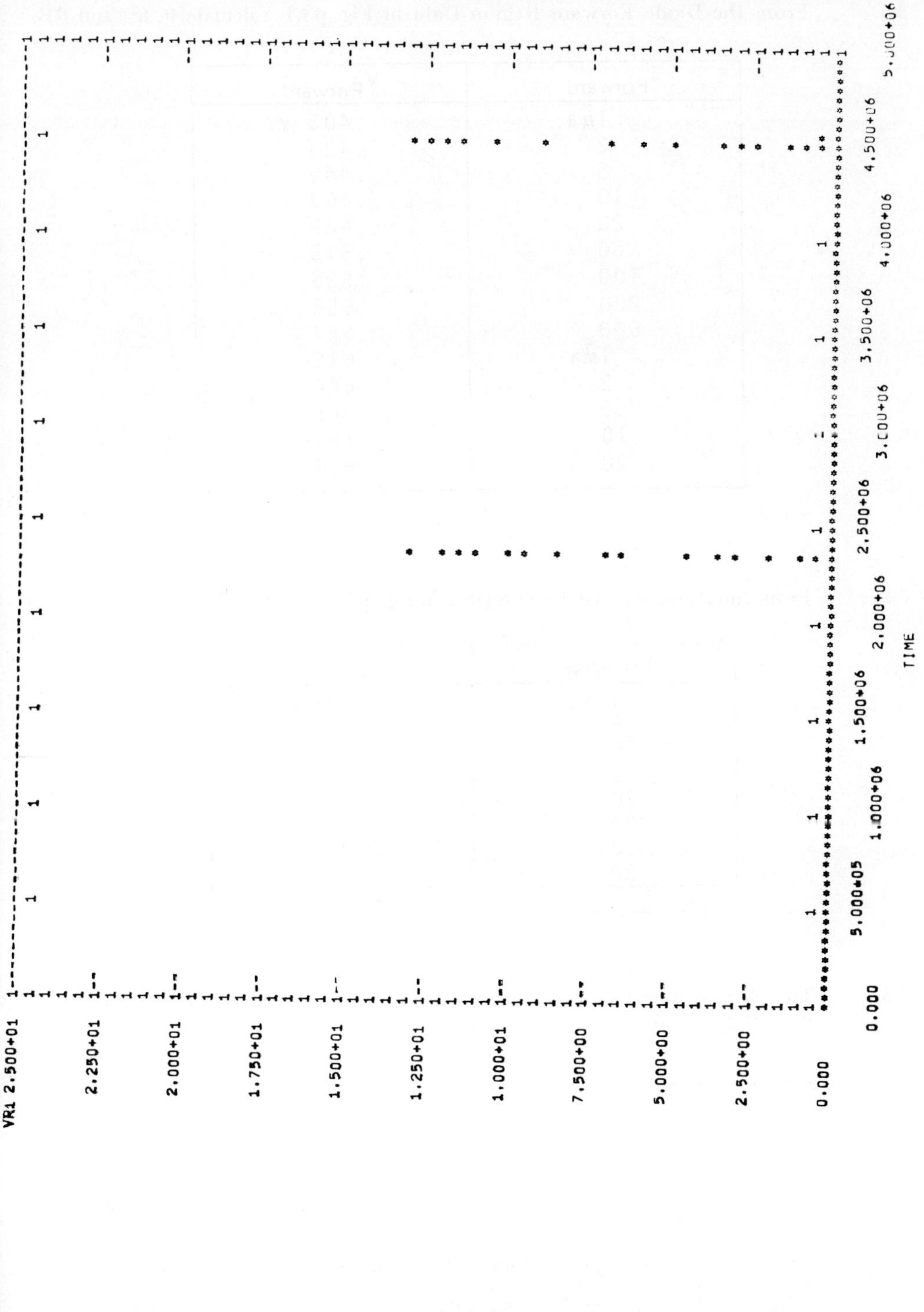

Fig. 3.140. Plot of VR1 vs. time.

PROBLEMS

3.1 From the Diode Forward Region Data in Fig. p3.1, calculate θ, IS, and RB.

$I_{Forward}$	$V_{Forward}$
1μa	.403 v.
2	.423
5	.449
10	.469
20	.489
50	.515
100	.536
200	.556
500	.587
1ma	.612
2	.644
5	.706
10	.786
20	.920

Fig. p3.1

3.2 From the Diode Reverse Region Data in Fig. p3.2, calculate RS.

$V_{Reverse}$	$I_{Reverse}$
1 v.	.5 na
5	4.5
10	6.4
15	8.0
20	9.4
25	10.9
30	12.4
35	14

Fig. p3.2

3.3 The storage time waveform for a high speed diode is shown in Fig. p3.3. If $0 = 28.0_J$ calculate KD.

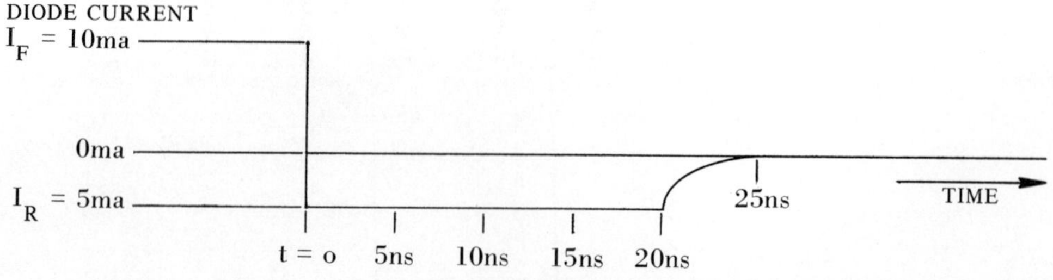

Fig. p3.3

3.4 From the diode transition capacitance data in Fig. p3.4, obtain values for ϕ n, CO which yield a reasonable fit to the data.

$V_{Reverse}$	$C_{Transition}$
0 v.	8.91 pf
.1	8.50
.2	8.14
.5	7.31
1	6.37
2	5.26
5	3.82
10	2.90

Fig. p3.4

3.5 Develop a procedure for entering transition capacitance in the diode model using the table feature of SCEPTRE. Use the data given in Prob. 3.4.

3.6 Calculate α_N for each set of values of I_c and I_B in Fig. p3.6. Show how the calculated values of α_N are entered as a tabular function of emitter current under "FUNCTIONS" in the MODEL DESCRIPTION.

I_C	I_B
.05 ma	.0025 ma
.1	.0045
.2	.0083
.5	.0192
1.0	.0357
2.0	.0645
5.0	.1510
10.0	.2940
20.0	.5700
50.0	1.5100
100.0	3.3300

Fig. p3.6

3.7 From the C_{ob} data in Fig. p3.7, obtain values for ϕ_c, M , COC. The stray capacitance C_{sc} = .7pf.

$V_{Reverse}$	C_{ob}
0 v.	5.06 pf
.1	4.90
.2	4.76
.5	4.43
1.0	4.05
2.0	3.59
5.0	2.96
10.0	2.52

Fig. p3.7

3.8 Develop a method for obtaining RC, using the SCEPTRE program:

Given all other D.C. device parameters (Θ_E, Θ_C, IES, ICS, α_N(JE), α_I (JC), RB, R1, R2) and $V_{CE(SAT)}$ at a specific $I_{C(SAT)}$ and $I_{B(SAT)}$

Write a SCEPTRE program to obtain RC for the 2N2369, given $V_{CE(SAT)} = .176_V$ with $I_{C(SAT)} = 10$ma, $I_{B(SAT)} = 1$ ma.

3.9 From the data in Fig. p3.9:

a. Determine T_E for the value of t_r.
b. Show how each value of T_E is entered as a function of emitter current under "FUNCTIONS" in the MODEL DESCRIPTION.
c. Is the method for determining T_E valid at low current levels? What criterion must be established to insure the validity of T_E ?

$I_{C_{Peak}}$	$B_o = B_{N_1} + B_{N_2}$	t_r 27% - 73%
200 ma	40	10.2 ns
100	45	9.2
50	50	13.2
20	55	17.9
10	50	19.9
5	45	21.15
2	40	21.3
1	38	22.6

Fig. p3.9

3.10 From the Zener Reverse Region Data in Fig. p3.10:

a. Calculate RZ.
b. Using the calculated value of RZ and the data in Fig. p3.10, show how the values for JZ are obtained and entered as a tabular function of voltage under "FUNCTIONS" in the MODEL DESCRIPTION.

	V_Z	I_Z
FORWARD REGION	+ .87 v	20.0 ma
	.78	10.0
	.75	5.0
	.72	2.0
	.68	1.0
	.67	.5
	.64	.2
	.62	.1
	0.0	0.0

Fig. p3.10

	V_Z	I_Z
	-1.0	0.0
	-5.0	0.0
	-10.0	.05
	-11.0	.07
REVERSE	-11.5	.25
REGION	-11.7	1.25
	-12.0	4.0
	-12.2	8.0
	-12.4	18.0
	-12.6	38.0
	-12.8	58.0
	-13.0	78.0

Fig. p3.10 (cont'd)

3.11 From the JFET drain characteristic data in Fig. p3.11, obtain values for V_P, G_c, ϕ_B.

V_{GS}	V_{DS}	I_{Drain}	V_{GS}	V_{DS}	I_{Drain}
0 v.	-.5v.	2.268 ma	+2v.	-.25 v.	.483 ma
	-1.0	4.121		-.5	.898
	-1.5	5.621		-1.0	1.532
	-2.0	6.805		-1.5	1.920
	-2.5	7.703		-1.75	2.026
	-3.0	8.337		-2.0	2.075
	-3.5	8.725		-2.1	2.080
	-3.75	8.831			
	-4.0	8.880	+3v.	-.25 v.	.224 ma
	-4.05	8.884		-.5	.388
	-4.1	8.885		-.75	.494
	-5	8.904		-1.0	.543
	-10	9.005		-1.1	.547
+1 v.	-.5 v.	1.499 ma			
	-1.0	2.683			
	-1.5	3.582			
	-2.0	4.216			
	-2.5	4.603			
	-2.75	4.709			
	-3.0	4.759			
	-3.1	4.763			

Fig. p3.11

3.12 From the JFET drain characteristic data in Fig. p3.11, obtain the values for RO, I_T, K.

3.13 From the C_{iss} and C_{rss} data of Fig. p3.13, obtain values for η_G, η_D, CG, CD, ϕ_B.

V_{DS}	V_{GS}	C_{iss}	C_{rss}	$V_{GD} = V_{DS} - V_{GS}$
-12.0	0	13.35	2.87	-12
-12.0	.1	12.86	2.86	-12.1
-12.0	.2	12.43	2.85	-12.2
-12.0	.5	11.42	2.83	-12.5
-12.0	1.0	10.28	2.79	-13.0
-12.0	2.0	8.91	2.72	-14.0
-12.0	5.0	7.02	2.52	-17.0
- .1	0.0	—	8.00	- .1
- .2	0.0	—	7.70	- .2
- .5	0.0	—	6.99	- .5
- 1.0	0.0	—	6.19	- 1.0
- 2.0	0.0	—	5.23	- 2.0
- 5.0	0.0	—	3.99	- 5.0

Fig. p3.13

3.14 From the MOSFET drain characteristic data in Fig. p3.14, obtain values for V_T, B.

V_{GS}	V_{DS}	I_{Drain}
-4 v	-.2 v	.081 ma
	-.4	.144
	-.6	.189
	-.8	.216
	-1.0	.225
-5 v	-.8 v	.575 ma
	-1.0	.675
	-1.5	.845
	-1.75	.885
	-2.0	.900
-6 v	-1.0 v	1.125 ma
	-2.0	1.800
	-2.5	1.970
	-2.75	2.010
	-3.0	2.025
-7 v	-2.0 v	2.700 ma
	-3.0	3.370
	-3.5	3.540
	-3.75	3.590
	-4.0	3.600

V_{GS}	V_{DS}	I_{Drain}
-8 v	-3.0 v	4.725 ma
	-4.0	5.400
	-4.5	5.570
	-4.75	5.60
	-5.0	5.625
-9 v	-4.0 v	7.200 ma
	-5.0	7.87
	-5.5	8.05
	-5.75	8.09
	-6.0	8.100
-10 v	-5.0 v	10.12 ma
	-6.0	10.80
	-6.5	10.97
	-6.75	11.00
	-7.0	11.025
	-12	11.525

Fig. p3.14

3.15 From the MOSFET drain characteristic in Fig. p3.14, obtain values for RO, I_T, K.

3.16 From the η versus V_{BB} data in Fig. p3.16, determine a value for the constant "a."

V_{BB}	η
5 v	.650
10	.637
15	.625
20	.612
25	.600

Fig. p3.16

3.17 From the data in Fig. p3.17, calculate R_{BB} for each value of V_{BB} and determine a value for the constant b.

V_{BB}	I_{B2}
5 v	.714 ma
10	1.39
15	2.03
20	2.63
25	3.21

Fig. p3.17

3.18 From the UJT data in Fig. p3.18, calculate N, IE1, α.

TEST DATA

V_{BB} = VK = 10 Volts

η_{VK} = .65

$R_{BB_{VK}}$ = 7 Kilohms

Θ = 35 Volts^{-1}

IS = 8 x 10^{-10} ma

V_{EB1} at I_E = 20ma = 2.0 Volts

V_{EB1} at I_E = 50ma = 2.4 Volts

I_{B2} at I_E = 20ma = 6.5 ma

I_{B2} at I_E = 50ma = 7.4 ma

Fig. p3.18

REFERENCES

1. Millman, J. and H. Taub, *Pulse, Digital and Switching Waveforms*, Chapter 20, McGraw–Hill Book Company, New York, 1965.

2. "SCEPTRE Support, Vol. I, Revised SCEPTRE User's Manual," *Air Force Weapons Laboratory Technical Report* TR–67–124.

3. Ebbers, J. J. and J. L. Moll, "Large Signal Behavior of Junction Transistors," *Proc. IRE*, Vol. 42, pp. 1761–1772.

4. Phillips, A. B., *Transistor Engineering*, Chapters 8, 9, 10, 16, McGraw–Hill Company, New York, 1961.

5. *Zener Diode Handbook*, First Edition Motorola Semiconductor Products, Inc., 1967.

6. Cordwell, W., "Radiation Device Modeling for the SCEPTRE Program," *IBM Report 69–825–2371*, (June, 1969).

7. Daniel, M. E., "Unijunction Transistor Model for Computer Circuit Analysis," *Development Report SC–DR–68–358*, Sandia Laboratories, (June, 1968).

8. *General Electric Transistor Manual*, 1968.

4

ELECTRONIC CIRCUIT EXAMPLES

4.1 INTRODUCTION

Throughout this book the reader is referred to Chapter 4 to observe and study various types of examples. The technique of illustrating the subtle and important points by using examples, rather than the use of lengthy discussions, has proven to be far more effective, especially with the new user. Many important points which have purposely not been mentioned, in what might normally be considered their context, are discussed and illustrated with examples in this chapter. Therefore, in order for the user to get the complete picture, it is necessary for him to carefully study every example, even if some are in areas which normally are not in his specialty or interest.

It is also important to point out that these examples, for the most part, represent the unusual, rather than the usual cases. In fact, it is for this reason that most of them were selected for inclusion in this chapter. The reader should bear in mind that in most practical cases the circuits are coded in a routine fashion, such as has already been illustrated in previous chapters of the text. In the examples to follow most of the circuit diagrams are drawn only in the SCEPTRE form to avoid the repetition of two figures showing essentially the same information.

4.2 SENSE/DIGIT CIRCUITRY

This first example contains no unusual aspects but does serve as a launching point since it contains such a wide variety of elements. Also, the output plots are quite excellent and show the power of good graphical data. The circuit diagram is shown in Fig. 4.1.

For this diagram, a correct listing is:

```
MODEL DESCRIPTION
MODEL 2N2369 (TEMP) (B-C-E)
ELEMENTS
RB, B-1 = 0.02
RC, C-2 = 2.0E-3
RE, 3-E = 0.5E-3
JE, 1-3 = DIODE Q(5.5E-12,37.5)
JN, 2-1 = 0.98 * JE
JC, 1-2 = DIODE Q (5.0E-9,23.6)
JI, 3-1 = 0.10 *JC
CE, 1-3 = Q1(2.8,1.5,VCE,.31,11.67,JE,5.5E-12)
CC, 1-2 = Q1(2.9,1.5,VCC,0.1,75.,JC,5.0E-9)
FUNCTIONS
Q1 (A,B,C,D,E,F,G) = (A/(B-C)**D+E*(F+G))
MODEL SENSE LINE (TEMP) (A-B)
PARAMETERS ARE FOR LINE LENGTH OF 1KBITS
ELEMENTS
R1, C-D = 2.0E-2
L1, A-C = 1
L2, D-B = 1
CIRCUIT DESCRIPTION
    SENSE/DIGIT CIRCUITRY
ELEMENTS
E1, 3-6 = 12
```

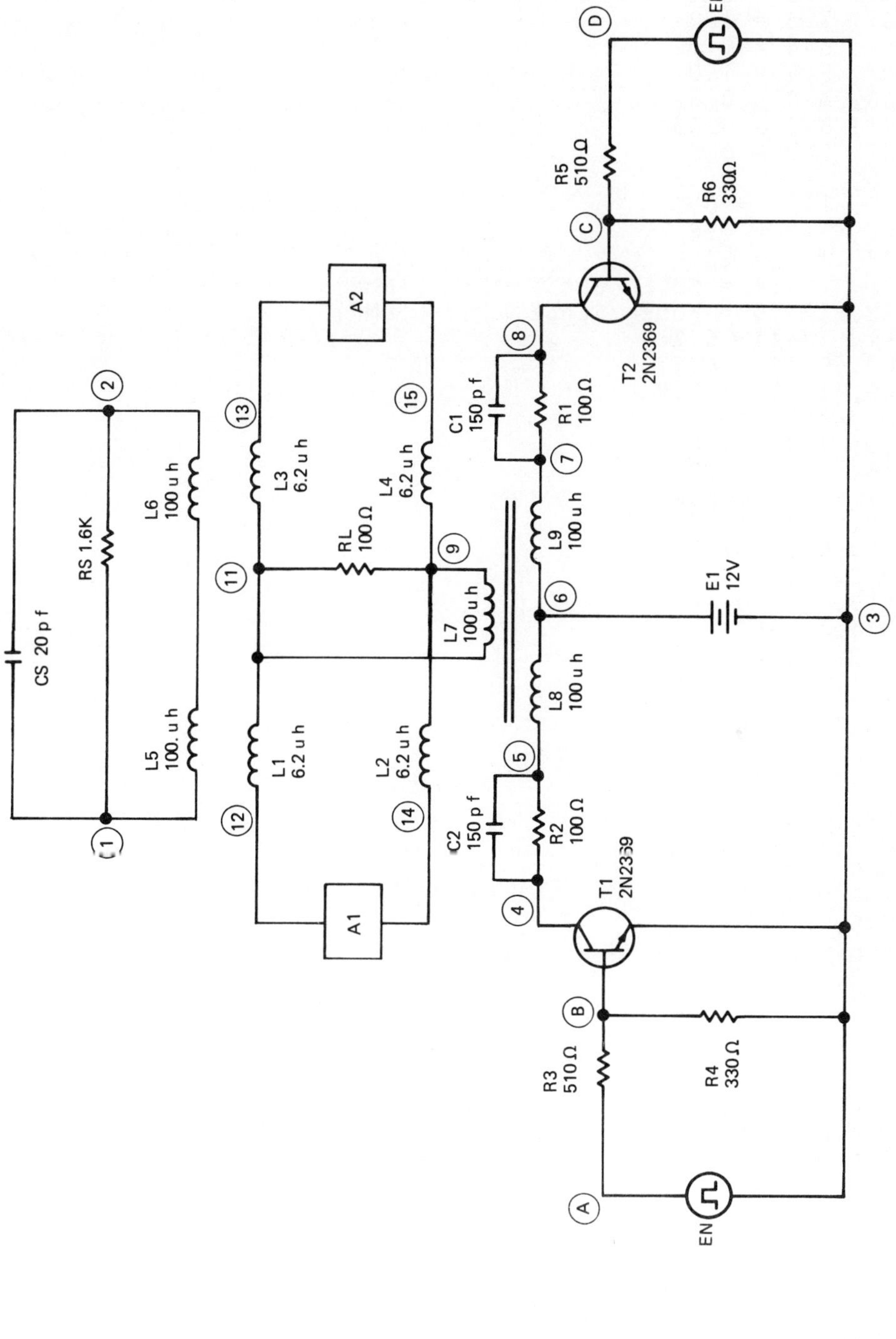

Fig. 4.1. Sense/digit circuitry.

```
EP, 3-D = TABLE 1(TIME)
EN, 3-A = TABLE 2 (TIME)
R1, 7-8 = 0.1
C1, 7-8 = 150
R2, 5-4 = 0.1
C2, 5-4 = 150
R3, A-B = 0.51
R4, B-3 = 0.33
R5, D-C = 0.51
R6, C-3 = 0.33
RL, 11-9 = 0.1
RS, 1-2 = 1.6
CS, 1-2 = 20
A1, 12-14 = MODEL SENSE LINE (TEMP)
A2, 13-15 = MODEL SENSE LINE (TEMP)
L1, 12-11 = 6.2
L2, 9-14 = 6.2
L3, 11-13 = 6.2
L4, 15-9 = 6.2
L5, 1-0 = 100
L6, 0-2 = 100
L7, 11-9 = 100
L8, 5-6 = 100
L9, 6-7 = 100
M12, L1-L2 = 6.1
M13, L1-L3 = 6.1
M14, L1-L4 = 6.1
M23, L2-L3 = 6.1
M24, L2-L4 = 6.1
M34, L3-L4 = 6.1
M15, L1-L5 = 24
M16, L1-L6 = 24
M25, L2-L5 = 24
M26, L2-L6 = 24
M35, L3-L5 = 24
M36, L3-L6 = 24
M45, L4-L5 = 24
M46, L4-L6 = 24
M56, L5-L6 = 99
M78, L7-L8 = 99
M79, L7-L9 = 99
M89, L8-L9 = 99
T1, B-4-3 = MODEL 2N2369 (TEMP)
T2, C-8-3 = MODEL 2N2369 (TEMP)
DEFINED PARAMETERS
PIDIF = X1 (IL1+IL3)
PVA1 = X2 (VRL+VL1+VL2)
PVA2 = X3 (VL1A1+VR1A1+VL2A1)
OUTPUTS
VRL, IL1,PIDIF(IDIF),PVA1(VA1),VRS,IL5,IL7,VL9,IR1,IL8,
VL8,PLOT
VL1A1,VR1A1,VL2A1,PVA2,VL1,IL2,VL2,PLOT
INITIAL CONDITIONS
VCCT1 = -12, VCCT2 = -12
```

```
FUNCTIONS
TABLE 1
0,0,5,5,55,5,60,0,65,0
TABLE 2
0,0,100,0,105,5,155,5,160,0,165,0
RUN CONTROLS
STOP TIME = 50, MAXIMUM PRINT POINTS = 200
RERUN DESCRIPTION
ELEMENTS
R1A1 = 0.018
END
```

From Fig. 4.2. it is seen that the frequency, damping ratio, rise time, etc., can all be calculated directly from the output plot for maximum convenience. Notice the use of the Defined Parameters to obtain various output quantities. The reader should study each listing carefully before going on to the next example.

4.3 TRANSFORMER COUPLED AMPLIFIER

A SCEPTRE schematic for a transformer coupled amplifier is shown in Fig. 4.3. This example illustrates a number of points which are discussed in Chapter 2 but are emphasized here.

One correct listing for the circuit in question is:

```
MODEL DESCRIPTION (INITIAL)
MODEL 2N9999AA (PERM) (B-E-C)
ELEMENTS
CE, B-E = EQUATION 1 (5., 40., TABLE 1(VCE))
CC, B-C = EQUATION 1 (10., 400., TABLE 2 (VCC))
J1, B-E = DIODE TABLE 1
J2, B-C = DIODE TABLE 2
J3, C-B = P1 * J1
DEFINED PARAMETERS
P1 = .98
OUTPUTS
VCE, VCC, J1, PLOT
FUNCTIONS
EQUATION 1 (A, B, C) = (A+B*C)
DIODE TABLE 1
0, 0, .3, 0, .65, .05, .7, .6, .72, 1.4, .73, 2, .74,
3.4, .77, 10, .8, 22
DIODE TABLE 2
0, 0, .58, 0, .62, .4, .64, 1, .66, 2, .67, 3, .69, 7,
.7, 12
CIRCUIT DESCRIPTION
EXAMPLE 2 NETWORK
ELEMENTS
E1, 1-2 = TABLE 1 (TIME)
DERIVATIVE E1 = TABLE DE1
E2, 1-4 = 20
CZ, 2-3 = 1E3
```

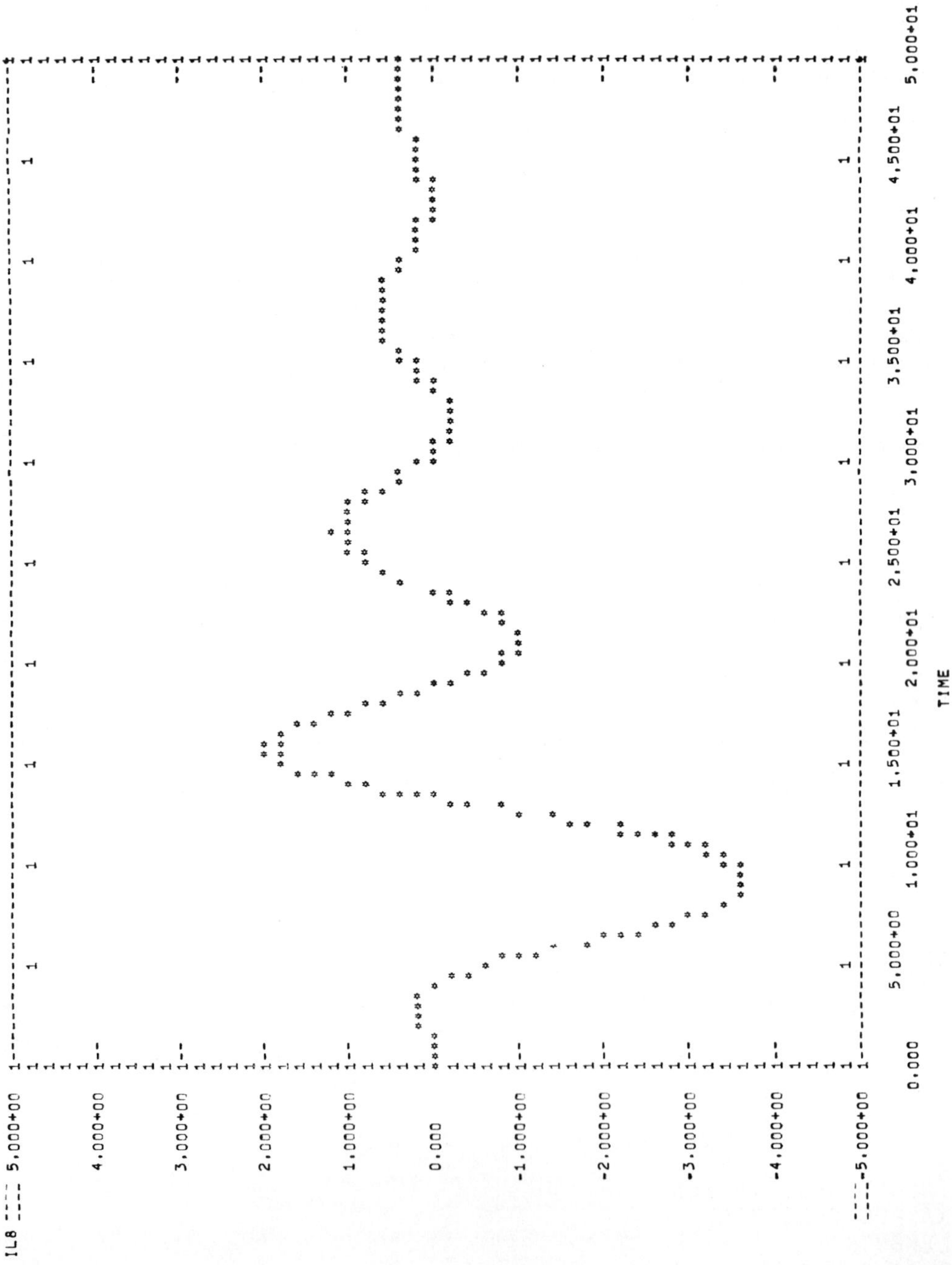

Fig. 4.2 Plot of IL8 vs TIME

```
CX, 5-6 = 1E3
R1, 4-3 = 30
R2, 3-1 = 20
R3, 5-1 = 2
R4, 4-6 = 240
R5, 4-7 = 3.3
R6, 9-1 = 1.8
T1, 3-5-4 = MODEL 2N9999AA (PERM)
T2, 6-1-8 = MODEL 2N9999AA (PERM, CHANGE P1 = .975)
L1, 7-8 = 100
L2, 9-1 = 900
M, L1-L2 = 299.7
OUTPUTS
VR6, VL1, VL2, PLOT
ICZ, ICX, IR3, IR2, IR1, J3T1, J2T1, CCT1, ICCT1
FUNCTIONS
TABLE 1
0, 0, 50, .5, 100, .5
TABLE DE1
0, .01, 50, .01, 50, 0, 100, 0
RUN CONTROLS
STOP TIME = 500, COMPUTER TIME LIMIT = 5
RUN INITIAL CONDITIONS
MAXIMUM PRINT POINTS = 3000
END
```

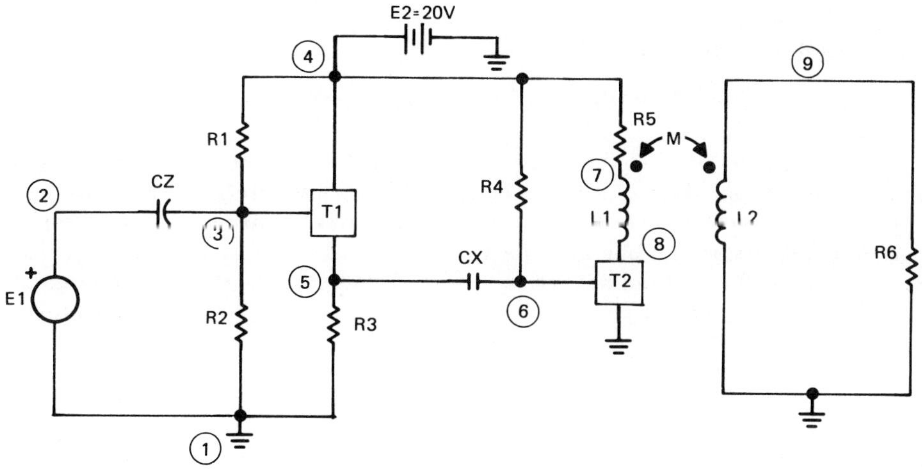

Fig. 4.3. Transformer coupled amplifier.

Discussing the points in the order in which they appear, notice first the term (INITIAL) after MODEL DESCRIPTION. This indicates that the 2N9999AA model is the first model to be stored permanently on the computer tape.

Next observe the use of a table in an equation. According to Sec. 2.9.3, TABLE 1 (VCE) and DIODE TABLE 1 are the same tables. TABLE 1 (VCE) could have been listed as DIODE TABLE 1 but not vice versa. The reason for the latter statement is seen from TABLE 2.4. If it is necessary to compute initial conditions, any J that is a function of the voltage across itself must always be specified as a primary or secondary dependent source. In this case

```
J1, B-E = DIODE TABLE 1
```

is the correct format, even though TABLE 1(VCE) seems to contain the same information.

A good general rule to follow is to always use the format DIODE TABLE, when the table in question is to be a function of the voltage across a current generator. Therefore, a better way to list CE is:

```
CE, B-E = EQUATION 1 (5., 40., DIODE TABLE 1)
```

This listing avoids any confusion, although both of the CE listings yield identical results as long as the current generator, J1, is set equal to DIODE TABLE 1. Actually, the best and most direct way to list CE is:

```
CE, B-E = X1(5.+40.*J1)
```

While discussing tables, notice that the CIRCUIT DESCRIPTION also contains a TABLE1. However, this is a completely different table and has no relation to the table listed under MODEL DESCRIPTION.

The next point of interest is the listing:

```
DERIVATIVE E1 = TABLE DE1
```

It is not obvious why such a listing is needed. However, careful study of the circuit diagram indicates that there is a loop consisting of four capacitors in series with input voltage, E1. According to Sec. 2.16.3 a source derivative is required in such cases. The important point here is that one must look carefully to discover such potential problem areas. However, if the user did not see this error, the program aborts and an appropriate diagnostic message is shown. The user then realizes the situation and reruns the problem correctly.

The listing:

```
T2, 6-1-8 = MODEL 2N9999AA (PERM, CHANGE P1 = .975)
```

indicates that the parameter P1 is changed to .975 from whatever it is in the permanently stored model. The model change holds only for this particular run and in any associated reruns. A final point that should be explained in the listing of this example is the entry MAXIMUM PRINT POINTS = 3000 which appears under RUN CONTROLS. In the absence of this command, SCEPTRE uses its preset maximum of 1000 print points. That is, the printed output format is presented at no more than one thousand solution steps regardless of how many solution steps were actually taken. The relationship that controls the number of output print points is:

$$\frac{\text{Number of steps taken}}{\text{X}} \leq \text{Maximum print points allowed.}$$

where X is the smallest possible integer which makes this relation true. If, for example, the transient solution of this run had required 2109 steps and the default maximum of 1000 was allowed to stand, then the results of 703 (or every third step since here X = 3) steps are printed for output. The fact that the preset maximum was countermanded in this case allows for all 2109 steps to be printed out. Many times this command is used in the reverse fashion; that is, to reduce the number of output print points and hence the amount of output print-out data. The other listings in this example are straightforward, but each should be studied by the reader to ensure complete understanding.

4.4 ANTILOG AMPLIFIER EXAMPLE

Figure 4.4 is the circuit diagram for an Antilog Amplifier. Fig. 4.5 shows the equivalent circuit for the system. This example illustrates several points, the main one being the conservation of computer running time.

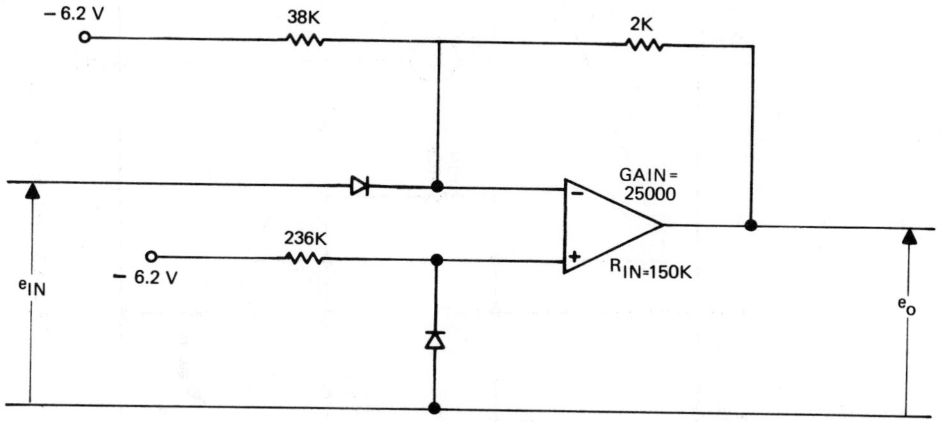

Fig. 4.4. Antilog amplifier.

The following is a listing that one might make for such an analysis.

```
CIRCUIT DESCRIPTION
RUN 1, ANTILOG AMPL
ELEMENTS
E1, 4-5 = Q1(TABLE 1 (TIME))
E2, 6-4 = 6.2
E3, 7-4 = 6.2
R1, 8-6 = 38
R2, 8-3 = 2
R3, 2-7 = 236
R4, 4-3 = 10
C1, 4-10 = 10
```

```
RS1, 4-10 = 2000
RB1, 10-2 = 0.01
JD1, 4-10 = DIODE EQUATION (1.E-9, 14.0)
C2, 5-11 = 10
RS2, 5-11 = 2000
RB2, 11-1 = .01
```

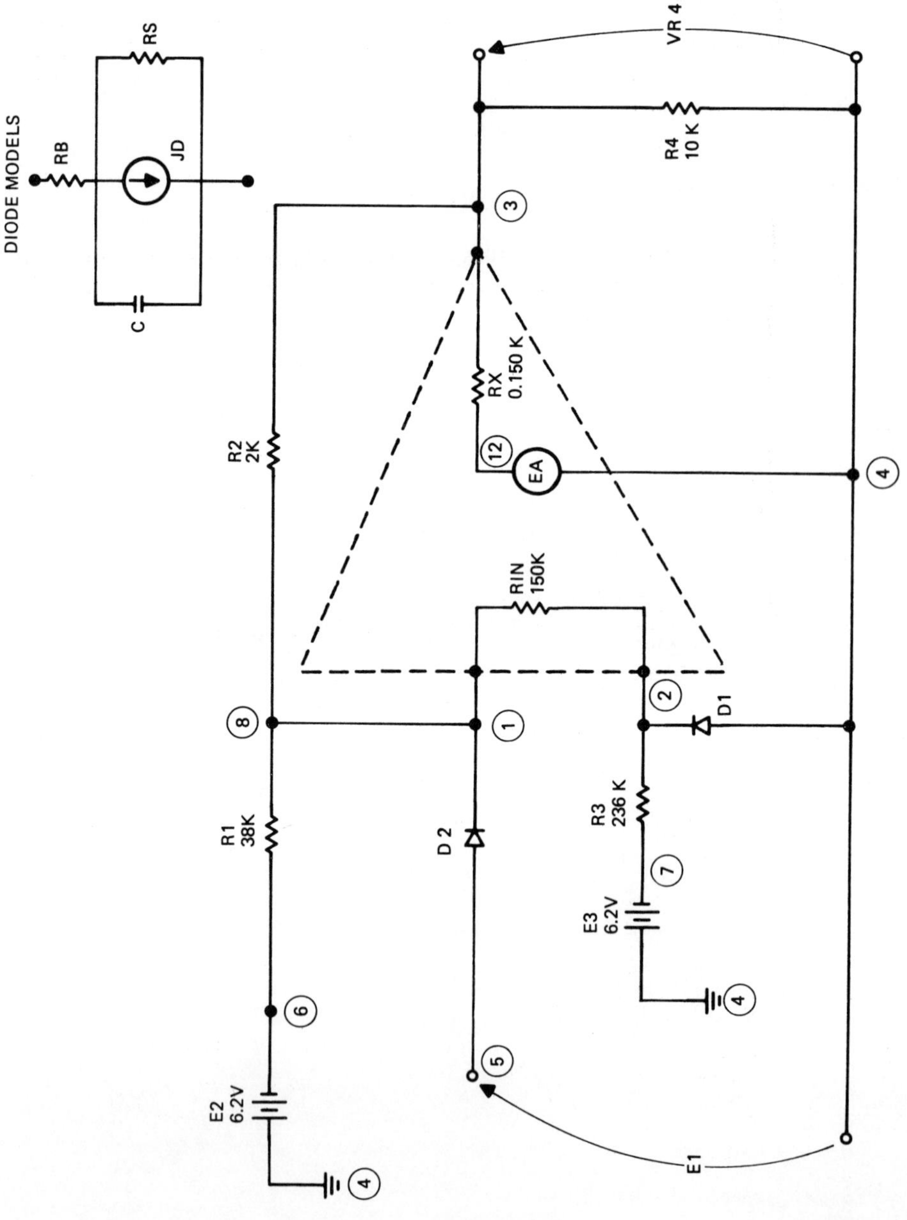

Fig. 4.5. Antilog amplifier.

```
JD2, 5-11 = DIODE EQUATION (1.E-9, 14.0)
RIN, 2-1 = 150
RX, 3-12 = 0.150
EA, 4-12 = 25E3 * VRIN
OUTPUTS
E1, VRIN, VR4, IRB1, IRB2, PLOT
FUNCTIONS
TABLE 1
0, 2.4, 1.E10, 240
Q1(A) = (-.125*ALOG(1.42E5/A)+.6)
RUN CONTROLS
RUN INITIAL CONDITIONS
STOP TIME = 1.E10
END
```

It is seen in this case that the diodes are entered under CIRCUIT DESCRIPTION rather than under MODEL DESCRIPTION. The format of EA is required as shown (see linearly dependent sources under Table 2.4). In this circuit the variable is to vary from 2.4 to 240. The output is desired over this range. However, this is not basically a transient problem. The user actually wants only to determine whether the output voltage is the antilog of the input voltage. Therefore, to minimize the capacitive and transient effects, TABLE 1 indicates that the variable changes from 2.4 to 240 in 1.E10 nanoseconds. This time duration is long enough for the output curve to closely represent the required steady-state type of transfer function.

The preceding discussion might seem logical; however, the careful observer should notice the long problem stop time in comparison to the small system time constants. This would result in excessive computer cost. A more economical approach would be to replace TABLE 1 with the Defined Parameter P1, where P1 is set equal to the table ordinate value at TIME = 0; i.e., P1 is set equal to 2.4. The master run is then made for initial conditions only. Under reruns, the value of P1 is changed over the required range in the desired smallest increments. This approach does not yield a convenient plot of the output voltage as a function of input voltage, but the initial conditions values could readily be plotted manually for the reruns, and certainly much computer time is saved.

The user should always keep an open mind on the type of approach to be used. Many times a little thought can save many dollars. Too often users do their thinking after the run has aborted or after the computer cost bill has been received.

4.5 SATURABLE REACTOR EXAMPLE

Two examples are illustrated in this section. One is a simplified reactor with a single valued Ø – NI curve. The second example is the more common case of a hysteresis loop. In this latter case, a FORTRAN subprogram is required. The subprogram suitable for these nonlinear inductors and transformers is developed in detail and has wide application.

4.5.1 Single Valued Ø – NI Curve

Figure 4.6 is a simple circuit to illustrate a saturable reactor with a single valued Ø – NI curve. Figure 4.7a indicates the relationship between Ø and NI, while Fig. 4.7b shows $\frac{dØ}{dNI}$ versus NI.

Since $L = N\frac{dØ}{dI}$ then $L = N^2\frac{dØ}{dNI}$ Let $E_1 = 5\sin(2.51E{-}6 * TIME)$ and N = 10.

One correct listing for the circuit is:

```
CIRCUIT DESCRIPTION
SIMPLIFIED SATURABLE REACTOR
ELEMENTS
E1, 1-2 = X1 (5. * SIN (2.51E-6*TIME))
R1, 3-1 = 1
L1, 2-3 = Q1(100.,TABLE 1(PNI))
DEFINED PARAMETERS
PNI = X2(10.*IL1)
OUTPUTS
VR1,PNI,L1,PLOT
XSTPSZ,PLOT
FUNCTIONS
TABLE 1 = -10,400, -2,400, -2,1E4, -.4,1E4, -.4,4E4,
.4,4E4, .4,1E4, 2,1E4, 2,400, 10,400
Q1(A,B) = (A*B)
RUN CONTROLS
STOP TIME = 5E6
MINIMUM STEP SIZE = 1
END
```

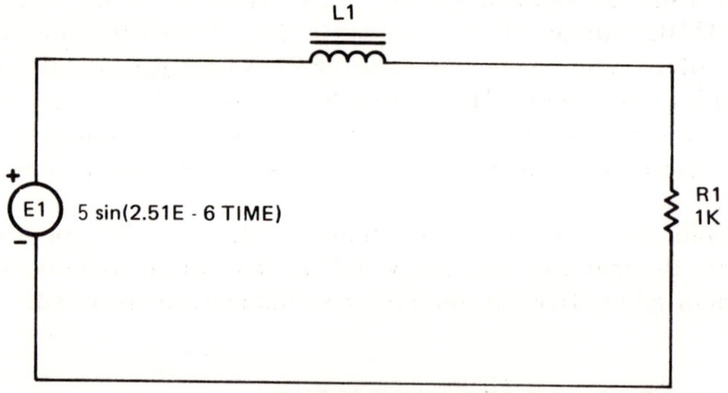

Fig. 4.6.

From the preceding listing is is seen that L1 is set equal to one hundred times (TABLE 1), where TABLE 1 is from Fig. 4.7b. The independent variable of the TABLE is the Defined Parameter PNI, which is the current times the number of turns. From Fig. 4.7b, if the ampere – turns are known, then

$\frac{dØ}{dNI}$ is known, and hence L is known ($L = 10^2 * \frac{dØ}{dNI}$).

The MINIMUM STEP SIZE = 1 statement is necessary because of the large STOP TIME (see Sec. 2.10.4).

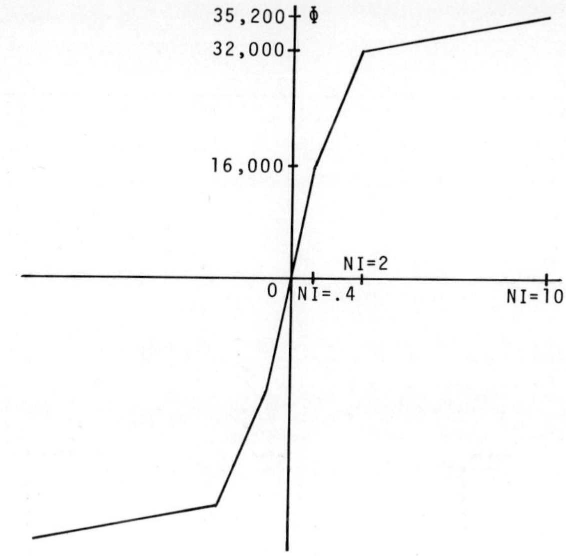

Fig. 4.7. A

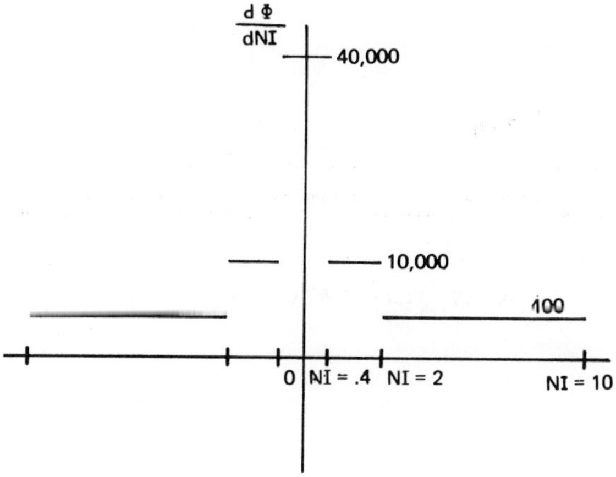

Fig. 4.7. B

Figure 4.8 is the output plot of the voltage across R1 as a function of time. Notice that VR1 is very small as long as the inductance is very high (4 henries). It begins to rise sharply as L1 decreases to 1 henry and then to .04 henry.

The step size is also outputted (XSTPSZ) for curiosity to see how close it came to the minimum allowed. The data shows that the smallest step taken was 39. Therefore without the step size change, the problem would not have run since the automatic step size minimum would have been, (STOP TIME)*(10^{-5}) = (5E6) (10^{-5}) = 50, and 39 < 50. The example in Sec. 4.11 discusses the step size philosophy more fully.

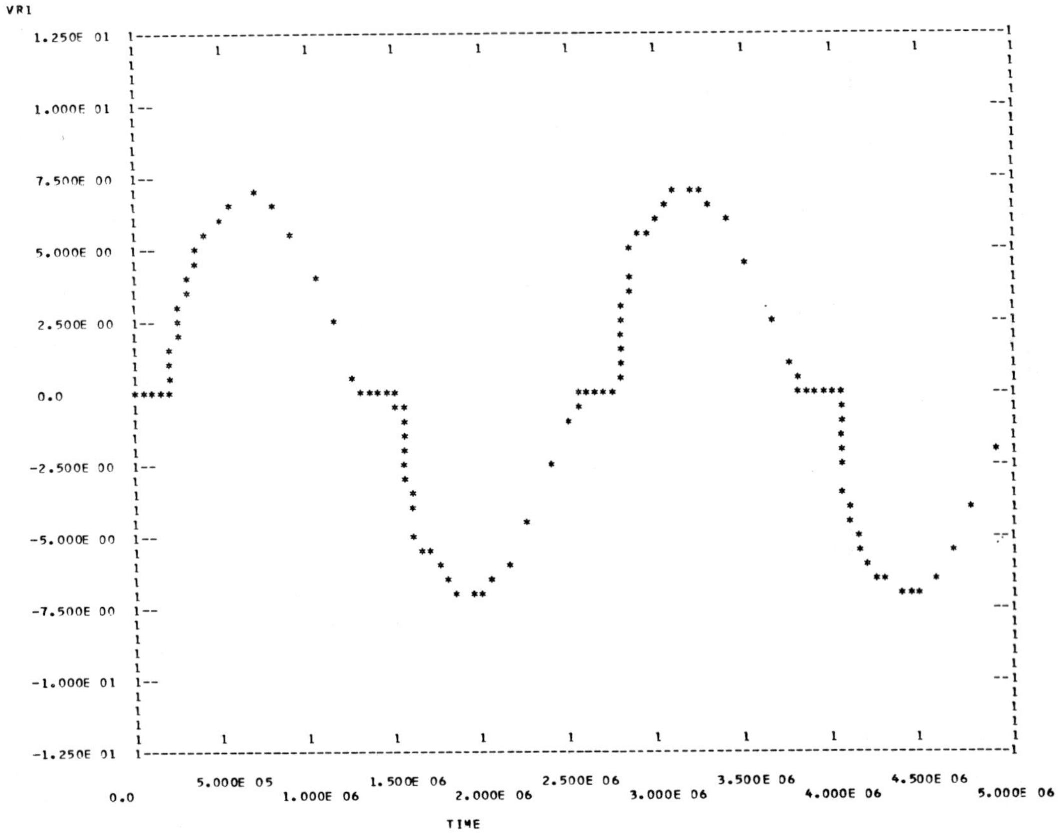

Fig. 4.8 Plot of VRI vs TIME

4.5.2 Inductor Hysteresis Example

In order to illustrate the principle, a very simple example is shown in Fig. 4.9. This problem differs from the example of Sec. 4.5.1 in that the hysteresis loop of Fig. 4.10 is used to control the inductance of the saturable reactor.

A special, and fairly involved, subprogram is required to simulate the hysteresis loop; however, this subprogram is general in nature and is applicable to a wide variety of inductor and transformer problems.

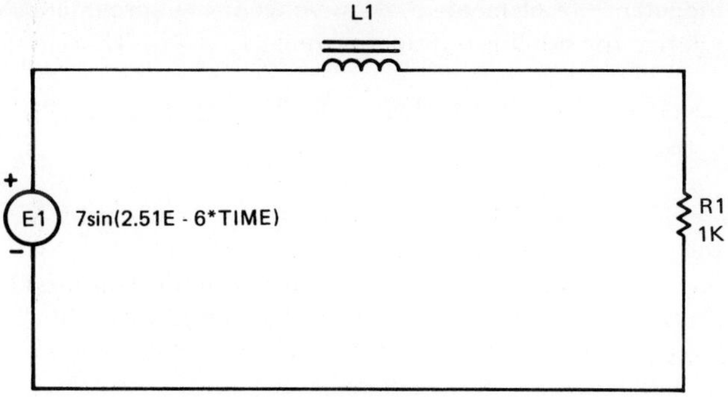

Fig. 4.9.

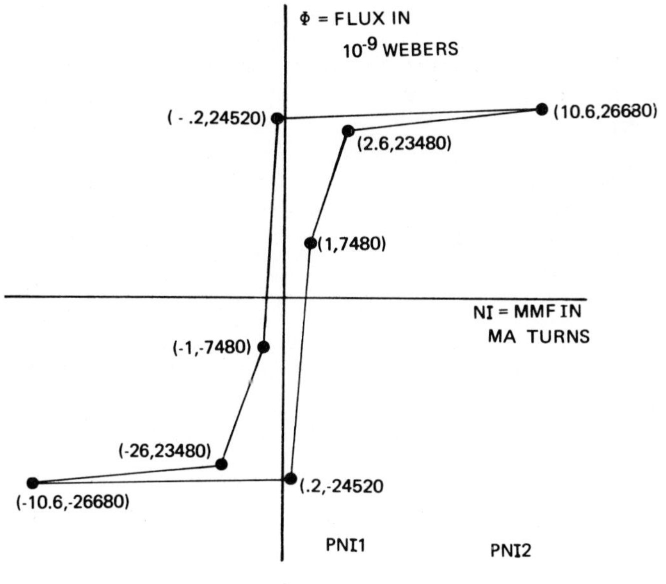

Fig. 4.10.

4.5.3 Subprogram FL for Hysteresis Loop Simulation

The subprogram is designed to simulate the hysteresis loop (major and minor) characteristics of nonlinear inductors in magnetic circuit applications. A typical hysteresis loop is shown in Fig. 4.11 where the flux (PHI) is the dependent variable and the ampere-turns (PNI) is the independent variable. Basically, FL operates by determining the appropriate values of flux and the derivative of flux from the hysteresis loop and using these values to compute the inductance of elements that reference the subprogram. The usual method of operation is to define the nonlinear inductor under ELEMENTS as:

L1, NODE–NODE = FL(PSAT, PNI, PNI1, PNI2, PLSAT, PHI1, PHI2, PL1, PSTATE), where the arguments are defined as:

PSAT = slope of flux vs. PNI in the reversible region of the hysteresis loop. This is entered by the user as either a constant or a variable.
PNI = net ampere turns that control the hysteresis loop. This usually is a product or sum of products that is entered as an EQUATION or EXPRESSION.
PNI1, PNI2 = constant boundary points of both irreversible regions.
PLSAT = self inductance L1 in the reversible region. It is usually, but not necessarily constant and is computed as the product of the number of turns squared and PSAT.

PHI1, PHI2 = flux versus PNI for the entire hysteresis loop. These are usually, but not necessarily input as tabular function. PHI1 must describe the single valued regions above PNI2 and below –PNI2 as well as the lower curve in the double valued region of the loop. PHI2 must also describe the single valued regions above PNI2 and below –PNI2 as well as the upper curve in the double valued region of the loop.

PL1 = self inductance L1 in the irreversible regions. It is equal to the product of the number of turns squared and dØ/dNI. The user must supply dØ/dNI in either tabular or closed form.

PSTATE = the initial state of the core. PSTATE = 0 for the lower reversible region, PSTATE = 1 for the irreversible region on the right side of the loop, PSTATE = 2 for the irreversible region on the left side of the loop and PSTATE = 3 for the upper reversible region.

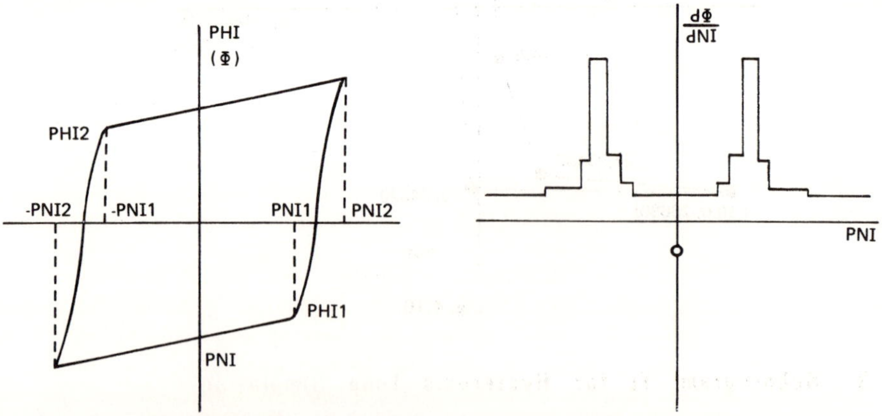

(A) HYSTERESIS LOOP I VS N

(B) DERIVATIVE OF (A) WITH RESPECT TO NET AMPERE TURNS

Fig. 4.11.

These arguments are all entered under DEFINED PARAMETERS. A typical partial listing, along with some appropriate commentary in parentheses follows:

DEFINED PARAMETERS

PSAT = NUMBER (if a constant is desired)

PNI = X1 (IL1*PN)

PNI1 = NUMBER

PNI2 = NUMBER

PLSAT = X2 (PN*PN*PSAT) (computes inductance as PN^2*dØ/dNI in reversible regions)

PHI1 = TABLE 1 (PNI)

PHI2 = TABLE 2 (PNI)

PL1 = Q1 (PN, TABLE 3 (PNI)) (computes inductance as PN^2*dØ/dNI in irreversible regions)

PSTATE = NUMBER (0, 1, 2 or 3)

PN = NUMBER (number of turns)

Under FUNCTIONS:

TABLE 1 (tabular function of flux including the lower curve of the double valued region)

TABLE 2 (tabular function of flux including the upper curve of the double valued region)

TABLE 3 (dØ/dNI,Fig. 4.11b)

Q1 (A, B) = (A*A*B)

Subprogram FL is shown in Fig. 4.12. This particular version is compatible with the IBM S/360 computer. Slight changes in the language are necessary for use with the 7094.

If the flux (PHI) is desired as output, several steps are required. First introduce PHI = NUMBER under DEFINED PARAMETERS. The particular "number" that is used is immaterial since it will be internally replaced by whatever value subprogram FL computes as soon as the analysis begins. In addition, the user must add PHI to the argument list of the FL reference (element L1 in this section) and also to the argument list of subprogram FL itself. Finally, PHI must be requested under OUTPUTS in the same way as any other Defined Parameter.

In some cases, the user may wish to use this subprogram in a run in which the initial conditions are to be computed by SCEPTRE. Under these conditions, Defined Parameter PNI (which is the independent variable of the hysteresis loop) must be requested for output, for algorithmic reasons.

Fig. 4.12. Subprogram FL.

```
      FUNCTION FL (PSAT, PNI, PNI1, PNI2, PLSAT, PHI1,
      PHI2, PL1, PSTATE)
      IMPLICIT REAL * 8(A-H, O-Z)
      COMMON/CNTRL S/TIME, XSTOPT, XMXERT, XIR, XTISSS,
      XMNISS, XMXISS,  XMNAIE,  XMXAIE, XIC, XRERNO,
      XMXPAS, XMNRIE, XMXRIE. XMNISO, XMXICP XICRER,
      XICAER
      XMXOTP, XXDIMS, XYDIMS, XICPAS, XNOPRQ,
      XNTNDX, XNHEAD, XNDFEQ, XERT, XNRF
      RN, XCNTNU, XSTPNO, XPASNO  XRUNNO, XSTPSZ,
      XTRANS, SKIP(4)
      IF(XPASNO . EQ. 1 .)GO TO 1000
      PDNI = PNI-OLDNI
      IF(XSTPNO .NE. COUNT) GO TO 10
      OLDNI = PNI
      OLDPHI = PHI
      COUNT = COUNT + 1.
      IF (DABS(PNI) .LE. PNI1) GO TO 100
10    IF (DABS(PNI) .GE. PNI2) GO TO 105
      IF (PDNI .EQ. 0.) GO TO 99
      IF ((PNI .LT. PNI2) .AND. (PNI .GT. PNI1))
      GO TO 50
C     LEFT SIDE OF LOOP
      IF (PDNI .GT. 0.) GO TO 100
      IF (FL .NE. PLSAT) GO TO 102
      IF (PHI-PHI2) 100, 102, 102
C     RIGHT SIDE OF LOOP
50    IF (PDNI .LT. 0.) GO TO 100
      IF (FL .NE. PLSAT) GO TO 101
      IF (PHI-PHI1) 101, 101, 100
99    PHI = PHI
      FL = FL
      JCOUNT = 0
      GO TO 2000
100   IF (JCOUNT .NE. 0)GO TO 1011
      BSAT = OLDPHI-PSAT * OLDNI
      PHI = PSAT * PNI + BSAT
      FL = PLSAT
      JCOUNT = 1
      GO TO 2000
1011  PHI = PSAT * PNI + BSAT
      FL = PLSAT
      GO TO 2000
105   PHI = PHI1
      FL = PLSAT
      JCOUNT = 0
      GO TO 2000
101   PHI = PHI1
      FL = PL1
      JCOUNT = 0
      GO TO 2000
```

```
102   PHI = PHI2
      FL = PL1
      JCOUNT = 0
      GO TO 2000
103   PHI = PHI1
      GO TO 1
104   PHI = PHI2
1     BSAT = PHI-PSAT * PNI
      FL = PLSAT
      JCOUNT = 1
      GO TO 2000
C     INITIALIZATION OF HYSTERESIS LOOP
1000  PDNI = 0.
      COUNT = 1.
      OLDNI = PNI
      OLDPHI = PHI
      IF (PSTATE .EQ. 0.) GO TO 103
      IF (PSTATE .EQ. 1.) GO TO 101
      IF (PSTATE .EQ. 2.) GO TO 102
      IF (PSTATE .EQ. 3.) GO TO 104
2000  RETURN
      END
```

4.5.4 Hysteresis Example Solution

From the preceding information, the following program is listed as a correct listing for the solution to the problem of Figs. 4.9 and 4.10.

```
CIRCUIT DESCRIPTION
SATURABLE REACTOR WITH HYSTERESIS
LOOP
ELEMENTS
E1, 1-2 = X1(1.*DSIN (2.51E-6*TIME))
R1, 3-1 = 1
L1, 2-3 = FL(PSAT,PNI,PNI1,PNI2,
PLSAT,PHI1,PHI2,PL1,PSTATE,PHI)
DEFINED PARAMETERS
PSAT = 200
PNI = X9(10.*IL1)
PNI1 = .2
PNI2 = 10.6
PLSAT = 20000
PHI1 = TABLE 1 (PNI)
PHI2 = TABLE 2 (PNI)
PL1 = Q1 (TABLE 3 (PNI))
PSTATE = 0
PHI = 0
OUTPUTS
VR1,PNI,L1,PLOT
XSTPSZ,PLOT
```

* The term DSIN is used because problem was run on S/360 version.

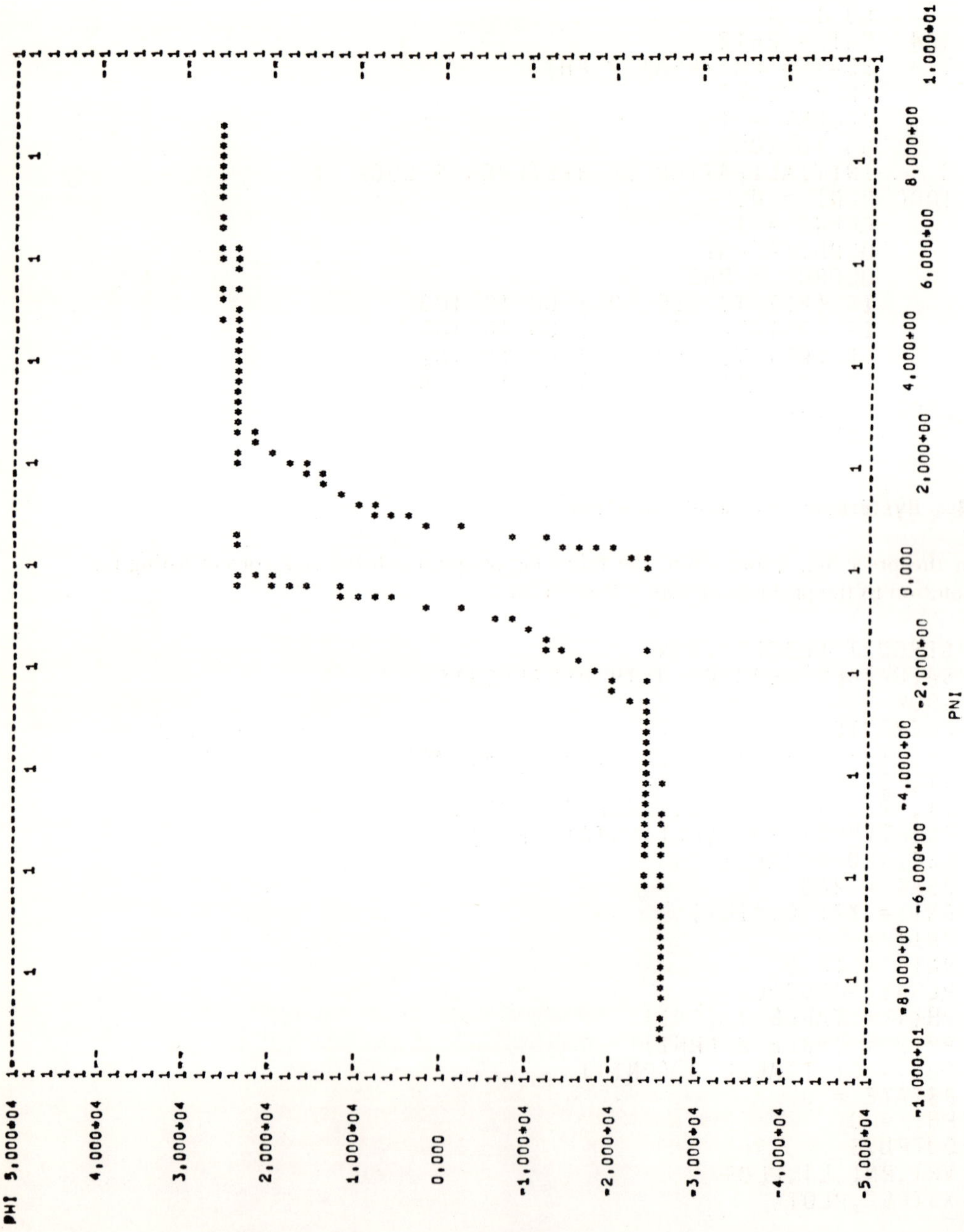

Fig. 4.13 Plot of PHI vs PNI

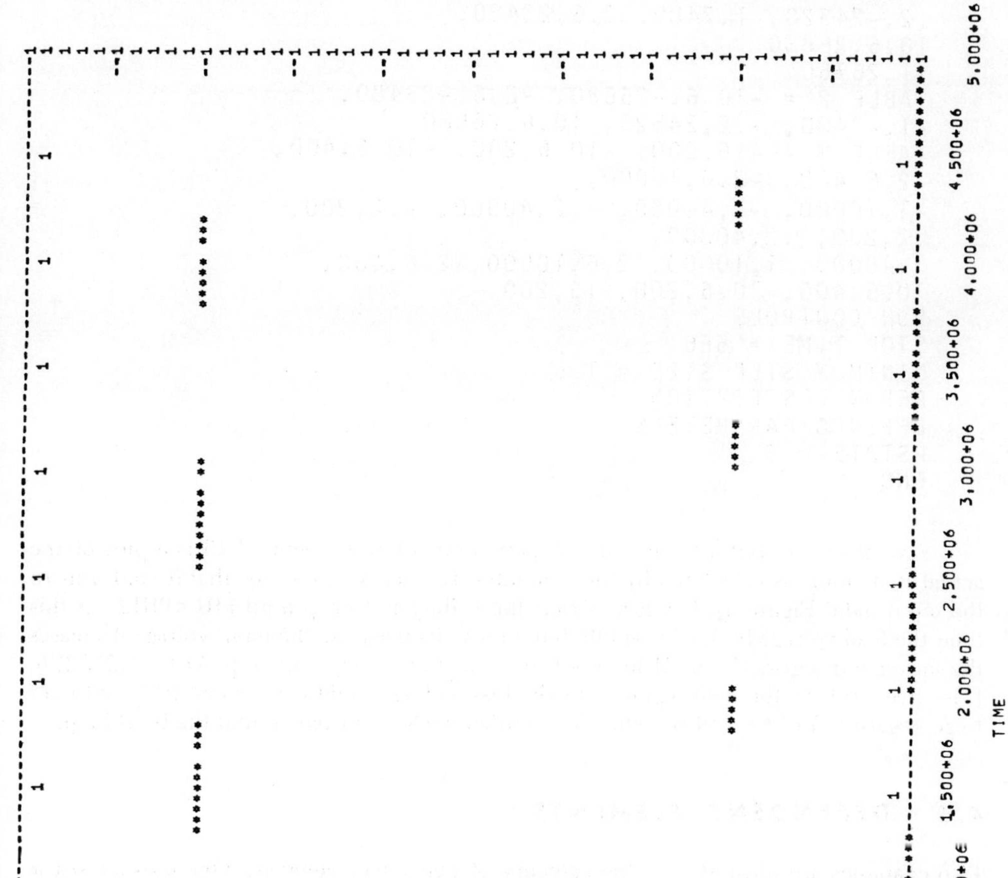

Fig. 4.14 Plot of L1 vs TIME

```
PHI, PLOT(PNI)
FUNCTIONS
Q1(A) = (100.*A)
TABLE 1 = -10.6, -26680,
.2,-24520, 1,7480, 2.6,23480,
10.6,26680
11,26760
TABLE 2 = -10.6,-26680, -2.6,-23480,
-1,-7480, -.2,24520, 10.6,26680
TABLE 3 = -15,200, -10.6,200, -10.6,400,
-2.6,400, -2.6,10000,
-1,10000, -1,40000, -.2,40000, -.2,200,
.2,200, .2,40000,
1,40000, 1,10000, 2.6,10000, 2.6,400,
10.6,400, 10.6,200, 15,200
RUN CONTROLS
STOP TIME = 5E6
MINIMUM STEP SIZE = 1
RERUN DESCRIPTION
DEFINED PARAMETERS
PSTATE = 3
END
```

Several of the output plots are of particular interest. Figure 4.13 is a plot of the actual B–H loop as calculated by the computer. It is reassuring to see that it conforms to the input data. Figure 4.14 indicates that the L does not drop until PHI <PHI2. At this time the load voltage begins to rapidly build up to its peak. As the input voltage decreases the movement across the B–H loop is from right to left across the top. At t = 1.2732E6, L is once again in the high region, and the load voltage rapidly decreases. PNI continues to go negative until t = 1.9E6. After that another cycle is started around the B–H loop.

4.6 DEPENDENT ELEMENTS

Two examples are illustrated in the category of dependent elements. One uses a resistor as a function of the voltage across itself, and the other uses a resistor as a function of the current through itself. Observe that the latter requires a special approach to get around a program restriction.

4.6.1 Voltage Dependent Resistor

The simple circuit of Fig. 4.15 is used to show that the SCEPTRE program can accurately handle a R = F(VR) in many applications. An acceptable listing for the circuit is:

```
CIRCUIT DESCRIPTION
ELEMENTS
E, 1-2 = 1
R, 2-3 = X1(1./VR)
C, 3-1 = 1
DEFINED PARAMETERS
PTRU = X2 (1.-1./(TIME+1.))
```

```
PABERR = X3 (VC-PTRU)
PRELER = X4 ((VC-PTRU)/PTRU)
OUTPUTS
VR,PTRU,PABERR,PRELER,R,PLOT
RUN CONTROLS
STOP TIME = 5
END
```

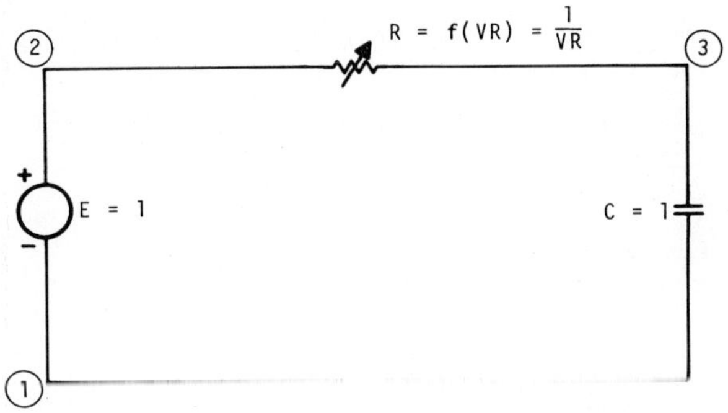

Fig. 4.15.

To illustrate the versatility of SCEPTRE, the DEFINED PARAMETER mode is used to check the actual results versus the theoretical results.

For this circuit, the closed form solution is:

$$VC(t) = 1 - \frac{1}{t+1}.$$

This is listed above as PTRU. The absolute error, listed as PABERR, is the difference between the VC and PTRU; i.e, PABERR is the difference between the computed results and the theoretical results.

The relative error, PRELER, is defined as the absolute error divided by PTRU. The plot of PABERR in Fig. 4.16 indicates that the maximum absolute error is about $2.5\text{x}10^{-5}$. The fact that the errors are so small (computer round off) proves that the formulation can accurately handle a nonlinear resistor if a computational delay is avoided (see the example in Sec. 4.11).

4.6.2 Current Dependent Resistor

In Table 2.4, footnote 1 states that no initial conditions can be computed if any resistor current is used as an independent variable or argument. This example is to show that the user should exercise his ingenuity in solving computer analysis problems. Sometimes clever approaches do not succeed, but the following example illustrates a case where the computer is "outsmarted."

In Fig. 4.17, R2 is a function of the current through itself, as defined by:

$$R2 = 1 + .1\ IR2$$

No initial conditions can be computed if this statement is entered directly into the program. However,

$$R2 = 1 + \frac{.1\ VR2}{R2}$$

is the same equation, but does not violate the program algorithms.

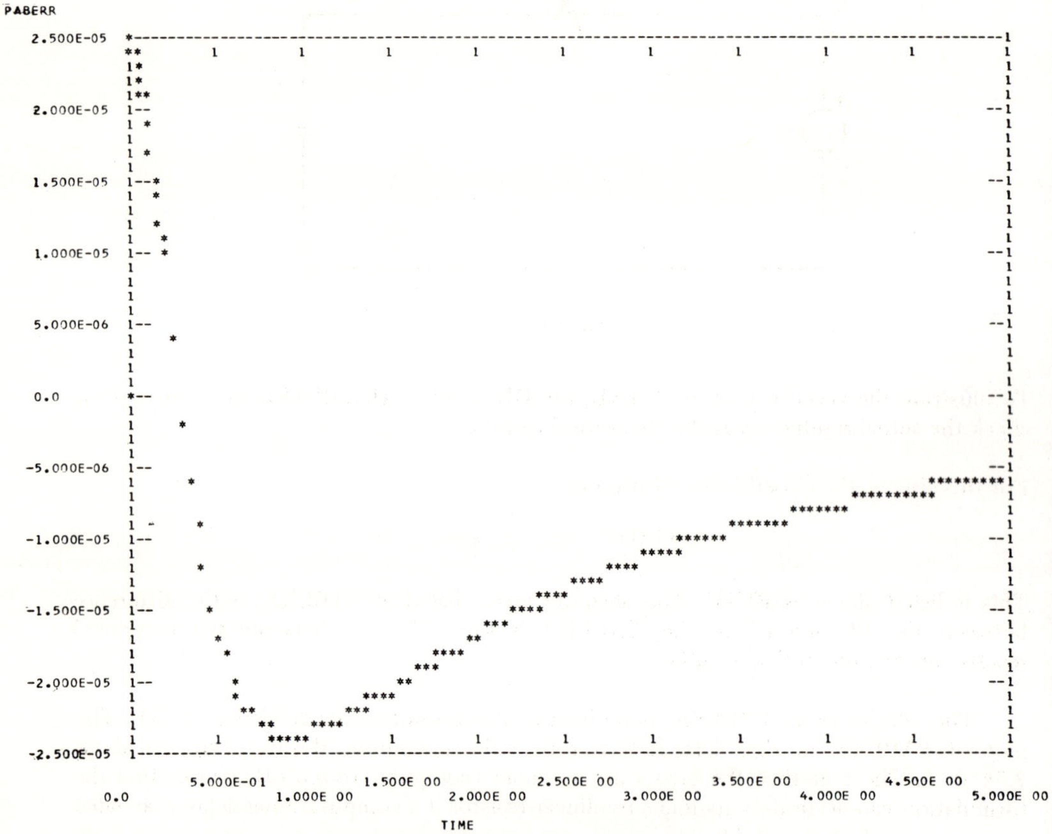

Fig. 4.16 Plot of PABERR vs TIME

A correct listing for this circuit is:

```
CIRCUIT DESCRIPTION
ELEMENTS
E1, 1-2 = TABLE 1
R1, 3-1 = 1
R2, 2-3 = Q1 (1.0,.1,VR2,R2)
C1, 3-1 = 1
OUTPUTS
VC1,VR2,VR1,IR2,E1,IC1,R2,PLOT
FUNCTIONS
TABLE 1
0,1,10,10,10,0,20,0
Q1(A,B,C,D) = (A+B*(C/D))
RUN CONTROLS
RUN INITIAL CONDITIONS
STOP TIME = 20
END
```

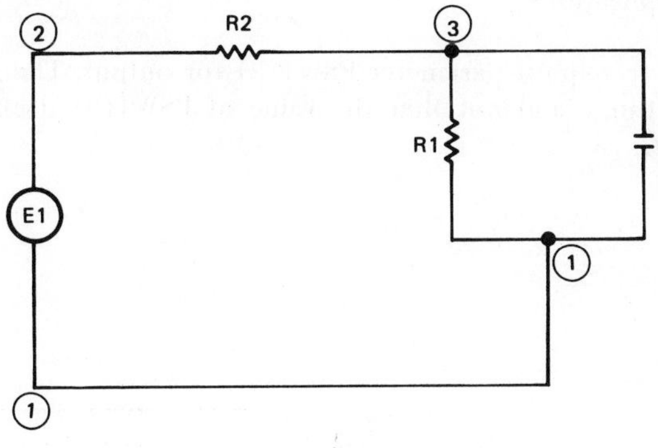

Fig. 4.17.

The program works as shown and the correct initial conditions are computed.

4.7 FSKIP EXAMPLE[2]

A very useful subprogram called FSKIP has been developed which has numerous applications and can result in tremendous savings in computer running time. Figure 4.18 illustrates the type of problems for which FSKIP most logically applies. The solid line represents the input waveform in a given circuit and the dotted line represents the slowest output voltage of interest. It is clear that no useful information is obtained between t = 300 and t = 3000.

Since the computer step size is governed by the system time constants, much computer time is wasted as the output proceeds from 300 to 3000. It is true that SCEPTRE automatically increases the STEP SIZE as the transients die out, so that larger steps are taken; however, the FSKIP subprogram eliminates this waste completely.

The subprogram is designed to monitor at least two specified state variable derivatives and when the system has "settled down" by a prescribed amount, it immediately jumps ahead in time. In this case, it skips immediately from 300 to 3000.

In order to make efficient use of this subroutine, the user must do the following:

1) Supply FSKIP (given below) in the appropriate place in the starter deck.

2) Define three parameters under DEFINED PARAMETERS:
PTGO = X, where X is some time greater than zero, to insure against premature operation of the subroutine.
PSKPTO = XX, where XX is the time to which the user desires the program to be advanced.
PSWTCH = FSKIP (TIME, PTGO, PSKPTO, DCI, DC7). This is effectively an entry that calls the subprogram. The last two arguments are the derivatives of the selected state variables. It is suggested that the input capacitances to the first and last transistor stages be chosen for most applications.

3) Be sure to request parameter PSWTCH for output. The reason for this is alogrithmic and not that the value of PSWTCH itself is inherently valuable.

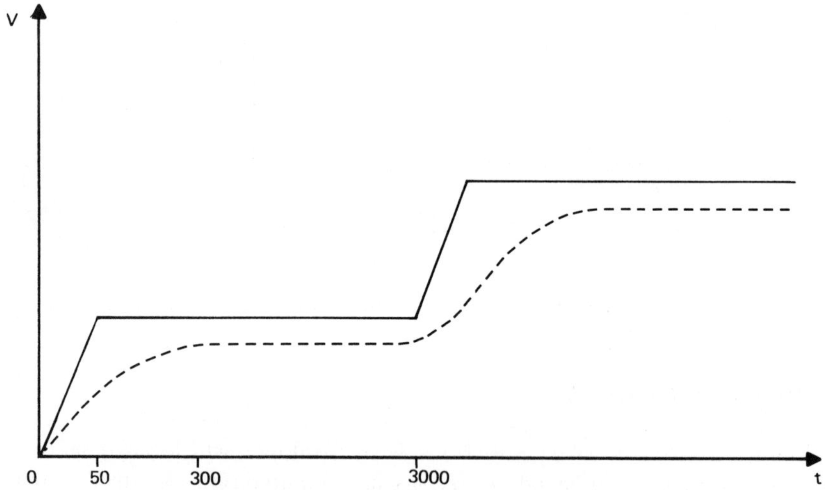

Fig. 4.18. Typical FSKIP voltages.

The subprogram FSKIP as it should be inserted into the SCEPTRE deck is:

```
1                   8                               16
$IBFTC              FSKIP                           FULIST
                    FUNCTION FSKIP (TIME, PTGO, PSKPTO, D, E)
                    DATA Z/0./
                    IF (TIME .EQ. 0.) Z = 0.
                    IF (TIME-PTGO) 4, 2, 2
2                   IF (Z) 1, 1, 4
```

```
1         IF (ABS(D) - .00005) 3, 3, 4
3         IF (ABS (E) - .00005) 5, 5, 4
5         TIME = PSKPTO
          Z = 1.
4         FSKIP = 1.
          RETURN
          END
```

There are a number of modifications that can easily be made to this version of FSKIP. For example, the number of state variable derivatives to be monitored can be increased by lengthening the PSWTCH call and adding the appropriate IF statements to the subroutine. Also, in some applications, the user may wish to tighten or loosen the .00005 figure that controls FSKIP operation. Finally, multiple skip intervals could be accommodated by more significant subprogram changes, but the basic principles given here are valid.

Figure 4.19(a) shows a circuit diagram and Fig. 4.19(b) is the input waveform.

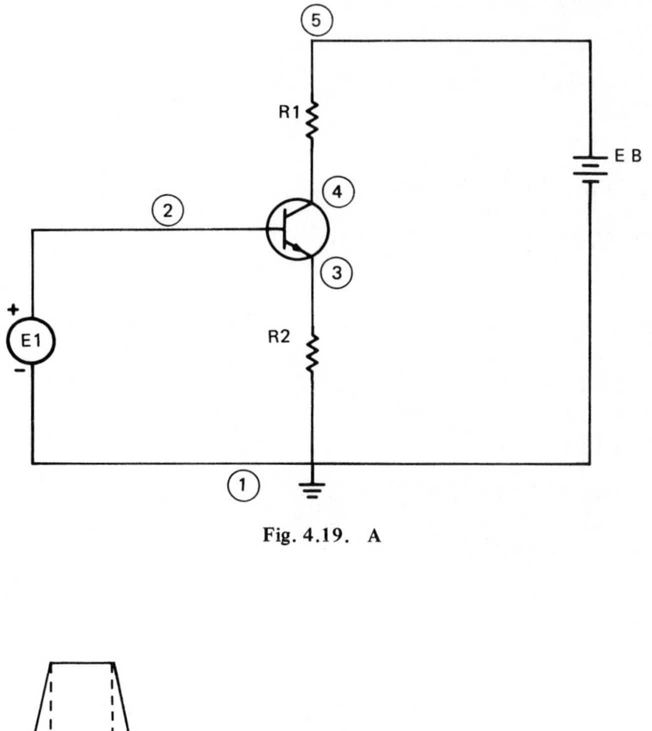

Fig. 4.19. A

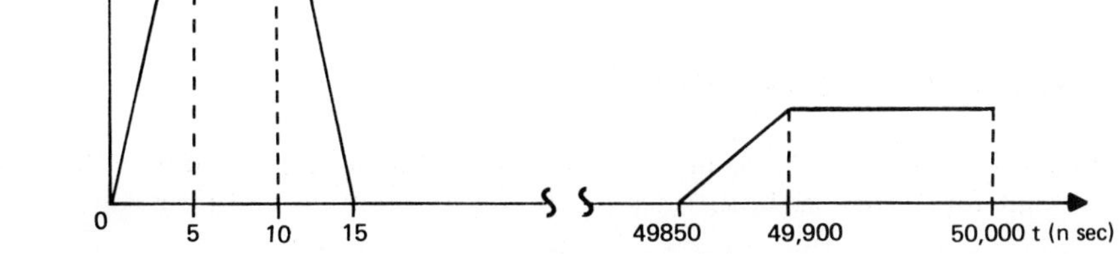

Fig. 4.19. B

The listing for this circuit is:

```
CIRCUIT DESCRIPTION
PULSE AMPLIFIER
ELEMENTS
R1, 5-4 = 1
R2, 3-1 = .1
E1, 1-2=TABLE 1 (TIME)
EB, 1-5 = 20
T1, 2-3-4 = MODEL 2N22XX
DEFINED PARAMETERS
P1 = X1(EB-VR1)
PTGO = 15.
PSKPTO = 49850
PSWTCH = FSKIP (TIME,PTGO,PSKPTO,DCET1,DCCT1)
OUTPUTS
P1(VOUT),VR2,E1,PSWTCH,PLOT
FUNCTIONS
TABLE 1
0,0,5,10,10,10,15,0,49850,0,49900,
1,50000,1
RUN CONTROLS
STARTING STEP SIZE = .1
RUN INITIAL CONDITIONS
STOP TIME = 50000
MINIMUM STEP SIZE = 1.E-10
END
```

The subprogram as used in this example is:

```
      FUNCTION FSKIP (TIME,PTGO,PSKPTO,D,E)
      DATA Z/0. /
      IF (TIME-PTGO) 4,2,2
2     IF (Z) 1,1,4
1     IF (ABS (D)-.001) 3,3,4
3     IF (ABS (E)-.001) 5,5,4
5     TIME = PSKPTO
      Z = 1.
4     FSKIP = 1.
      RETURN
      END
```

Notice the capacitors which are chosen to be monitored and also observe that in the subprogram the error is loosened from .0005 to a value of .001. A plot of the output is shown in Fig. 4.20.

4.8 FLIP/FLOP EXAMPLE[2]

The unaided DC solution of a symmetrical flip-flop, that is not biased by either input or load, is "half on." That is, both sides of the flip-flop are in the linear region between

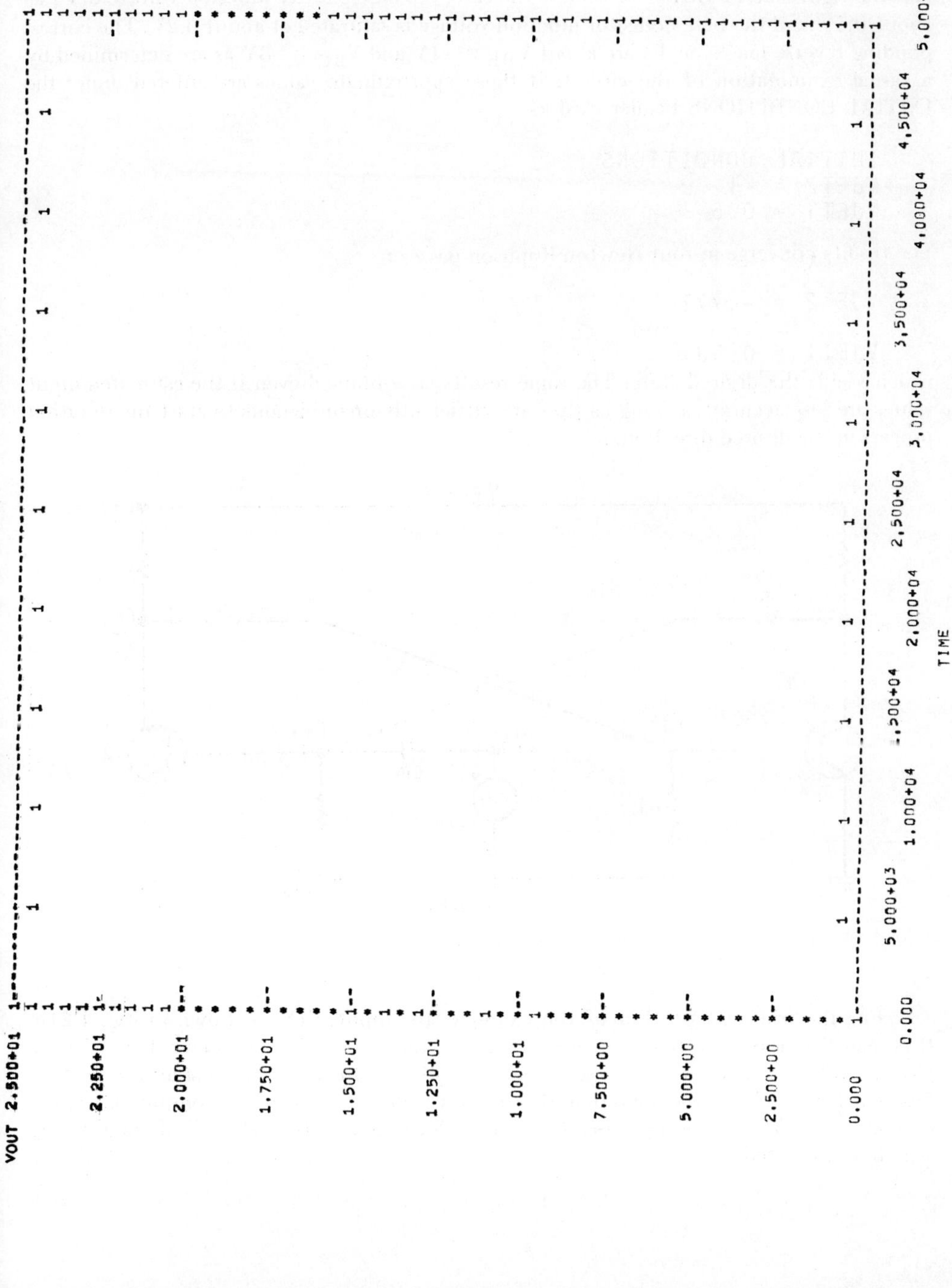

Fig. 4.20 Plot of VOUT vs TIME

cutoff and saturation. Mathematically, this is quite proper, but the practical results should leave the circuit in one of its two possible bi-stable states. This is accomplished in SCEPTRE by the insertion of initial condition values (for the DC run) that approximate those which hold if the circuit is in the desired state. Consider the flip-flop in Fig. 4.21a and assume that the final result of the DC computation finds the left side (T_1)ON and the right side (T_2)OFF. If such is the case, the base-emitter junction voltage ofT_1 is about 0.5V and the base-collector junction voltage is saturated at about 0.2V. The corresponding reverse biases on T_2 are about $V_{BE} = -1V$ and $V_{BC} = -8V$ as are determined by a casual examination of the circuit. If these approximate values are entered under the INITIAL CONDITIONS header card as

```
INITIAL CONDITIONS
VJET2 = -1
VJET1 = 0.5
```

The results converge in four Newton Raphson passes to

```
VJET2 = -.771
                    and
VJET1 = 0.531
```

which yields the desired state. The same results are obtained even if the estimated input values are less accurate as long as they are sufficiently unambiguous to start the iteration process in the desired direction.

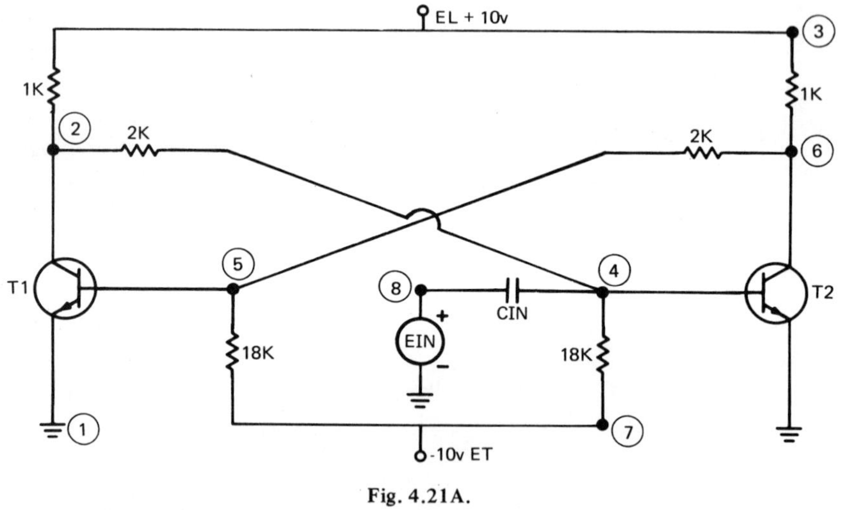

Fig. 4.21A.

For the transient portion of the example, the input, EIN, is shown in Fig. 4.21b. It is seen that the input remains zero for 5ns and then rises to a value of 5v. Fig. 4.22 is a plot of VCCT2 versus time. It is seen that the voltage value remains constant for the first 5ns which reconfirms the fact that the initial conditions, which are computed, are the correct ones. After the input rises to 5v, the plot indicates the "states" of the flip-flop switch, as expected.

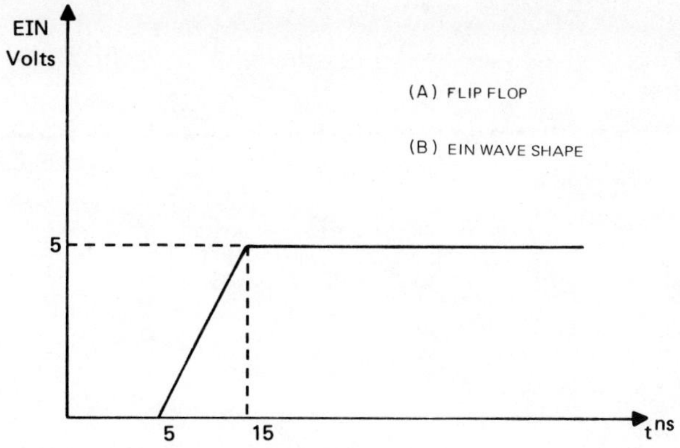

Fig. 4.21B.

A complete listing for Fig. 4.21 is:

```
CIRCUIT DESCRIPTION
BISTABLE CIRCUIT
ELEMENTS
T1, 5-2-1 = MODEL 2NNPN*
T2, 4-6-1 = MODEL 2NNPN
RL1, 3-2 = 1
RL2, 3-6 = 1
RB2, 2-4 = 2
RB1, 6-5 = 2
RT1, 5-7 = 18
RT2, 4-7 = 18
EL, 1-3 = 10
ET, 7-1 = 10
EIN, 1-8 = TABLE 1(TIME)
CIN, 8-4 = 1E4
DEFINED PARAMETERS
P1 = X1(10.-VRL1)
P2 = X2(10.-VRL2)
OUTPUTS
VCET1,VCCT1,VCET2,VCCT2,ICIN,PLOT
P1,P2,PLOT
INITIAL CONDITIONS
VJET1 = .5
VJET2 = -1
FUNCTIONS
TABLE 1 = 0,0, 5,0, 15,5, 100,5
RUN CONTROLS
RUN INITIAL CONDITIONS
STOP TIME = 200
END
```

* Assuming, of course, that the Models have been previously stored.

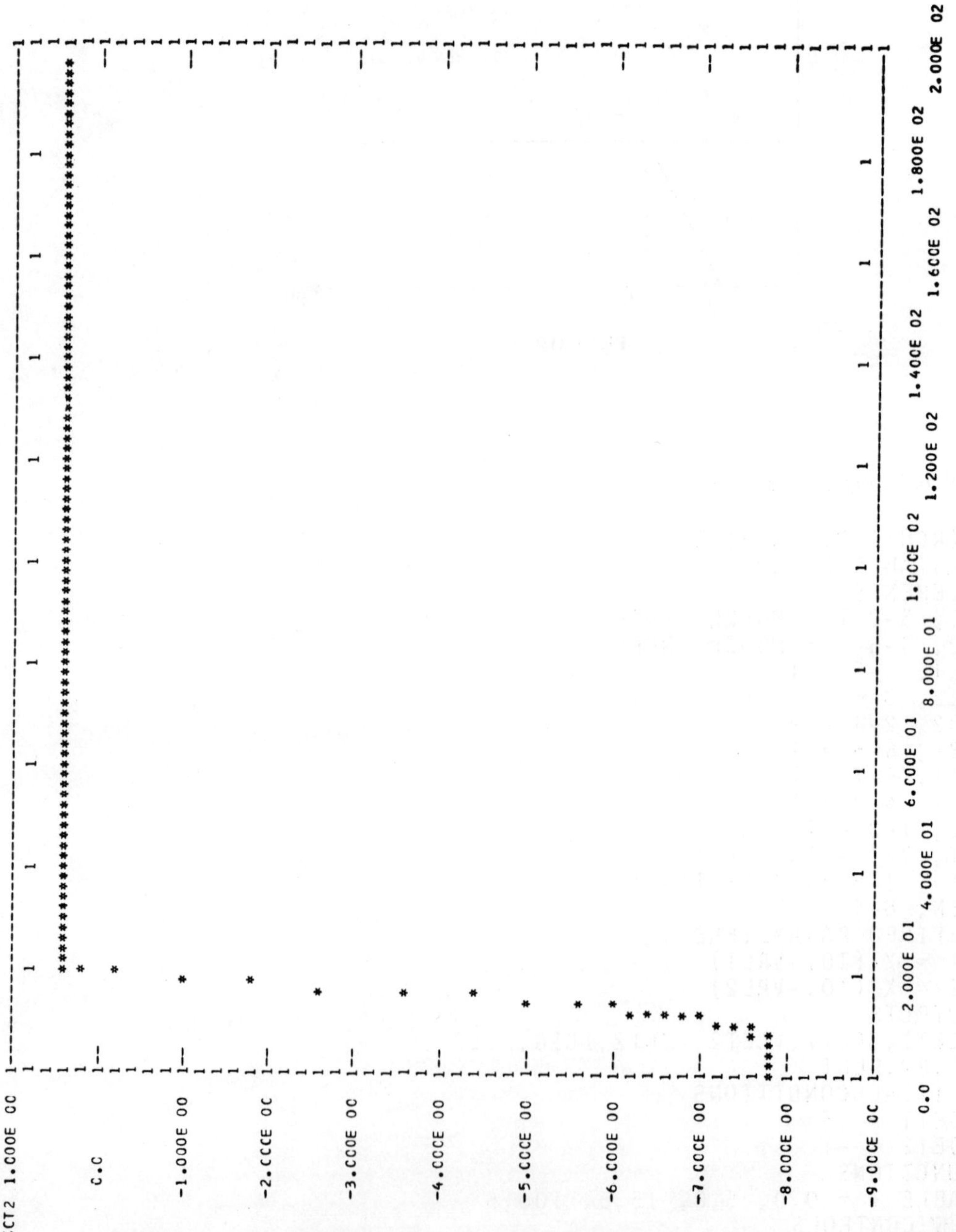

Fig. 4.22 Plot of VCCT2 vs TIME

4.9 A-C FILTER EXAMPLE

While SCEPTRE is not primarily used to solve a-c networks, this example illustrates that some such problems are handled readily. Figure 4.23 is a circuit where the input voltage, El, is the sum of two frequencies. It can be shown that there should be no attenuation and a -90° phase shift for the ω_1 component and a very high attenuation to the ω_2 signal.

With $\omega_1 = 1$ and $\omega_2 = 20$, the correct listing follows:

```
CIRCUIT DESCRIPTION
ELEMENTS
E1, 1-2 = X1 (SIN(TIME)+20.*SIN(20.*TIME))
L1, 2-3 = 1
L2, 3-4 = 1
C, 3-1 = 1
RO, 4-1 = 1
OUTPUTS
E1,VL1,VC,VL2,VRO,PLOT
RUN CONTROLS
STOP TIME = 20
END
```

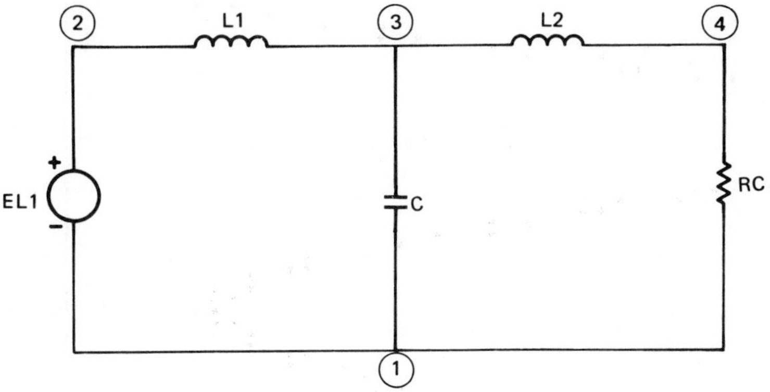

Fig. 4.23.

Figure 4.24 is the output plot. At first glance it appears that there is no phase shift in the output, but a check at the end of the 3rd period ($t = 18.840$) reveals the 90° lag. While a true a-c formulation would have been more efficient than SCEPTRE, it would not have shown the transient effects which have about "died out" at $t = 20$. In many a-c applications, the user is vitally interested in the possibility of clipping, by either saturation or cut-off. Also, in many applications, the turn-on transient is important in a-c circuits. The general purpose a-c type programs do not yield any of this type information, since they assume linear conditions. Therefore, many engineers use SCEPTRE for a-c circuit analysis in a number of applications.

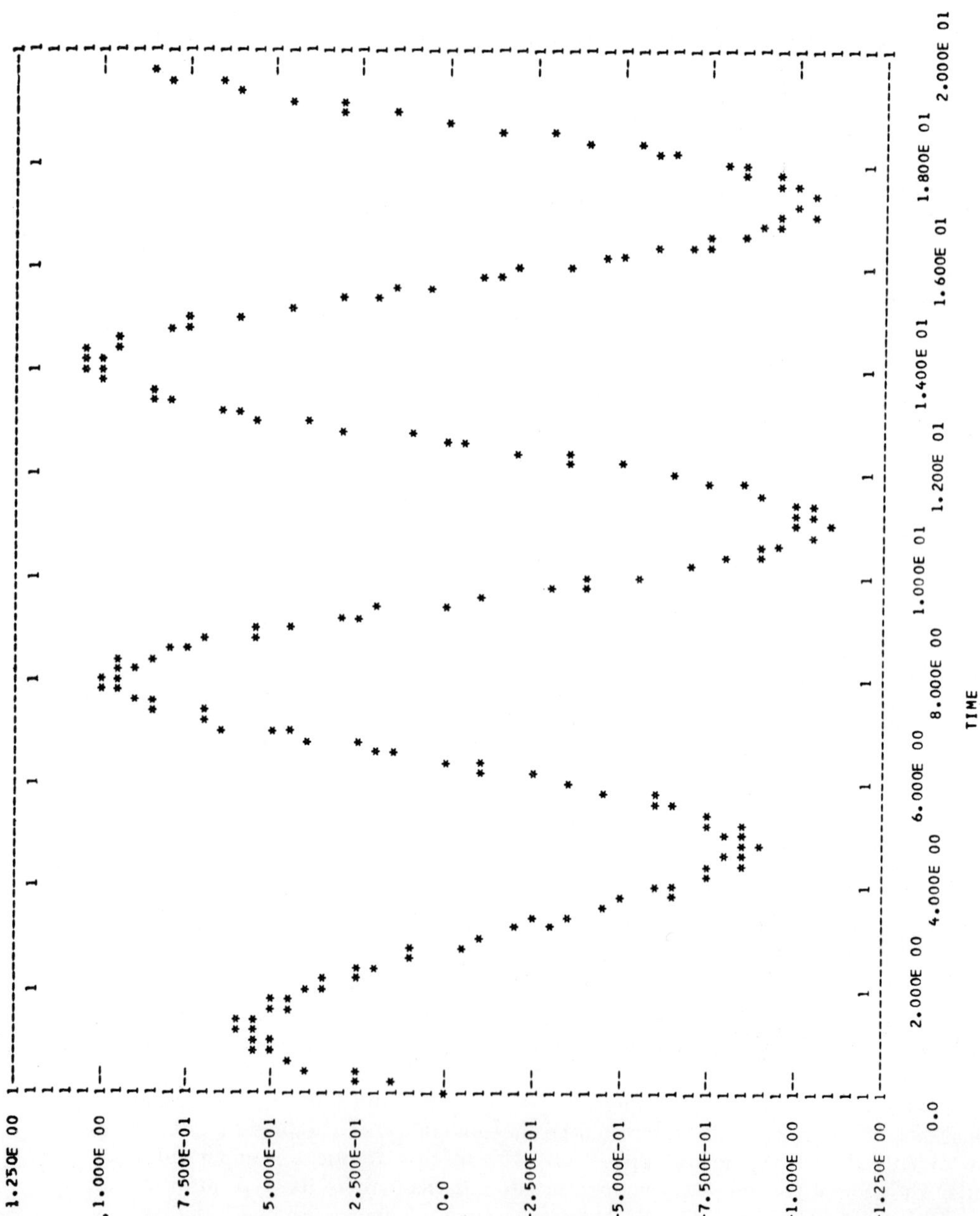

Fig. 4.24 Plot of VL2 vs TIME

4.10 RERUN EXAMPLE

Figure 4.25 is a diagram of a two-stage amplifier which is biased such that both stages are OFF. The object in this example is to use a series of reruns with a successively increasing noise voltage at the input, E1. The master run is made with a noise spike that peaks at one volt. It is desired to automatically terminate the run whatever the voltage levels are such to turn T2 ON. The reruns are set up with increasingly higher noise levels.

Using the STOP IF feature, the program automatically continues through the reruns until T2 is turned on. This condition is designated by the program statement

STOP IF(VCET2. GT. 0.)

In other words, when the base to emitter junction of T2 is positive, the transistor is just "turning-on." *

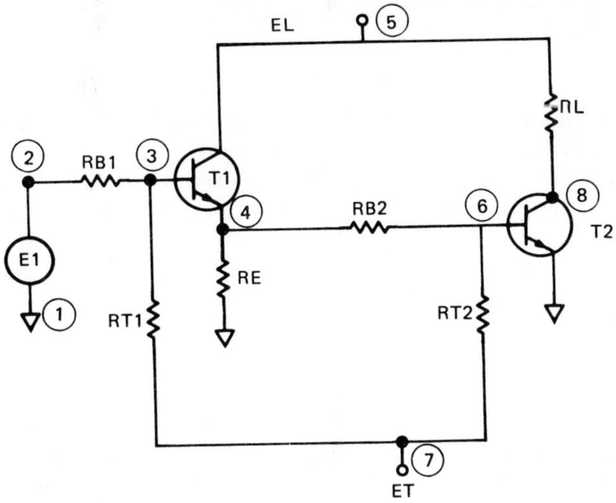

Fig. 4.25. Two stage amplifier

It is important to remember the difference between the STOP IF and the TERMINATE IF statements. In this example, VCET2 does not go positive until t = 267 in the second rerun. At that point, the run is completely terminated and the results are printed out. If the TERMINATE IF statement is used, the second rerun terminates, but the program then goes on to the third rerun. A correct listing for the circuit of Fig. 4.25 is:

```
CIRCUIT DESCRIPTION
ELEMENTS
E1, 1-2 = TABLE 1(TIME)
RB1, 2-3 = 2
RT1, 3-7 = 18
RE, 4-1 = 1
RB2, 4-6 = 2
RL, 5-8 = 1.5
RT2, 6-7 = 18
ET, 7-1 = 5
```

* Please do not confuse this term with the present-day teenage jargon.

```
EL, 1-5 = 10
T1, 3-5-4 = MODEL 2NNPN
T2, 6-8-1 = MODEL 2NNPN
OUTPUTS
VCET1,VCCT1,VCET2,VCCT2,PLOT
FUNCTIONS
TABLE 1 = 0,0, 250,1, 500,0, 600,0
RUN CONTROLS
RUN INITIAL CONDITIONS
STOP TIME = 1000
STOP IF (VCET2 .GT. 0.)
RERUN DESCRIPTION (3)
FUNCTIONS
TABLE 1 = 0,         0,          0,      0
          250,     1.5,          2,    2.5
          500,       0,          0,      0
          600,       0,          0,      0
END
```

Figure 4.26 shows the value of VCET2 during the second rerun. It is seen that the program does stop exactly when VCET2 goes positive.

The user should exercise his ingenuity in using the RERUN and STOP IF features. There are many practical ways to incorporate these features to save computer time and yield excellent results.

4.11 BOOBY PRIZE

This example is intended to incorporate the type of hidden problems which constitute a majority of all of the errors made in SCEPTRE programming. The reader is advised to study carefully the following listing, as it applies to Fig. 4.27, and see how many mistakes can be determined, before reading the explanation. The circuit looks quite simple, but watch out! One hint, there are seven errors in the listing and one additional problem which results in excessive running time. Do not become discouraged, it is inconceivable that more than two or three of these problems would be present, simultaneously, in any actual, practical circuit. In fact, in most cases, none are present. The problems are worth noting, however, because they do come up from time to time and knowing how to deal with them quickly helps to eliminate the trouble.

```
MODEL DESCRIPTION
MODEL 2N706A (B-E-C)
ELEMENTS
CE, 1-E = EQUATION 1(5.,70.,J1)
CC, 1-2 = EQUATION 1(8.,370.,J2)
RB, B-1 = .3
RC, C-2 = .015
J1, 1-E = DIODE EQUATION (1.E-7,35.)
J2, 1-2 = DIODE EQUATION (5.E-7,37.)
JA, E-1 = .1*J2
JB, 2-1 = P1*J1
DEFINED PARAMETERS
```

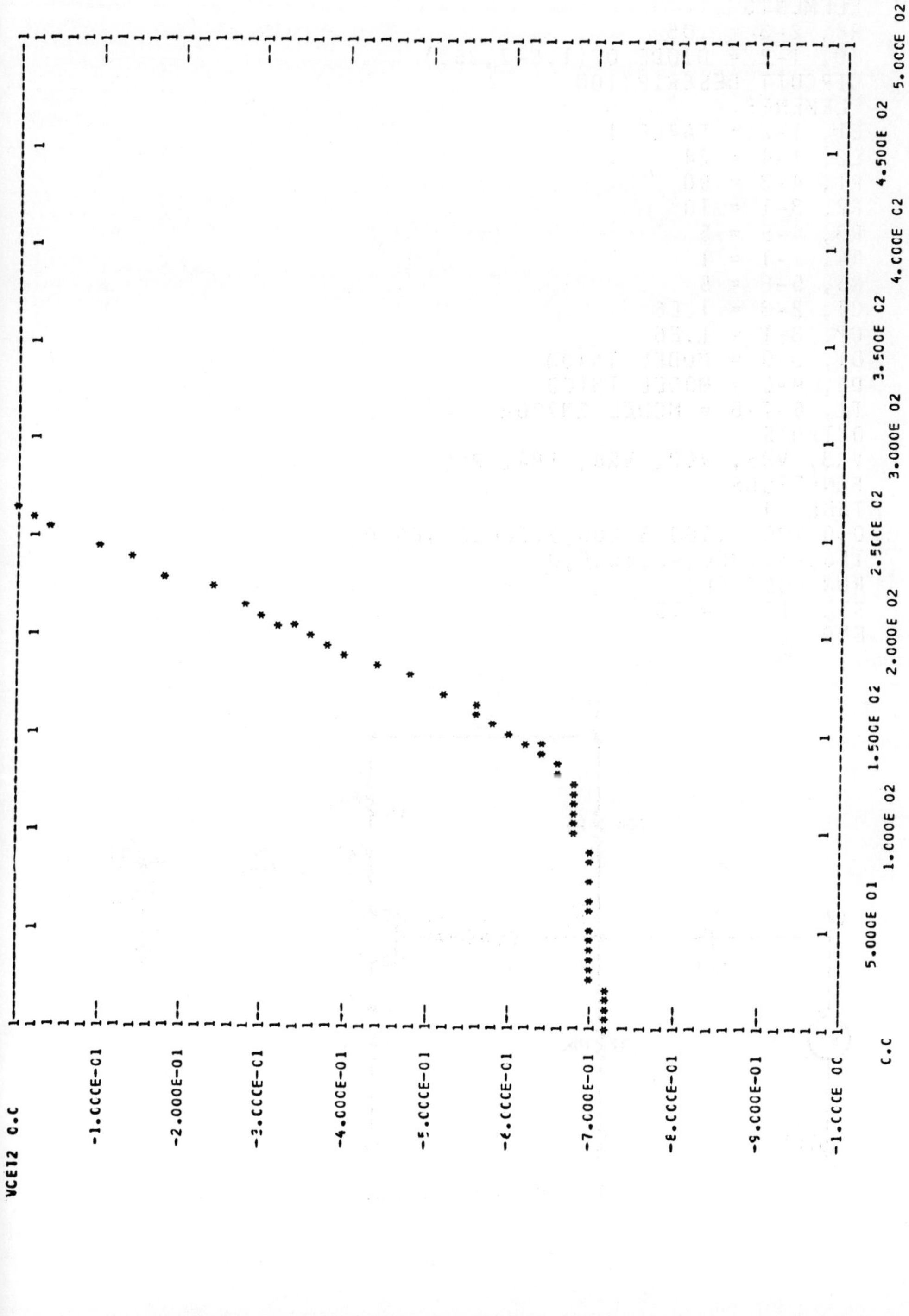

Fig. 4.26 Plot of VCET2 vs TIME

```
P1 = .98
FUNCTIONS
EQUATION 1(A,B,C) = (A+B*C)
MODEL 1N100(1-3)
ELEMENTS
RB, 2-3 = .05
JD, 1-2 = DIODE Q (1.E-7,35.)
CIRCUIT DESCRIPTION
ELEMENTS
E1, 1-2 = TABLE 1
E2, 1-4 = 24
R1, 4-3 = 90
R2, 3-1 = 10
R3, 4-5 = 5
R4, 7-1 = 1
R5, 5-8 = 5
C1, 2-3 = 1.E6
C2, 8-1 = 1.E6
D1, 3-9 = MODEL 1N100
D2, 9-6 = MODEL 1N100
T1, 6-7-5 = MODEL 2N706A
OUTPUTS
VR3, VR5, VC2, VRB, IR4, PLOT
FUNCTIONS
TABLE 1
0,0,100,0,100,5,200,5,200,0,1E6,0,
1E6,-5,10E6,-5,10E6,0
RUN CONTROLS
STOP TIME = 10E6
END
```

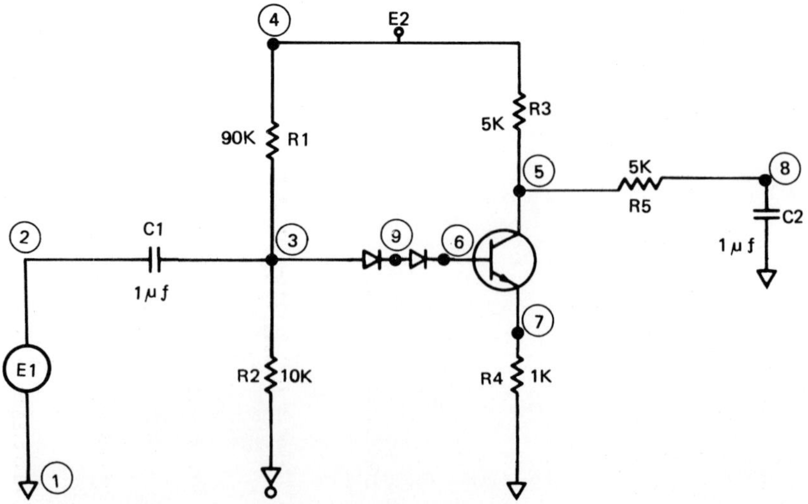

Fig. 4.27. Booby prize figure.

4.11.1 Simple Errors

The three simple errors are discussed first, and once pointed out, are obvious. The term VRB is called for output. This is incorrect, the term should be VRBT1. The statement, RUN INITIAL CONDITIONS, should have been entered in order for the quiescent initial d-c conditions to be established before applying the input wave form.

The input wave form is plotted in Fig. 4.28. It is taken directly from TABLE 1 in the SCEPTRE listing. Notice that the final "point" listed is point C (10E6,0). The problem here is that this leaves the input signal unbounded. The computer assumes that the input continues upward indefinitely at t = 10E6 (see Sec. 2.9.3). To correct this problem, one more data point should be added to the table such as (11E6,0).

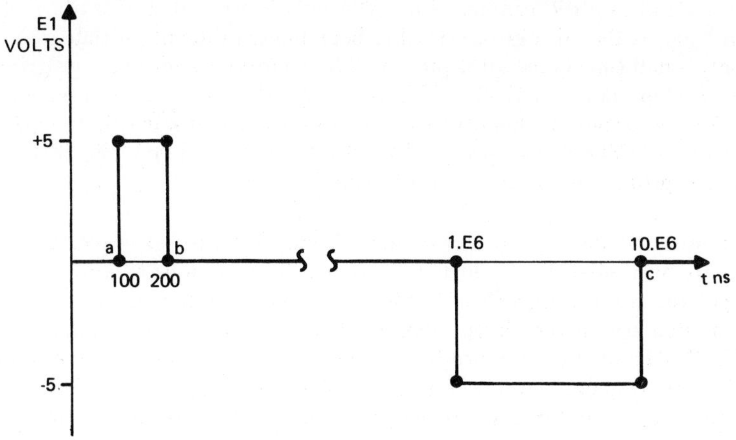

Fig. 4.28.

4.11.2 Illegal Cut-Set

This example contains a current source cut-set at node 9. Section 2.16.1 discusses this type of problem and Fig. 2.14 illustrates the existing condition at node 9. To correct for this situation, a large shunt resistor or a large resistor to ground must be added (Sec. 2.16.2).

While cut-sets are usually easy to eliminate, they do subtly appear quite often in large circuits. This is especially true in integrated circuits where the emitter of one transistor connects directly to the base of another. If any cut-sets are missed by the user, a diagnostic calls out the nodes involved. When eliminating cut-sets, the user should take care that the added resistor does not substantially affect the circuit operation. Adding resistors that are overly large is not recommended.

4.11.3 Step Size and Running Time

The diagnostic SMALLER MINIMUM STEP SIZE REQUIRED is probably the most common one of all. Table 2.5 summarizes the definitions for the step size as well as for the other important quantities. However, simply stating the definitions does not necessarily describe clearly the overall philosophy. Four different quantities are all integrally related. They are, minimum step size, starting step size, maximum integration passes and computer time limit.

One of the features of SCEPTRE is that it automatically adjusts the step size, consistent with the accuracy requirements. The only limitation is that the program is preset with a minimum value. As stated in Sec. 2.10.4, this minimum value can be changed to any number. An immediate question should now come to the reader's mind. "Why does not the user always put in a MINIMUM STEP SIZE = 1.E–39?" This is a logical question. If the user routinely made such an entry on every circuit listing, the program would, no doubt, never abort due to a smaller step size requirement. Also, and this is a very important point, SCEPTRE never uses a step size any smaller than required, so that such a listing does not increase the running time. The only reason, then, for not making such an entry each time, is that if a gross error has been made in the magnitude of an element or an extremely small time constant is present. The computer continues to "grind" away, even down to a step size of 1.E–39. This is costly. However, if the user has a TIME LIMIT listed for the program, this offers protection against running times exceeding the limit. In addition, SCEPTRE has a preset MAXIMUM INTEGRATION PASSES = 20,000. This also provides protection against excessive running times.

The user must decide the correct course of action. If, for example, the user has a circuit with extremely small time constants and an extremely long running time, and it is necessary to get the results, regardless of the cost, it is absolutely correct to enter a very small minimum step size, a very large number of integration passes and a long computer time limit. SCEPTRE, in this case, produces the results as economically as possible under the adverse, but unavoidable, circuit conditions. The foregoing discussion serves to illustrate to the reader that the three quantities being discussed are all merely user protection devices, and if the users know exactly what they are doing, all three requirements may be completely "opened-up" and no program will abort due to these three conditions.

The fourth quantity, previously mentioned, but not yet discussed is the STARTING STEP SIZE. This quantity effects the results in two ways. The first is in the running time. When initially starting out at t = 0, SCEPTRE repeatedly and automatically reduces the starting step size in half, until it is able to satisfy its error criterion. If the pre-set value of 10^{-3}* (STOP TIME) is large compared to the first step size used, then computer time is wasted (though seldom of significance) while the program automatically reduces the value to the proper one. Incidentally, the step size that is taken can never go below the minimum step size. Therefore, if a diagnostic reads, SMALLER MINIMUM STEP SIZE REQUIRED at the very outset of the run, this indicates that the initial step aborted the run while the starting step size was reducing its value. The second way the starting step size may cause problems is if a spike occurs at the input at a time greater than zero; i.e. the first computer increment might just "step" over the spike and not realize it is there.

The preceding information is now applied to the example of Fig. 4.27. The listing given for the example indicates a STOP TIME = 10E6. This gives a STARTING STEP SIZE = 10E3. From Fig. 4.28 it is seen that the first spike is never detected by SCEPTRE. Point "b" in the figure is at 200 nsec, and the first step is well beyond that point. The easiest way around this problem is to start time equal to zero at point "a," then the spike is detected and SCEPTRE reduces its STARTING STEP SIZE to the required amount. However, to save further computer time, it is also suggested to enter the statement, STARTING STEP SIZE = 1, although this is not overly important. After running the problem, a diagnostic might appear stating that SMALLER MINIMUM STEP SIZE IS REQUIRED. The major cause in this example is not the circuit, but the long STOP TIME. The minimum pre-set step size is, $10^{-5}*(10E6) = 100$. From estimating the system time constants, the step size of 100 nsec is not small enough at the beginning or even during the transient run. Therefore, in addition to the statement concerning starting step size, another listing should be added, such as, MINIMUM STEP SIZE = 1.E−6. Also, because of the long STOP TIME, the third related parameter should be listed as:

```
MAXIMUM INTEGRATION PASSES = 50000.
```

This example has been designed to illustrate the foregoing problems. If a user is actually confronted with such a circuit, the suggested approach is again summarized as follows: This first correction is to start t = 0 at point "a" of Fig. 4.28. Then, because of the long STOP TIME, the experienced user quickly realizes that the MINIMUM STEP SIZE and MAXIMUM INTEGRATION PASSES need to be adjusted. These need not be careful adjustments but can be made arbitrarily large. In order to avoid lengthy computer time, if the circuit is not functioning properly, there are two precautions which can be taken. The first is to enter a COMPUTER TIME LIMIT which automatically terminates the run after a certain time. The second suggestion is to use the CONTINUE feature. The user could run the problem for, say 100 nsec, and then after inspecting the results to insure everything is satisfactory, run out the remaining 10E6 nsecs of time.

4.11.4 Computational Delay

Computational delay has been discussed previously in Sec. 1.3. In the example being developed, a computational delay occurs because the voltage dependent current generator in the diode model has no capacitor across it.

In most practical applications, the diode always has a shunt capacitance, but it is purposely omitted here to bring out an important point. There is nothing to prevent the user from representing the diode as a current generator without the shunt capacitance; in fact, some low frequency applications tempt the user to do just this. The difficulty with this practice is that the current generator is voltage dependent and the voltage in question must be across some element. If the current generator is dependent on a capacitor voltage, the current source is updated at the start of each solution step based on known internal state variables. If, however, the equivalent capacitor is removed and dependence is placed on the voltage across the current source itself or some shunt resistor, the current source may be updated based on the information from the previous solution increment. A computational delay is created and significant errors can result. For this reason the user is cautioned against removing the shunt capacity associated with diodes for all transient applications. This consideration does not hold for the initial conditions program.

The problem of computational delay arises in other applications. Every variable quantity in a circuit is a function of another quantity. Ideally, SCEPTRE would update every variable at the start of each time step by using the updated value of the independent variable. Unfortunately, the only quantities that have definitely been updated are the state variables and time. If the variable quantities depend upon any of these, everything is satisfactory, but if not, the best that the program can do is to use the value of the independent variable that existed at the previous time step. This can lead to error. The error introduced is proportional to the degree of nonlinearity of the functional dependence as well as the step size being taken.

There is no problem with the use of resistor currents or voltages as independent variables if they are entered according to Table 2.4. The validity of the use of other independent variables cannot be unambiguously stated; the status is topology dependent. The program always prints out a warning message if a computational delay occurs, provided that the proper expression, equation, or table formats are used.

4.11.5 Large Time Constant Spread

The last problem area to be discussed for this "loaded" example is one to do with time constants which are far apart in magnitude. The time constants within the transistors are in the nanosecond range, while the output filter has a time constant in the millisecond range. The problem here is not one of errors but of computer running time. SCEPTRE always slices up time small enough to "see" even the smallest circuit time constant. While this is sometimes a disadvantage, it is often necessary for accurate results if truly automatic step control is to be retained by the program without user intervention.

A long STOP TIME is necessary to accommodate the long filter time constant reaction; but on the other hand, the small time constant is forcing the computer to take very small steps. This results in many steps and lengthy computer running times. There are several solutions. The direct solution is to do nothing; but, as already stated, this is quite expensive. The second possibility is to increase the value of the transistor capacitors. If the impedance presented by the small capacitors is very large at the highest frequency of interest, these capacitors could be arbitrarily increased in magnitude without affecting the output signal. This change would reduce the computer time significantly. The third possibility is similar to the second. That is, instead of increasing the two transistor capacitors, the output filter capacitor could be reduced several orders of magnitude to reduce the spread between time constants, which is the goal.*

It is important to note that this third approach changes the time scale. In other words, if C2 is changed from 1uf to .01uf, the output time scale is changed from nanoseconds to (nsec x 10^2). If the circuit in question has many transistors, it is usually easier to reduce the size of the one large capacitor, rather than change all of the transistor capacitors, expecially if the transistors are stored models.

The fourth possible solution is more involved but has a wider range of significant applications. Consider the network of Fig. 4.29 which contains a time constant spread of one million. It clearly requires a problem duration of at least 4 x 10^6 nanoseconds to encompass the complete transient response of this network. Yet it is known, from Sec. 1.10 that the small time constant holds down the average step size that is taken, and this combination forces a large number of solution steps to be taken. When this circuit is solved with SCEPTRE as it stands, almost 2800 solution steps are required to progress to

---

* Keep in mind that it is not the actual value of the capacitors which is a problem, but rather it is the wide spread between them.

a problem time of 250,000 nanoseconds - still far short of that required for the complete transient solution. With a little thought, however, one recognizes that the small capacitor is effectively charged to its final value at about five time constants, or at t = 5 nanoseconds. Suppose that at that point in the problem, C1 is arbitrarily switched to a large value, on the order of that of C2 or larger. In theory, the inhibiting factor that the small time constant had on the solution step size has been removed; and, therefore, there should be a marked improvement in solution speed. In practice, this is just what happens as the entire transient solution is completed at $t = 4 \times 10^6$ nanoseconds at the cost of only 253 steps. The solution speed increased by a factor of more than 250. The error incurred by this technique amounts to only millivolts and even that is reduced if the switch is thrown a bit later in time.

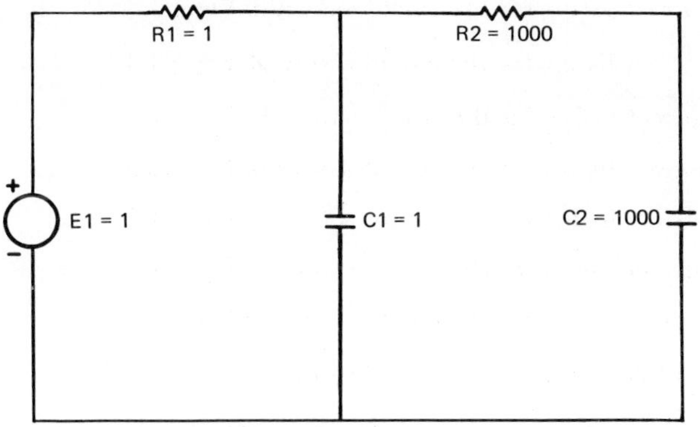

Fig. 4.29.

The switching process can be accomplished in a number of different ways, but probably the simplest is to enter the small capacitor as:

```
C1, NODE - NODE = X1(P1*1.)
```

where P1 is a Defined Parameter that references a time controlled table. This Table holds P1 = 1, until a given point in time (here t = 5), at which it is increased to some value in excess of one thousand. The same technique applies if the small capacitor was originally given as a variable instead of a constant. In this case the small capacitor could appear as:

```
C1, NODE - NODE = X1 (P1 * (5. + 70. * J1))
```

where P1 performs exactly as before. It is evident that no significant change in any network forcing function should be allowed after the element switch has occured unless, of course, a second switch is made in the Table that first restores P1 = 1.

The clever user has probably thought of a fifth approach which only applies with the type of input waveform in this particular example; that is, using the FSKIP subprogram between 200ns and 1.E6ns.

4.11.6 Summary

In the example just concluded, many important points are covered which should be studied very carefully. The reader will want to refer to this example many times when the various diagnostics, which are explained here in detail, appear.

PROBLEMS

4.1 What type of error would be caused by the omission of RUN INITIAL CONDITIONS in the Listing of Sec. 4.7?

4.2 Assume a transient run actually takes 3200 steps in which the user had specified MAXIMUM PRINT POINTS = 2000. At how many steps will the printed output be generated?

4.3 Write a correct listing for the circuit shown in Fig. p4.1.

4.4 Write a correct listing for the circuit shown in Fig. p4.2.

4.5 Write a correct listing for the circuit shown in Fig. p4.3.

4.6 Write a correct listing for the circuit shown in Fig. p4.4.

4.7 Write a correct listing for the circuit shown in Fig. p4.5.

4.8 Write a correct listing for the circuit shown in Fig. p4.6.

4.9 Write a correct listing for the circuit shown in Fig. p4.7.

4.10 Write a correct listing for the circuit shown in Fig. p4.8.

4.11 Write a correct listing for the circuit shown in Fig. p4.9

4.12 Write a correct listing for the circuit shown in Fig. p4.10.

4.13 A transient analysis is to be made on a network containing a resistor R12. Develop the additional coding necessary to compute the energy dissipated by R12 during the run.

(Hint: Read the discussion in Sec. 2.6 and the example in Sec. 5.2.2)

4.14 If the conventional system of units has been used in Prob. 4.13, what are the units of energy?

ONE SHOT MULTIVIBRATOR

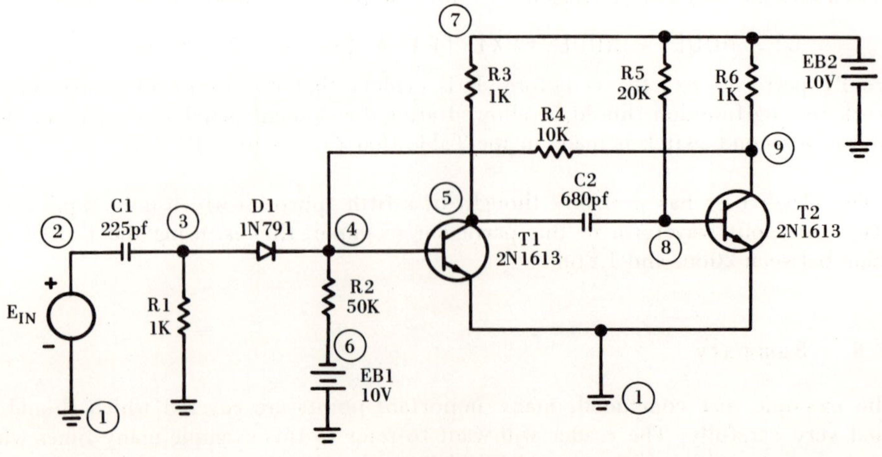

Fig. p4.1

COLPITTS OSCILLATOR

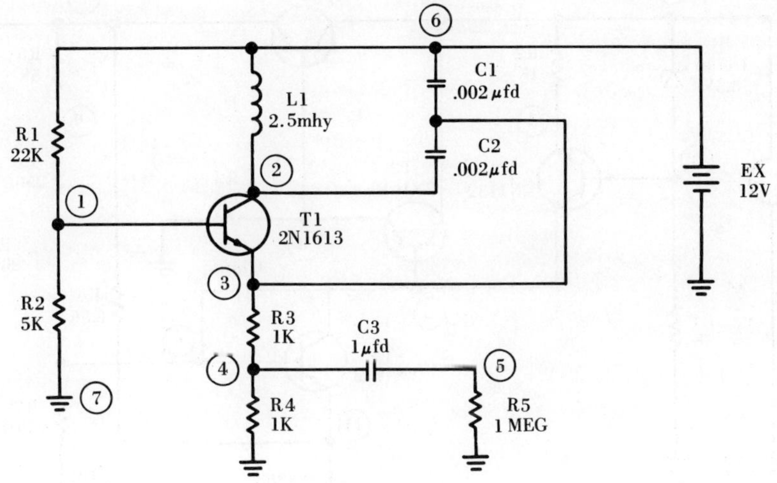

Fig. p4.2

ASTABLE MULTIVIBRATOR

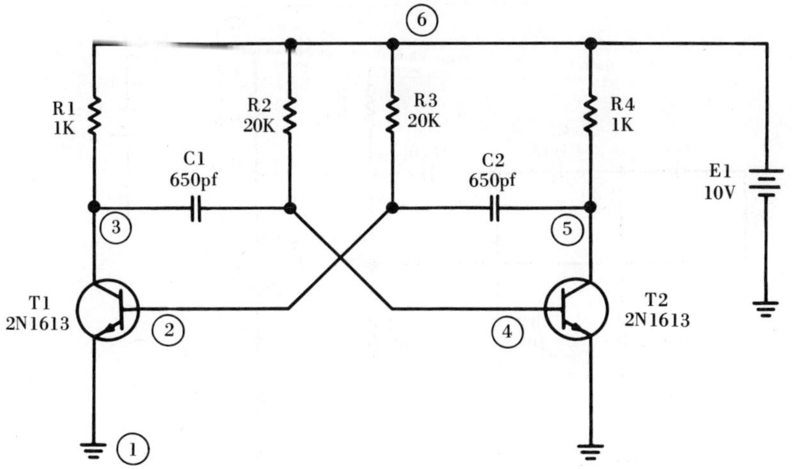

Fig. p4.3

POWER SUPPLY REGULATOR

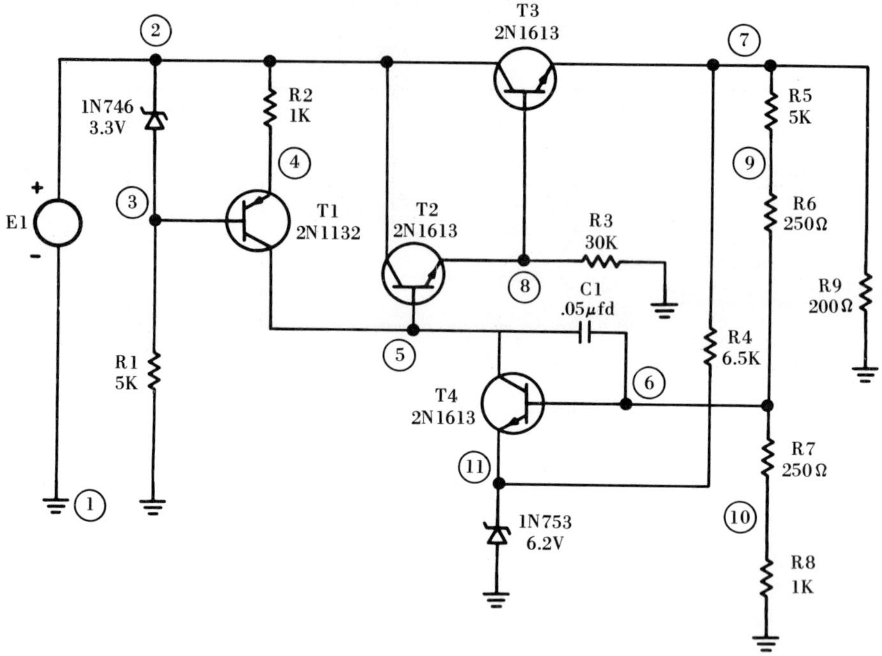

Fig. p4.4

PHASE SHIFT OSCILLATOR

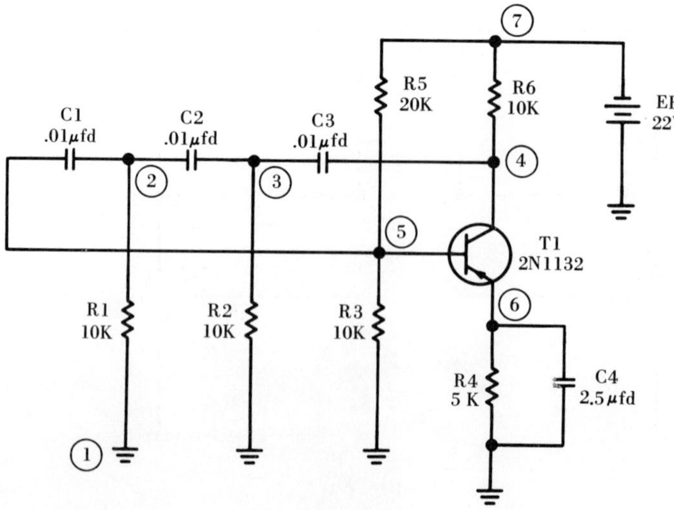

Fig. p4.5

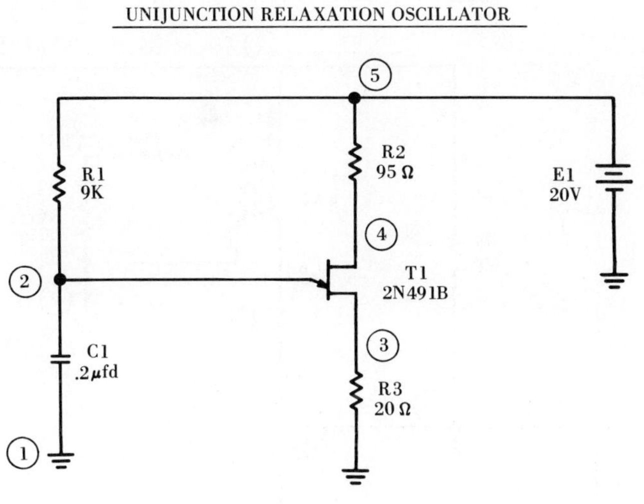

Fig. p4.6

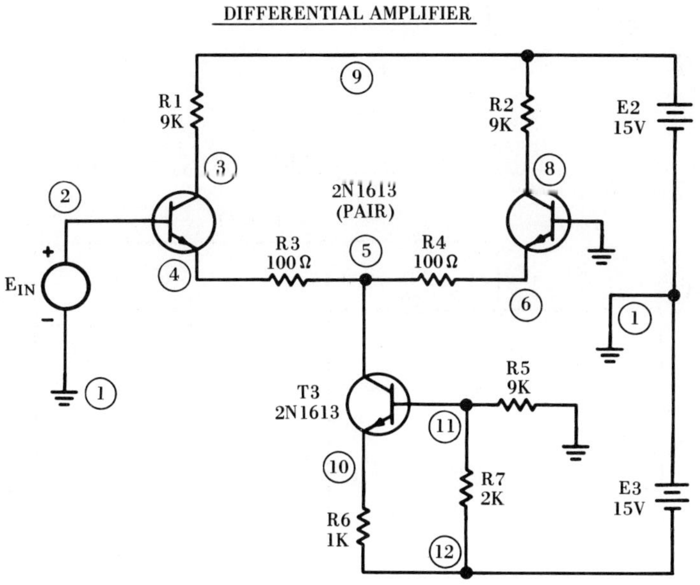

Fig. p4.7

TRANSFORMER COUPLED AMPLIFIER

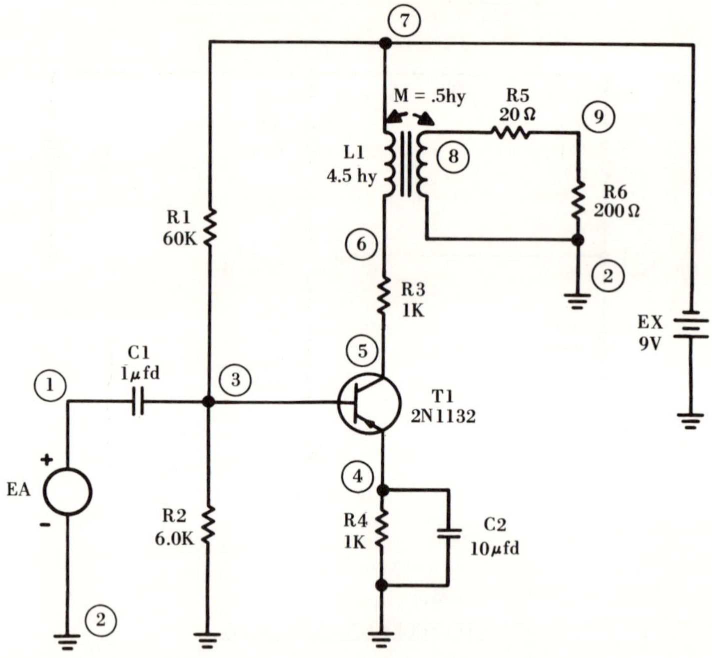

Fig. p4.8

LOW PASS ACTIVE FILTER

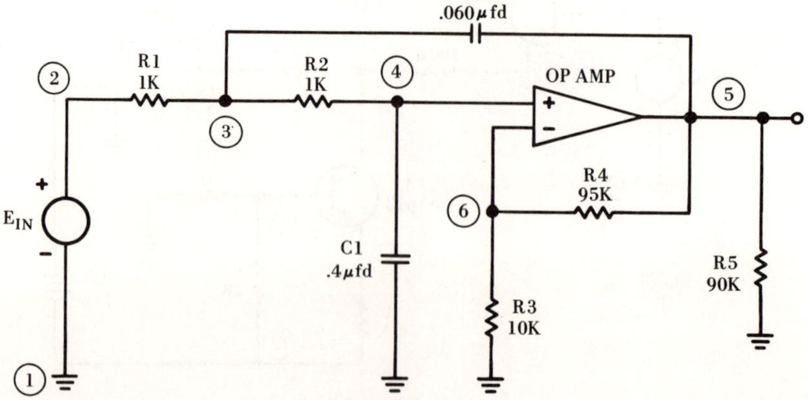

Fig. p4.9 (a)

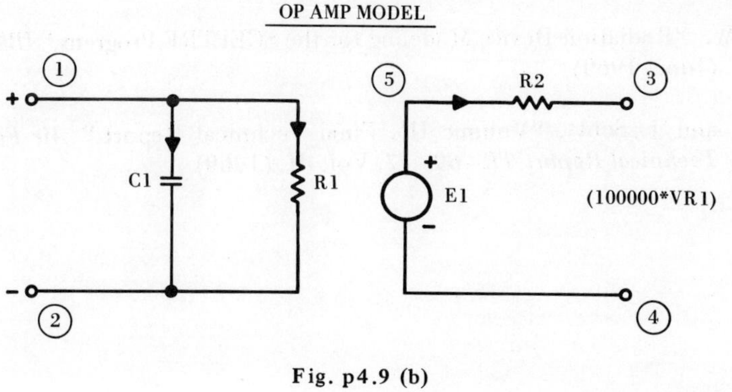

Fig. p4.9 (b)

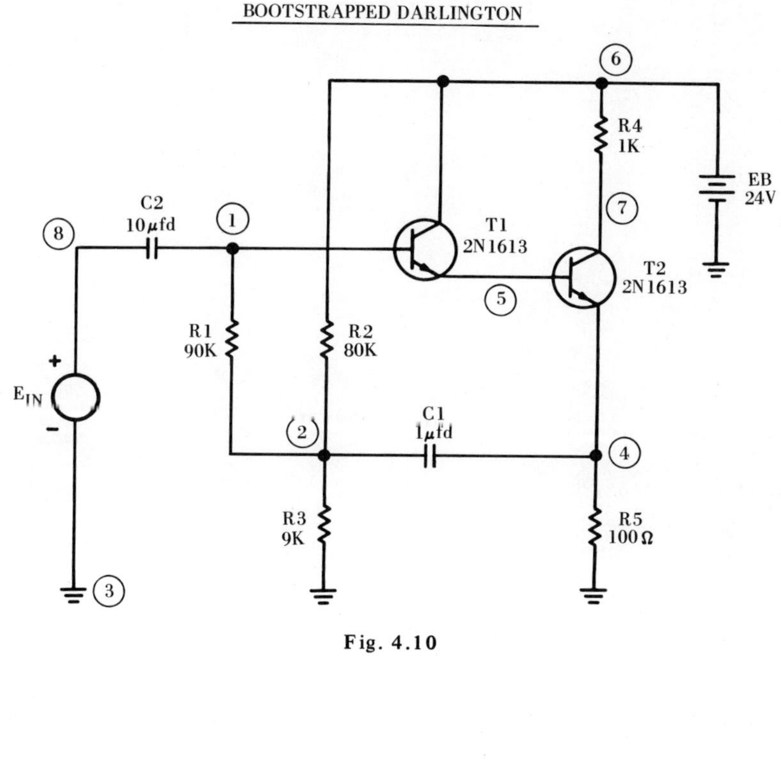

Fig. 4.10

REFERENCES

1. Cordwell, W., "Radiation Device Modeling for the SCEPTRE Program," *IBM Report 69–825–2371*, (June, 1969)

2. Sedore, S. and J. Sents, "Volume III, Final Technical Report," *Air Force Weapons Laboratory Technical Report TR–69–77*, Vol. III, (1969)

5

ANALYSIS OF AUTOMATIC CONTROL SYSTEMS

5.1 INTRODUCTION

The SCEPTRE circuit-analysis computer program is used to determine the transient response of a wide class of control systems. As a preliminary example of the class of problems for which the method is applicable, consider the system shown in Fig. 5.1.

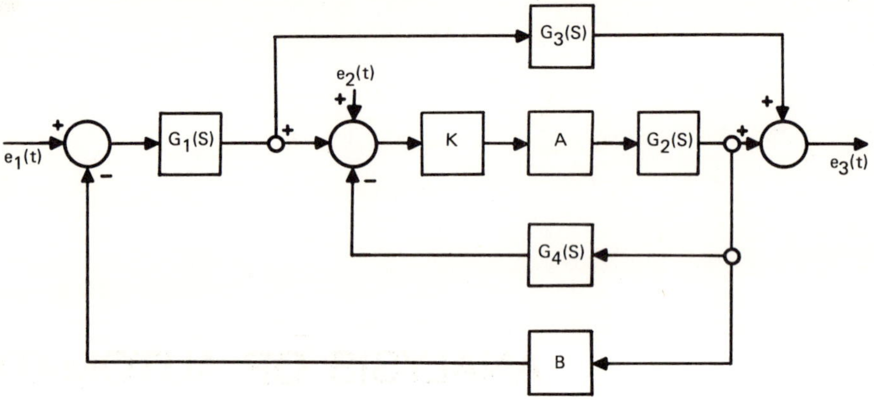

Fig. 5.1 A typical system block diagram

Blocks A and B represent non-linear electronic amplifier circuits consisting of transistors, resistors, capacitors, etc. Blocks $G_1(s)$, $G_2(s)$, $G_3(s)$, and $G_4(s)$ are transfer functions of linear sub-systems. K is a simple gain device. The objective in analyzing the system is to ascertain the effects on the output variable, $e_3(t)$, due to the nominal input signals, $e_1(t)$ and $e_2(t)$, and the effects caused by extraneous inputs to circuits A and B.

The system-analysis mode of SCEPTRE is accomplished entirely within the original configuration of the program; no sub-programs or other special constructs are required. The method embodies considerable flexibility in that the user may derive either a general transfer-function model, consisting of a ratio of n–th order polynomials in (s) having unspecified coefficients which can be degenerated into a lower-ordered function; or, if it is necessary to save computer time and SCEPTRE defined parameters, the user can derive a specific transfer-function model for any desired numerator and denominator polynomials.

An advantage of the SCEPTRE transfer-function model is that it is never necessary to compute derivatives of the input variable even though the transfer function has finite zeros. A simple SCEPTRE model for an ideal amplifier is also presented. This can be used to realize "pure" gain elements in the system block diagram. Finally, since the transfer-function model is derived as a four-terminal device, the summing junctions which appear in system specifications are realized by the proper interconnection of input and output terminals.

It is important to note at this point that it is not necessary for the user to understand precisely the following development in order to use SCEPTRE effectively for Control System analysis. The illustrative examples which are presented in this chapter are complete enough to show the user the procedures to follow in order to solve any particular problem. Section 5.2, however, does develop the foundation of this programming method for those interested in the theory.

5.2 TRANSFER FUNCTION MODELS

A linear time-invariant single-input, single-output system can always be described by a transfer function, G(s), of the form:

$$G(s) = \frac{E_o(s)}{E_{in}(s)} = \frac{\sum_{i=0}^{m} a_{m-i} s^{m-i}}{\sum_{i=0}^{n} b_{n-i} s^{n-i}}, \quad m<n \tag{5.1}$$

where $E_o(s)$ and $E_{in}(s)$ are the transforms of the output and input signals respectively, and the a_{m-i} and b_{n-i} are real constants. The restriction $m < n$ is typical for the transfer functions of most electrical networks, and is also satisfied for most electromechanical devices. *

In representing systems by interconnections of transfer functions, loading effects are not considered. Strictly speaking, such representations are valid only if each transfer function operates into an essentially infinite load impedance and from a zero source impedance. In practice, the utility of the transfer functions is such that the analyst is usually disposed to allow their use in non-idealized situations.

5.2.1 Model Derivation

Determining time response from s–domain specifications is accomplished either by the convolution integral or by solving differential equations. Since SCEPTRE contains a built-in capability for solving systems of first-order differential equations, the latter approach is employed, and the necessity of changing the basic program avoided. Converting G(s) into a set of first-order differential equations, while avoiding differentiation of the input function, $E_{in}(t)$, is accomplished by use of the state-variable formulation, as described in the following derivation.

Equation (5.1) can be factored into,

$$\frac{E(s)}{E_{in}(s)} = \frac{1}{\sum_{i=0}^{n} b_{n-i} s^{n-i}} \tag{5.2a}$$

and

$$\frac{E_o(s)}{E(s)} = \sum_{i=0}^{m} a_{m-i} S^{m-i} \tag{5.2b}$$

---

* The order of the numerator can be made equal to that of the denominator (m – n), but the possibility of computational delay is created. In addition, an error is usually introduced on the first solution pass.

which can be written as:

$$E_{in}(s) = \sum_{i=0}^{n} E(s)\, b_{n-i}\, s^{n-i} \tag{5.3}$$

$$E_o(s) = \sum_{i=0}^{m} a_{m-i} E(s) s^{m-i} \tag{5.4}$$

The inverse Laplace transform of the above equations, assuming zero initial conditions, results in:

$$e_{in}(t) = \sum_{i=0}^{n} b_{n-i} e^{(n-i)}(t) \tag{5.5}$$

and

$$e_o(t) = \sum_{i=0}^{m} a_{m-i} e^{(m-i)}(t) \tag{5.6}$$

where $e^{(i)}(t)$ is the i–th derivative of e(t) with respect to time. Eqs. (5.5) and (5.6) are written in state-variable form as:

$$e(t) = x_1$$

$$\dot{x}_1 = x_2$$

$$\dot{x}_{n-1} = x_n$$

$$\dot{x}_n = \frac{1}{b_n}\left[e_{in}(t) - \sum_{i=1}^{n} b_{n-i} x_{n-i+1}\right] \tag{5.7}$$

$$e_o(t) = \sum_{i=0}^{m} a_{m-i} x_{m-i+1}$$

Eqs. (5.7) represents a satisfactory time-domain representation of the given transfer function.

5.2.2 SCEPTRE Solution of First-Order Differential Equation

The procedure for using SCEPTRE to solve n first-order differential equations is quite straightforward. It involves the DEFINED PARAMETER mode described in Chapter 2. The simplest way to explain this method is by use of an example.

Assume that the user has the problem of solving the following set of first order, simultaneous differential equations that may be entirely independent of any electrical network:

$$\dot{X} = -6X + 5Y + 10$$

$$\dot{Y} = 5X - 7Y + 2Z$$

$$\dot{Z} = .2Y - .2Z - .5$$

with X(0) = 6, Y(0) = 5, Z(0) = 4 as initial conditions.

The user enters each of the derivatives under DEFINED PARAMETERS in explicit form. A proper sequence is:

```
CIRCUIT DESCRIPTION
EXAMPLE 5.2.2
DEFINED PARAMETERS
DPX = X1(-6.*PX + 5.*PY + 10.)
DPY = X2(5.*PX - 7.*PY + 2.*PZ)
DPZ = X3(.2*PY - .2*PZ - .5)
PX = 6
PY = 5
PZ = 4
OUTPUTS
PX(X), PY(Y), PZ(Z), XSTPSZ, PLOT
RUN CONTROLS
STOP TIME = 100
END
```

The defined parameter derivatives DPX, DPY and DPZ are created which are integrated into the defined parameters PX, PY and PZ. Notice that the initial conditions for PX, PY, and PZ are not entered under the INITIAL CONDITIONS sub-heading. Rather, these quantities are updated at each integration step to solve the state equations. The quantities X, Y, and Z are treated in the same manner as the state variables of the general transient problem and therefore, these equations are subjected to the same step size limitations in whatever integration routine is used. These quantities are explicitly labeled as X, Y, and Z because of the format used under OUTPUTS. The step size is output through the use of the internal name XSTPSZ (see Table 2.7).

5.2.3 SCEPTRE General Transfer Function Model

By mechanizing Eq. (5.7) within the framework of the SCEPTRE input language, a SCEPTRE model for the four-terminal device shown in Fig. 5.2 is written. The zero current source, J1 = 0, is artificially introduced to permit the voltage between nodes I and GI to be used as the input variable for the transfer function and to enable the transfer function to present infinite load impedance to the preceding stage. The voltage source E0 results in zero source impedance for the following state. These considerations allow transfer-function models to be interconnected either with other transfer-function models or with ordinary electrical circuits.

To illustrate the over-all procedure, let G(s) in Fig. 5.2 be the nth–order transfer function:

$$G(s) = \frac{a_m s^m + a_{m-1} s^{m-1} + \ldots + a_1 s + a_o}{b_n s^n + b_{n-1} s^{n-1} + \ldots + b_1 s + b_o} \tag{5.8}$$

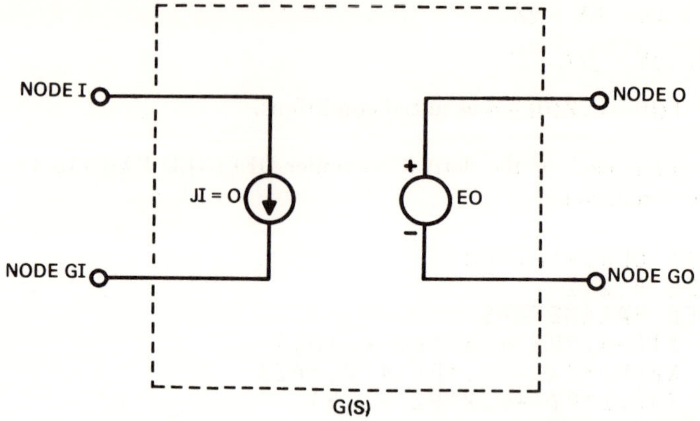

Fig. 5.2 The four-terminal transfer function model

The user must create the two elements shown in the figure, and he must supply three types of expressions. The voltage source E0 must be given as:

$$E0 = a_m P_{m+1} + a_{m-1} P_m + \ldots + a_1 P_2 + a_0 P_1 .^*$$

In addition, the nth derivative of the transfer function is supplied under DEFINED PARAMETERS as:

$$DP_n = \frac{(VJI - b_{n-1} P_n - \ldots\ldots\ldots - b_o P_1)}{b_n}$$

Finally, the remaining differential equations are supplied under DEFINED PARAMETERS in the repetitive form:

$$DP_{n-1} = P_n$$
$$DP_{n-2} = P_{n-1}$$

.
.
.
.

$$DP1 = P2$$

Note that n differential equations are needed, which is as expected to represent an nth–order transfer function.

It will shortly become clear that a significant amount of coding will be required from the user if the procedure implied above is followed for every transfer function in every problem. The stored model feature in SCEPTRE is a natural way to alleviate this

* For the special case m = n, this relation is revised to: $E0 = a_m DP_m + a_{m-1} P_m + \ldots + a_1 P_2 + a_o P_1$

situation. For this reason, observe the coding required to permanently store the following fifth-order version of Eq. (5.8).

$$G(s) = \frac{a_4 s^4 + a_3 s^3 + a_2 s^2 + a_1 s + a_o}{b_5 s^5 + b_4 s^4 + b_3 s^3 + b_2 s^2 + b_1 s\, b_o}$$

For this form, m = 4 and n = 5.

```
MODEL DESCRIPTION
MODEL 5 ORDER (PERM) (I - 0 - GI - GO)
ELEMENTS
JI, I-GI = 0
EO, GO-0 = XI (PA4*P5+PA3*P4+PA2*P3+PA1*P2*PA0*P1)
DEFINED PARAMETERS
DP5 = X2 ((VJI - PB4*P5-PB3*P4-PB2*P3-PB1*P2-PB0*P1)/PB5)
DP4 = X3 (P5)
DP3 = X4 (P4)
DP2 = X5 (P3)
DP1 = X6 (P2)
PA0 = 0, PA1 = 0, PA2 = 0, PA3 = 0, PA4 = 0
PB0 = 0, PB1 = 0, PB2 = 0, PB3 = 0, PB4 = 0, PB5 = 0
P1 = 0, P2 = 0, P3 = 0, P4 = 0, P5 = 0
END
```

The Defined Parameters are named PA0, . . . , PB0, ... to correspond with coefficients a_o, ... , b_o, ... Five other parameters, P1 through P5, serve as the dependent variables of the five differential equations. The next to last card in the above sequence establishes the initial conditions* of these differential equations. The model is general in that any of "a's" and "b's" can be changed from one problem to the next, and the model can be degenerated to represent any lower-ordered function by proper assignment of parameter values. The latter operation is not encouraged, however, since degeneration wastes computer time and can rapidly deplete the maximum number of Defined Parameters (100) that are allowed in any run. A better procedure is to permanently store a general model for each order transfer function of interest. Once stored, any transfer function can be called into use in any run with a single card.

5.2.4 Example: General Model Use

It is desired to use the general fifth-order model that was developed in the preceding section to represent the explicit fifth-order transfer function shown below:

$$G(s) = \frac{7s^4 + 2s^3 - 10s^2 + 6s + 1}{s^5 + 4s^4 + 16s^3 + 100s^2 + 200s + 1000}$$

If the transfer function is in a system with nodes numbered as shown in Fig. 5.3, the call on G under CIRCUIT DESCRIPTION is G, 2–3–7–8 = MODEL 5 ORDER (PERM,

* Failure to supply these initial conditions is not costly. The program automatically assumes zero values for these quantities and prints out an appropriate diagnostic message.

CHANGE PA4 = 7., PA3 = 2., PA2 = −10., PA1 = 6., PA0 = 1., PB5 = 1., PB4 = 4., PB3 = 16., PB2 = 100., PB1 = 200., PB0 = 1000.) The key point is that PAi and PBi are associated with a_i and b_i. It should be clear that any of the coefficients except b_5 may be left at zero by simply omitting reference to them in the call. Coefficient b_5 is excepted since if $b_5 = 0$, the function is not fifth-order, and the user is better advised to use a general fourth-order function.

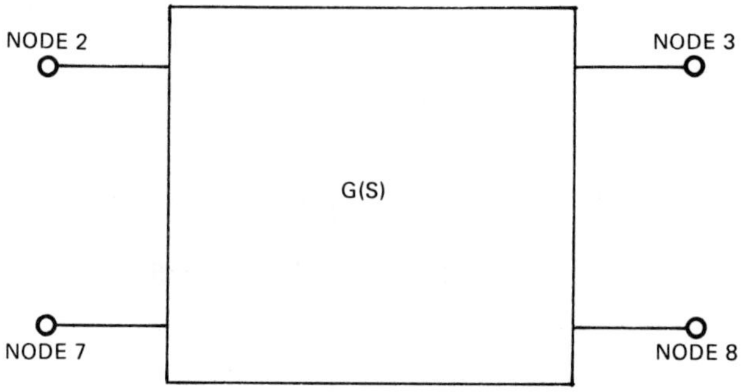

Fig. 5.3 Transfer function

Occasion may arise when a model applicable to one, and only one, transfer function is needed. This is accommodated by either storing the specific model (permanently or temporarily) or by simply describing it directly under CIRCUIT DESCRIPTION. To illustrate, a specific model for the transfer function below is to be developed:

$$G(s) = \frac{s + 7}{s^4 + 20s^3 + 4s + 1000}$$

Here m = 1 and n = 4. This time the model is not stored but is described directly under CIRCUIT DESCRIPTION. Again assuming the node nomenclature of Fig. 5.3, and following Section 5.2.3, the correct listing is:

```
CIRCUIT DESCRIPTION
ELEMENTS
JI, 2-7 = 0
EO, 8-3 = X1     (1.*P2 + 7. *P1)
DEFINED PARAMETERS
DP4 = X2 (VJI - 20.*P4-4.*P2-1000.*P1)
DP3 = X3 (P4)
DP2 = X4 (P3)
DP1 = X5 (P2)
```

Note that the coefficient of s^2 in the denominator (b_2) equals zero, and that this is reflected in the DP4 relation by the absence of any P3 term. The important point to be gleaned from this example is that it is specifically tailored to one specific transfer function and cannot be altered to any other G(s).

5.3 SUMMING AND AMPLIFICATION MODELS

5.3.1 Summing Junction Model

In addition to transfer functions, system analysis requires the use of summing junctions (see Fig. 5.4).

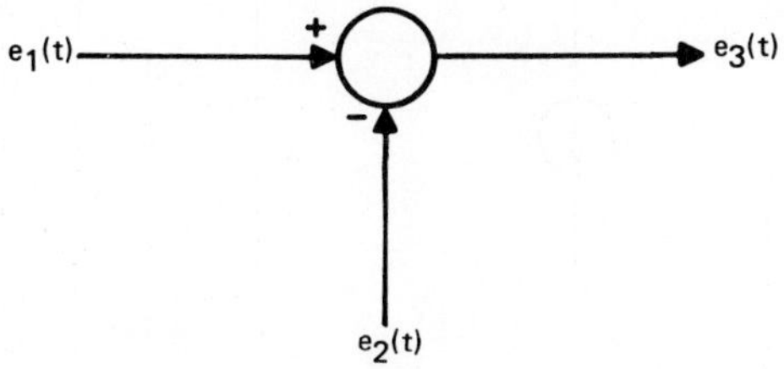

Fig. 5.4 Typical summing junction

Writing Kirchoff's voltage law for the voltage loop of Fig. 5.5 reveals that:

$$e_3(t) = e_1(t) - e_2(t)$$

as required. Correct polarity must be observed for all input voltages to insure that the signals add or subtract as desired. Summing junctions with more than two inputs are realized in an analogous manner. In the electrical-circuit used by SCEPTRE, the summing junction of Fig. 5.4 is realized by the network configuration shown in Fig. 5.5.

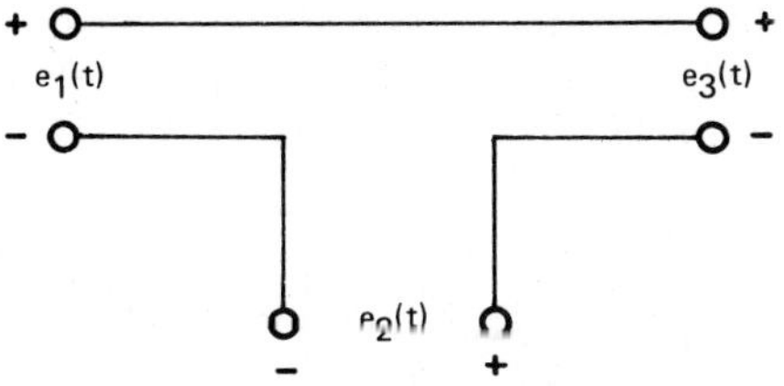

Fig. 5.5 Network summing junction

5.3.2 Amplifier Model

In system analysis one frequently encounters sub-systems which can be approximated by ideal amplifiers, (i.e., amplifiers having infinite input impedance, zero output impedance, and gain K for all input signals.) Although these can be derived from a general model with $n \geq 1$, by proper insertion of some Defined Parameters, an alternate approach, which uses only one parameter, is presented. Figure 5.6 shows an appropriate circuit model for an ideal amplifier having gain K = 100.

A suitable four-terminal SCEPTRE model is as follows:

```
MODEL DESCRIPTION
MODEL GAIN (TEMP) (I-O-GI-GO)
ELEMENTS
JI, I-GI = 0.0
```

```
EO, GO-O = X1(VJI*PA)
DEFINED PARAMETERS
PA = 100.
```

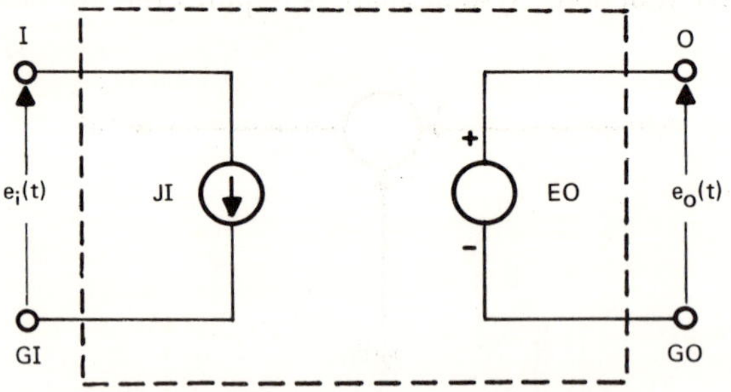

Fig. 5.6 Circuit model of an ideal amplifier

5.4 SCEPTRE SYSTEM-ANALYSIS APPLICATION

The remainder of this chapter consists of several practical examples of the SCEPTRE system-analysis concept.

5.4.1 Example: Linear Control System

In this example the results of an analog-computer study of an aircraft pitch-control problem are duplicated. The original analog computer study appeared in Control System Analysis and Synthesis[1] by D'Azzo and Houpis. The given system is shown in Fig. 5.7.

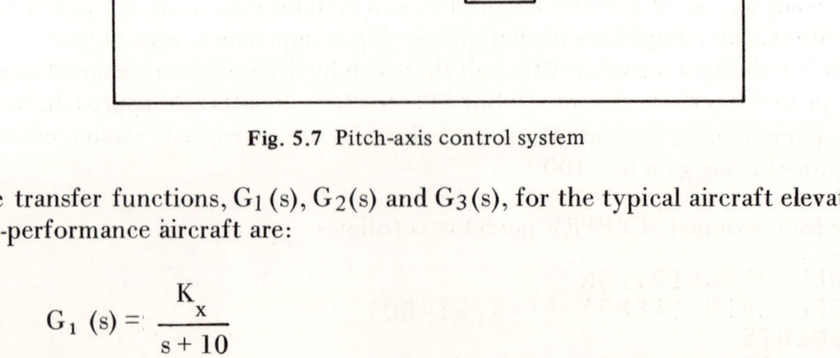

Fig. 5.7 Pitch-axis control system

The transfer functions, $G_1(s)$, $G_2(s)$ and $G_3(s)$, for the typical aircraft elevator servo in a high-performance aircraft are:

$$G_1(s) = \frac{K_x}{s + 10} \tag{5.9}$$

$$G_2(s) = \frac{10s + 4}{s^2 + 0.7s + 5} \tag{5.10}$$

$$G_3(s) = \frac{1}{s} \tag{5.11}$$

Using the values $K_x = 2.8$, $A_2 = 4.5$ and $A_1 = 13.0$, a block diagram suitable for SCEPTRE system and analysis is illustrated in Fig. 5.8.

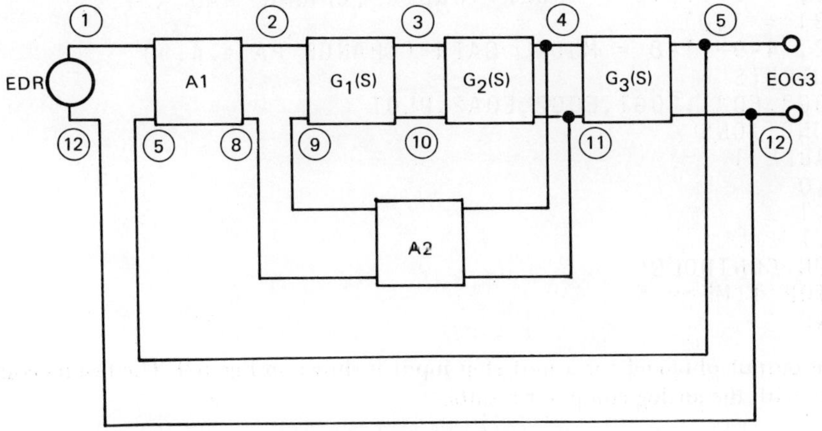

Fig. 5.8 Revised block diagram

The resulting SCEPTRE system-analysis program including model descriptions follows:

```
MODEL DESCRIPTION
MODEL GAIN (TEMP) (I-O-GI-GO)
ELEMENTS
JI, I-GI = 0.0
EO, GO-O = X1(VJI*PA)
DEFINED PARAMETERS
PA = 100.
MODEL 1ORDER (TEMP) (I-O-GI-GO)
ELEMENTS
JI, I-GI = 0
EO, GO-O = X1(PA0*P1)
DEFINED PARAMETERS
DP1 = X2((VJI-PB0*P1)/PB1)
PA0 = 0, PB0 = 0, PB1 = 0, P1 = 0
MODEL 2ORDER (TEMP) (I-O-GI-GO)
ELEMENTS
JI, I-GI = 0
EO, GO-O = X1(PA1*P2 + PA0*P1)
DEFINED PARAMETERS
DP1 = X3(P2)
DP2 = X2((VJI- PB1*P2- PB0*P1)/PB2)
PA0 = 0, PA1 = 0, PB0 = 0, PB1 = 0, PB2 = 0, P1 = 0
```

```
CIRCUIT DESCRIPTION
PITCH CONTROL
ELEMENTS
EDR, 12-1 = TABLE 1 (TIME)
A1, 1-2-5-8 = MODEL GAIN (CHANGE PA = 13.0)
G1, 2-3-9-10 = MODEL 1ORDER (CHANGE PAO = 2.8,
PBO = 10.0, PB1 = 1.0)
G2, 3-4-10-11 = MODEL 2ORDER (CHANGE PA1 = 10.0,
PAO = 4., PB2 = 1.0,PB1 = 0.7, PBO = 5.0)
G3, 4-5-11-12 = MODEL 1ORDER (CHANGE PAO = 1.,
PB1 = 1.0)
A2, 4-9-11-8 = MODEL GAIN (CHANGE PA = 4.5)
OUTPUTS
EOG3,EOA1,EOG1,EOG2,EOA2,PLOT
FUNCTIONS
TABLE 1
0,0
0,1
5,1
RUN CONTROLS
STOP TIME = 5
END
```

The output obtained for a unit step input is shown in Fig. 5.9. The results compare favorably with the analog computer results.

Fig. 5.9 Transient response

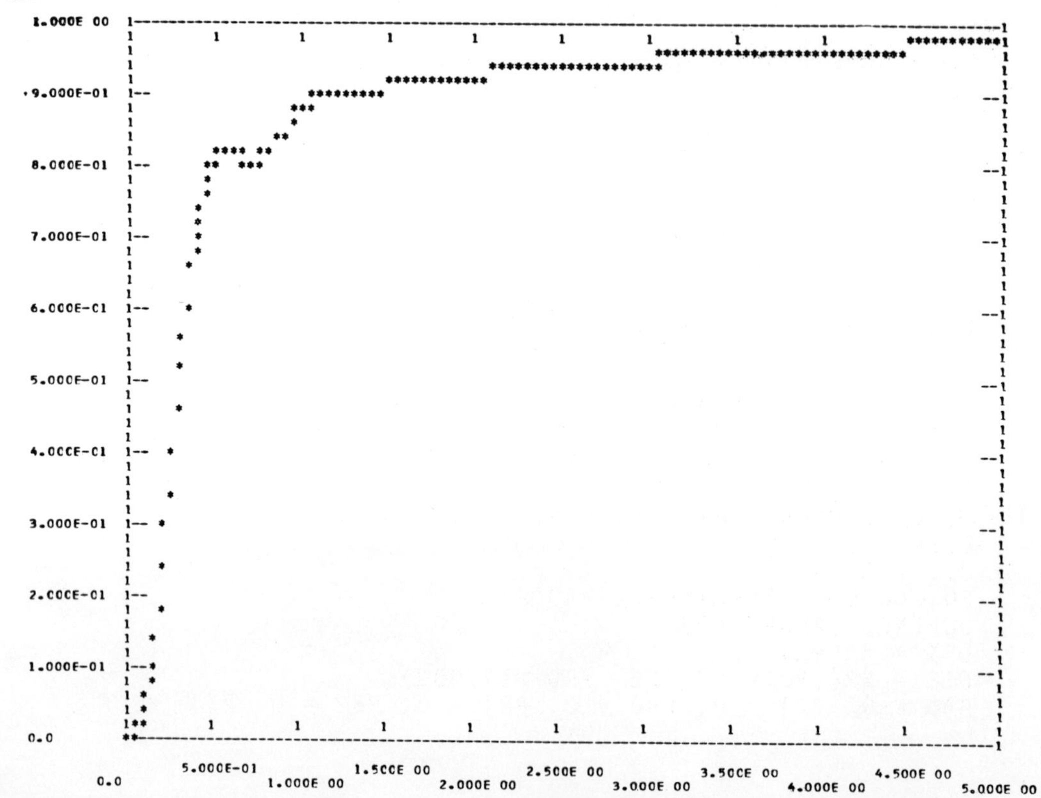

5.4.2 Example: Second-Order Nonlinear Control System

A block diagram of an instrument servo system containing a saturating amplifier is shown in Fig. 5.10. It is desired to obtain the output phase trajectory for a reference input of $\theta_R = 5$. Normally phase-plane techniques are used to solve this type of problem.

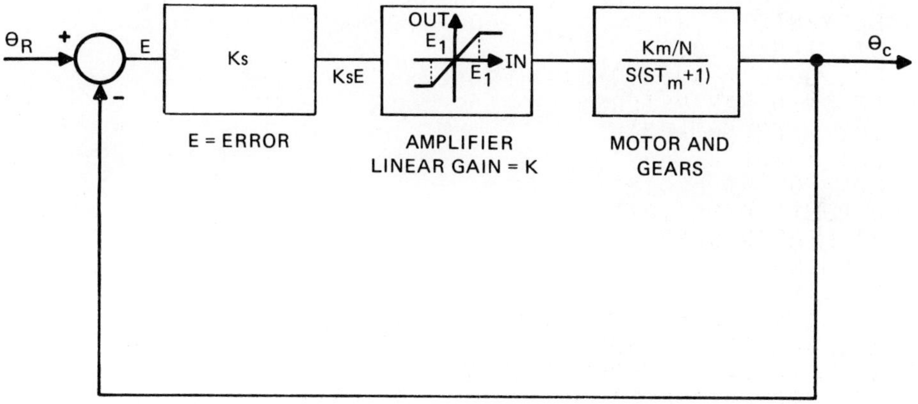

Fig. 5.10 Instrument servo system block diagram

System constants for this example:

Ks = 1

K = 1

Km/N = 3

Tm = 1

E_1 = 1 = The value of $|KsE|$ which causes saturation.

Substituting the appropriate constant values yields:

$$G1(s) = \frac{3}{s^2 + s} \tag{5.12}$$

This function is simulated by a specific version of Model 2 order. A system block diagram suitable for SCEPTRE system analysis is shown in Fig. 5.11.

The resulting SCEPTRE systems analysis program is listed below. A discussion and clarification of the saturating amplifier model and phase-plane output techniques follows the program listing.

```
MODEL DESCRIPTION
MODEL GAIN (TEMP) (I-O-GI-GO)
ELEMENTS
JI, I-GI = 0.0
```

```
EO, GO-O = X1(VJI)
MODEL GSAT (TEMP) (I-O-GI-GO)
ELEMENTS
JI, I-GI = 0.0
EO, GO-O = TABLE 1 (VJI)
FUNCTIONS
TABLE 1
-100,-1, -1,-1, 1,1, 100,1
MODEL 2 ORDER (TEMP) (I-O-GI-GO)
ELEMENTS
JI, I-GI = 0.0
EO, GO-O = XO(3.0*P1)
DEFINED PARAMETERS
DP2 = X2(VJI-P2)
DP1 = X3(P2)
P2 = 0, P1 = 0
CIRCUIT DESCRIPTION
NONLINEAR SERVO
ELEMENTS
EIN, 9-1 = 5
A1, 1-4-8-5 = MODEL GAIN
A2, 4-6-5-7 = MODEL GSAT
G1, 6-8-7-9 = MODEL 2ORDER
DEFINED PARAMETERS
PDEO = X4(3.0*P2G1)
OUTPUTS
PDEO (RATE), PLOT (EOG1 (ANGLE))
VJIA1,EOA1,EOA2,PLOT
RUN CONTROLS
STOP TIME = 10
END
```

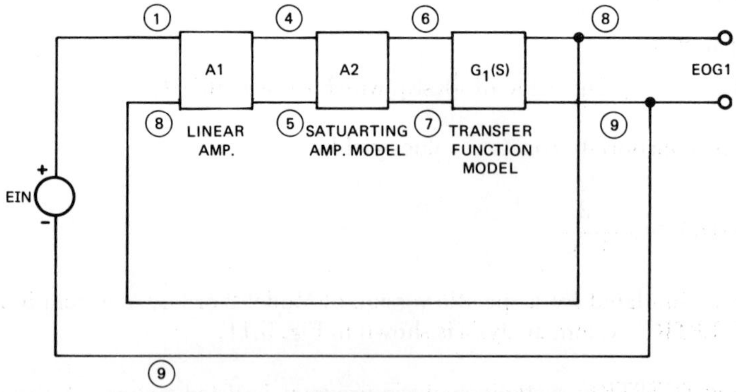

Fig. 5.11 Block diagram in SCEPTRE form

The saturating amplifier model shown in Fig. 5.12a indicates a SCEPTRE description of the amplifier characteristic shown in Fig. 5.12b. The system constants K = 1 and E1 = 1 of Fig. 5.10 indicate that the amplifier output voltage at saturation is 1 volt. The phase plane of the system is a plot of θc versus $\dot{\theta}c$. From Fig. 5.11 it is seen that θc is equal to EOG1.

Since E0 = PA0*P1:

$\dot{E}0$ = PAO*DP1

= PAO*P2

= $\dot{\theta}$c

Therefore

θc = 3*P2

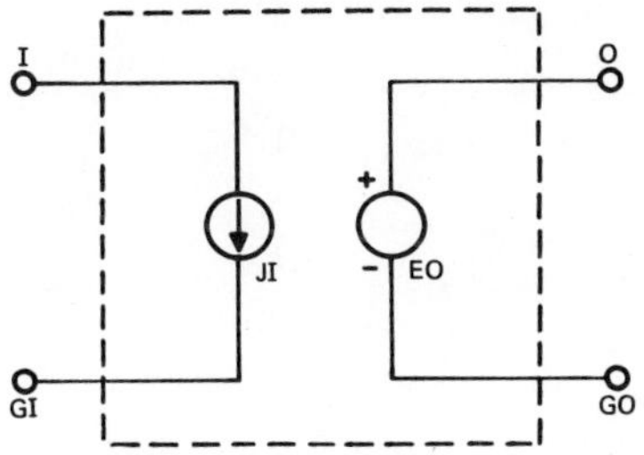

(A) SATURATING AMPLIFIER MODEL

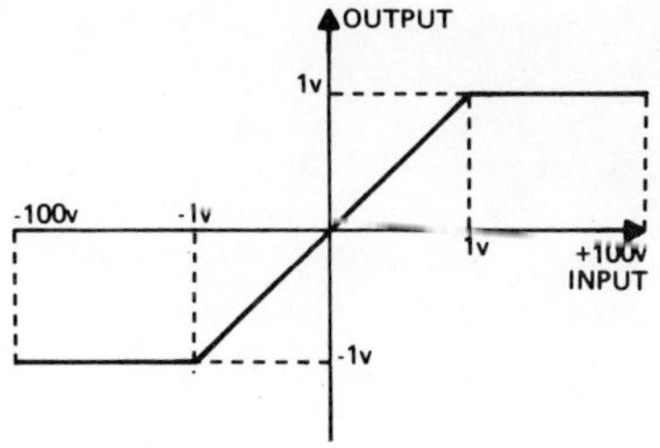

(B) SATURATING AMPLIFIER INPUT-OUTPUT CHARACTERISTIC

```
MODEL DESCRIPTION
MODEL GSAT (TEMP)   (I-O-GI-GO)
ELEMENTS
JI, I-GI = 0.0
EO, GO-O = TABLE 1 (VJI)
FUNCTIONS
TABLE 1
-100, -1, -1, -1, 1, 1, 100, 1
```

Fig. 5.12 SCEPTRE saturating amplifier model

A plot of $\dot{\theta}$c versus θc results if one first defines PDE0 = (PA0G1*P2G1) and then plots PDE0($\dot{\theta}$c) as a function of E0G1(θc). Notice that the output called for in the program renames the quantities to angle and rate for maximum clarity. The results of the SCEPTRE systems analysis are illustrated in Fig. 5.13, and the phase trajectory obtained by an analog computer simulation is illustrated in Fig. 5.14. These figures show excellent agreement between the two methods.

5.4.3 Example: Third-Order Nonlinear Control System

The system shown in Fig. 5.15 is a third-order plant controlled by an ON–OFF controller. The reader with experience in the Automatic Control field realizes immediately that a system of this type has a limit cycle oscillation. For that matter it is not difficult for

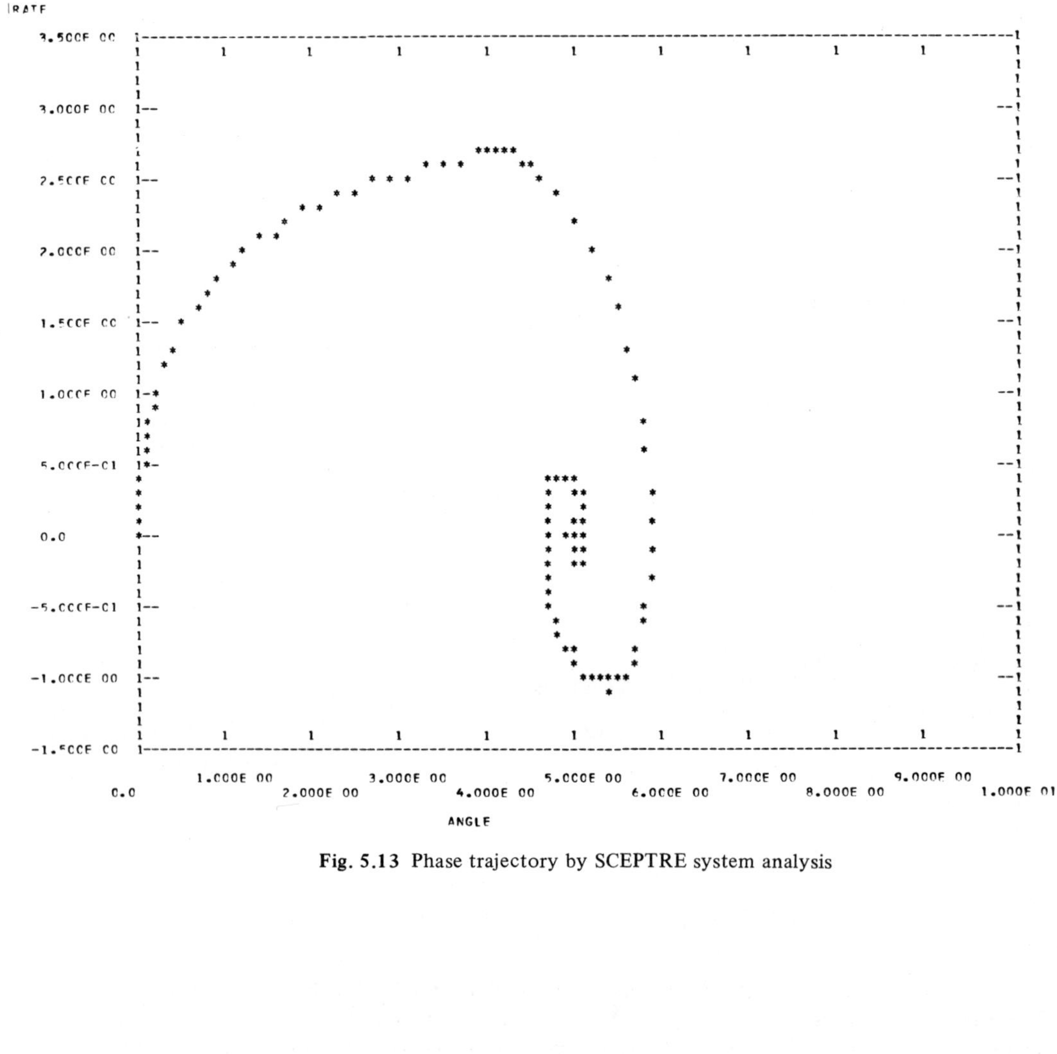

Fig. 5.13 Phase trajectory by SCEPTRE system analysis

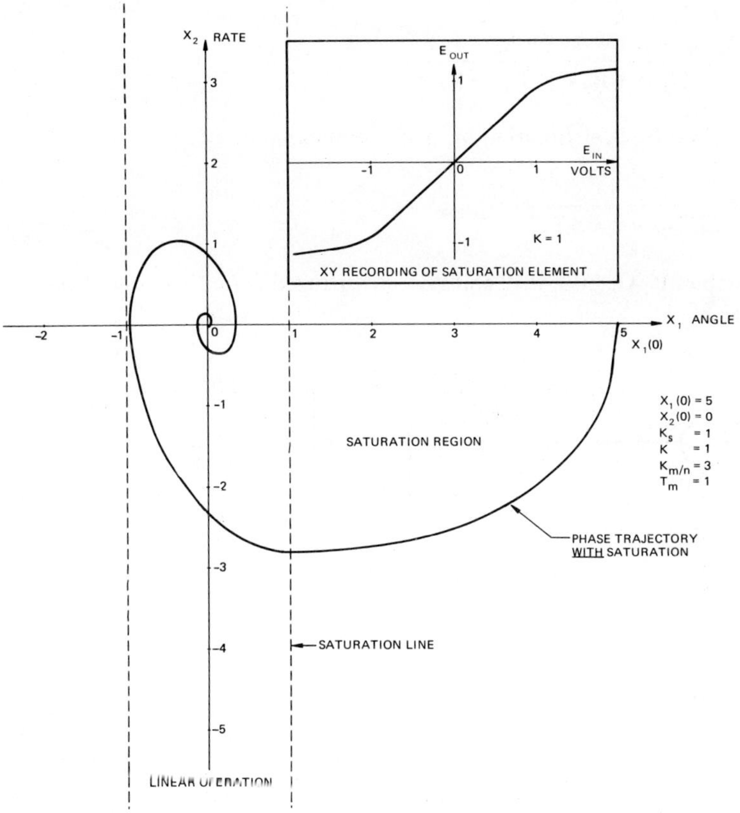

Fig. 5.14 XY recording of phase trajectories of servo system with saturation

anyone to realize, by inspection, that the system has no stable point since the controller theoretically has no zero position. The controller output is either +10 or −10. The control engineer normally applies the Describing Function technique to determine the frequency and the amplitude of the limit cycle. A third-order model is stored temporarily to accommodate the following general transfer function:

$$G(s) = \frac{A_2 s^2 + A_1 s + A_0}{B_3 s^3 + B_2 s^2 + B_1 s + B_0} \tag{5.13}$$

Since the desired transfer function in this system expands into:

$$G_1 s = \frac{1}{s^3 + 2s^2 + s} \tag{5.14}$$

only parameters PAO, PB3, PB2 and PB1 are changed from their stored zero values in the model call.

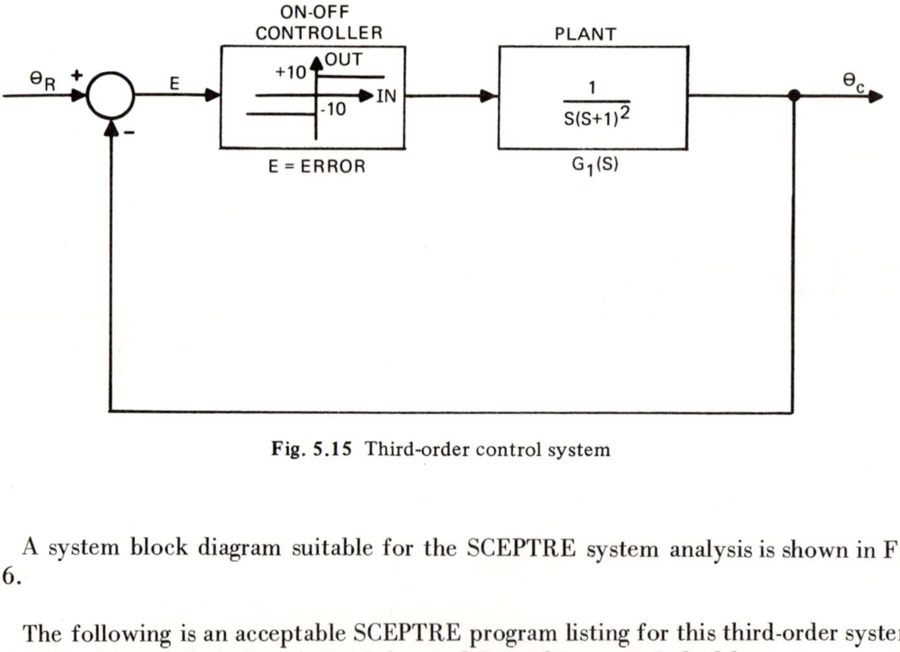

Fig. 5.15 Third-order control system

A system block diagram suitable for the SCEPTRE system analysis is shown in Fig. 5.16.

The following is an acceptable SCEPTRE program listing for this third-order system. The reader is reminded that if all of the models in this example had been permanently stored at some previous time, only the entries under CIRCUIT DESCRIPTION would be required.

```
MODEL DESCRIPTION
MODEL RELAY (TEMP) (I-0-GI-GO)
ELEMENTS
JI, I-GI = 0.0
EO, GO-0 = TABLE 1 (VJI)
FUNCTIONS
TABLE 1
-100,-10, -1.E-6,-10, 1.E-6,10, 100,10
```

```
MODEL 3ORDER (TEMP) (I-O-GI-GO)
ELEMENTS
JI, I-GI = 0.0
EO, GO-O = XO(PA2*P3 + PA1*P2 *PA0*P1)
DEFINED PARAMETERS
DP3 = X3((VJI-PB2*P3 - PB1*P2 - PB0*P1)/PB3)
DP2 = XY(P3)
DP1 = XZ(P2)
PA0 = 0, PB3 = 0, PB2 = 0, PB1 = 0, P3 = 0, P2 = 0
P1 = 0
PA1 = 0, PA2 = 0, PB0 = 0
CIRCUIT DESCRIPTION
THIRD ORDER NONLINEAR SYSTEM
ELEMENTS
EIN, 7-2 = 1
A1, 2-4-6-5 = MODEL RELAY
G1, 4-6-5-7 = MODEL 3ORDER (CHANGE PA0 = 1.0,
PB3 - 1.0, PB2 = 2.0, PB1 = 1.0)
OUTPUTS
EOG1,VJIA1,EOA1,PLOT
RUN CONTROLS
STOP TIME = 30
END
```

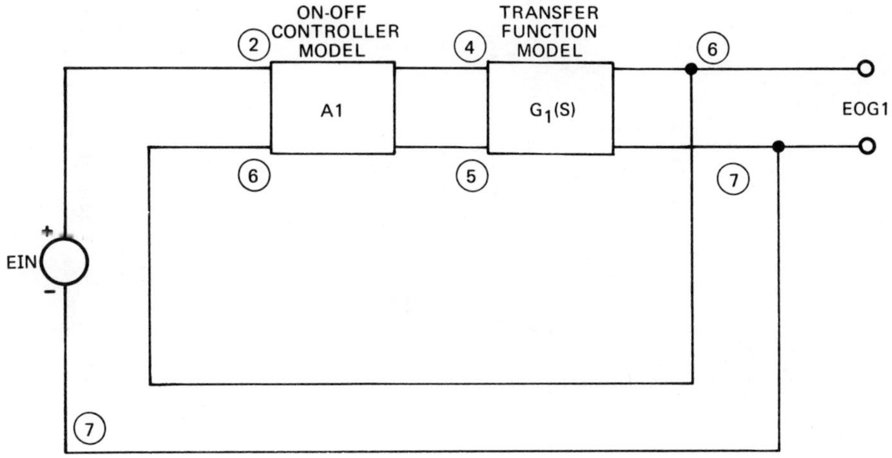

Fig. 5.16 System block diagram in SCEPTRE form

The "ON–OFF" controller (RELAY) model shown in Fig. 5.17a provides a SCEPTRE description of the device characteristics shown in Fig. 5.17b. An approximation to the ideal relay characteristic is required due to the computational problems involved at the transition point. At the zero point the relay state is indeterminate.

An arbitrarily large slope is used to overcome this problem, and the results obviously are not affected. The only complication coming from this approach is that the

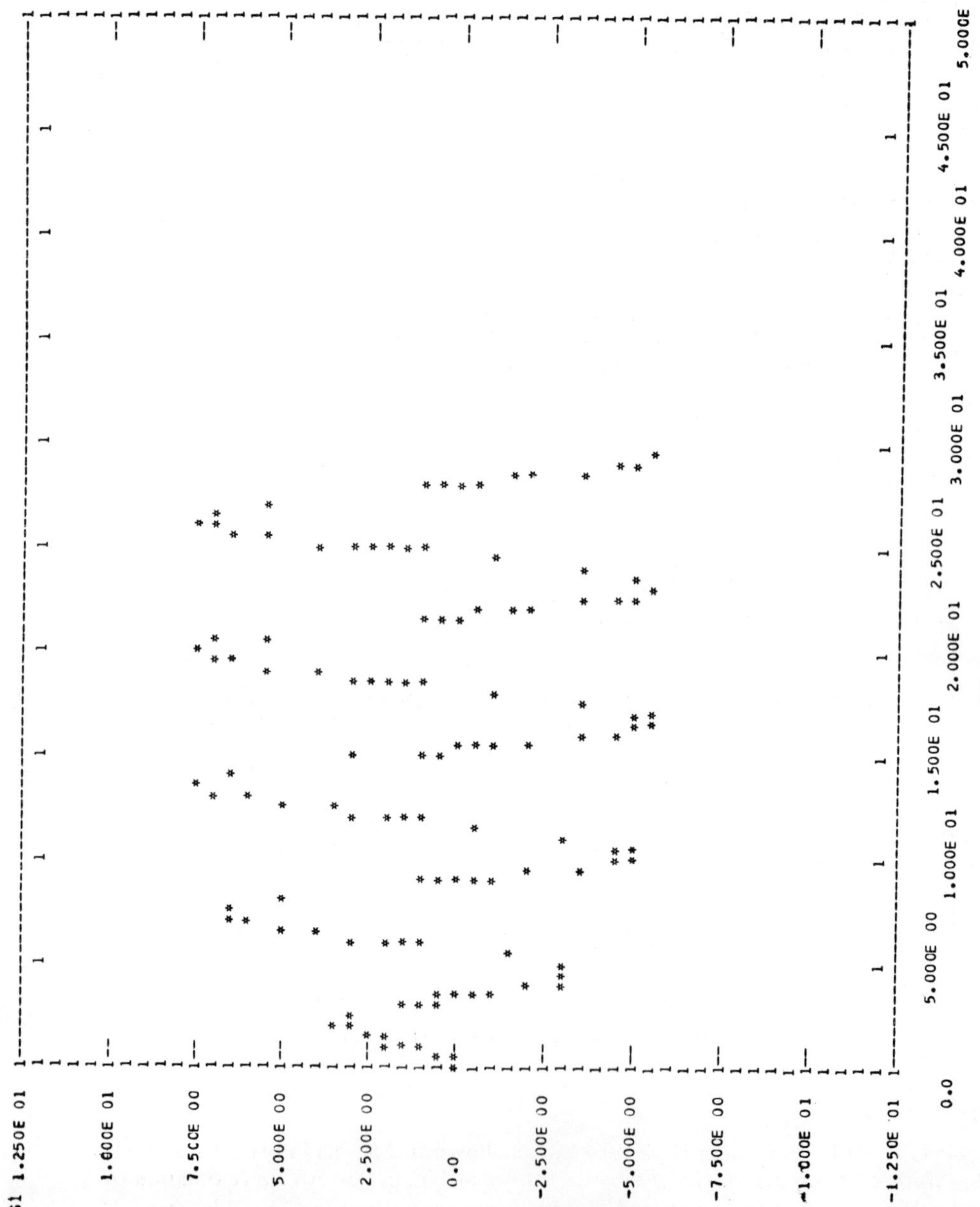

Fig. 5.18 SCEPTRE solution of the third-order system

system is potentially stable with zero input. Therefore, a small (or any) input must be applied as θ_R. Once again, the practical results are not affected.

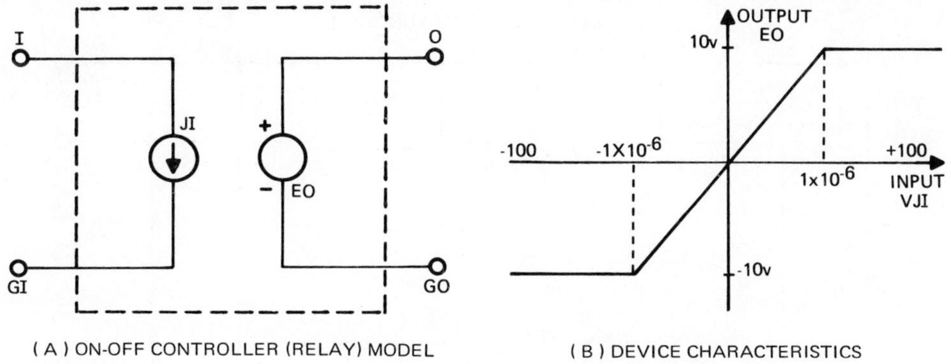


```
MODEL RELAY (TEMP) (I-O-GI-GO)
ELEMENTS
JI, I-GI = 0.0
EO, GO-O = TABLE 1 (VJI)
FUNCTIONS
TABLE 1 = -100, -10, -1.E-6, -10, 1.E-6, 10, 100, 10
```

Fig. 5.17 SCEPTRE relay model

The problem originally appeared in Control Systems Theory [2] by Elgerd and was solved using Describing Function techniques. The results indicate an amplitude of 6.38 volts and a frequency of .163 HZ for the limit cycle. The SCEPTRE results shown in Fig. 5.18 compare quite favorably. In fact, the SCEPTRE results are no doubt more accurate than those obtained using the approximate Describing Function technique.

5.4.4 Example: Sampled-Data Control System (Z Transform Analysis)

A unit step function is applied to the sampled-data control system illustrated in Fig. 5.19. The value of $G_1(s)$ is shown and the sampling period, T, is one second—with the sampling duration, t, being 0.2 second. The goal is to plot the output response for 5 seconds.

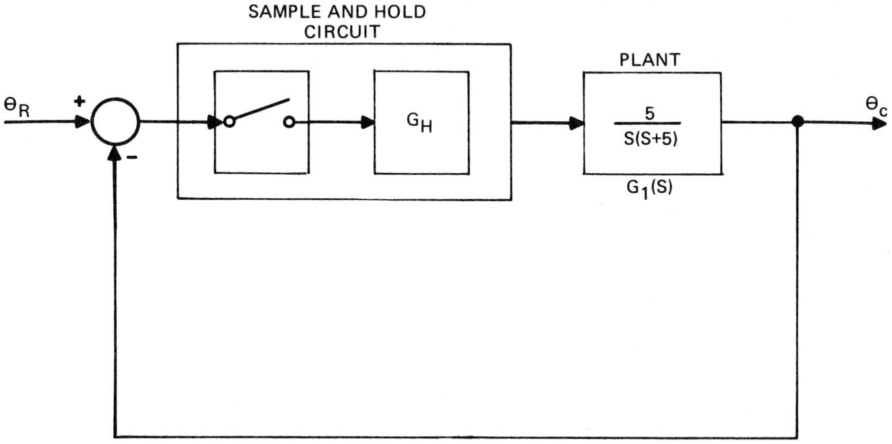

Fig. 5.19 Sampled-data control system

A system block diagram in SCEPTRE form is shown in Fig. 5.20.

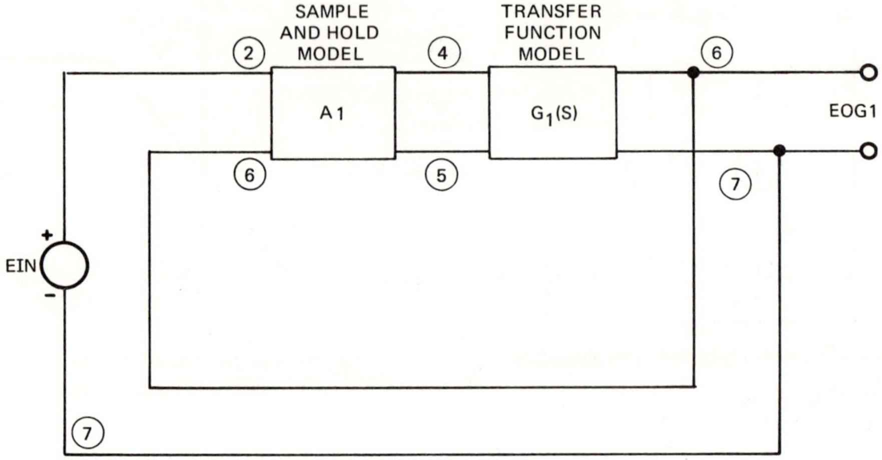

Fig. 5.20 Block diagram in SCEPTRE form

The resulting SCEPTRE system analysis program is listed below. Clarification of the MODEL SAMP follows the program listing.

```
MODEL DESCRIPTION
MODEL SAMP (TEMP) (I-0-GI-GO)
ELEMENTS
JI, I-GI = 0.0
E1, 1-2 = X1(VJI)
R1, 2-3 = TABLE 1 (TIME)
C1, 3-1 = 100E-6
EO, GO-0 = X2(VC1)
FUNCTIONS
TABLE 1
0,50, .2,50, .2,1.E6, 1,1.E6,
1,50, 1.2,50, 1.2,1.E6, 2,1.E6,
2,50,2.2,50, 2.2,1.E6, 3,1.E6,
3,50, 3.2,50, 3.2,1.E6, 4,1.E6,
4,50, 4.2,50, 4.2,1.E6, 5,1.E6
MODEL 2ORDER (TEMP) (I-0-GI-GO)
ELEMENTS
JI, I-GI = 0.0
EO, GO-0 = XO(5.*P1)
DEFINED PARAMETERS
DP2 = X2(VJI-5.*P2)
DP1 = XZ(P2)
P2 = 0, P1 = 0
CIRCUIT DESCRIPTION
SAMPLED DATA CONTROL SYSTEM
ELEMENTS
EIN, 7-2 = TABLE 1 (TIME)
A1, 2-4-6-5 = MODEL SAMP
```

```
G1, 4-6-5-7 = MODEL 2ORDER
OUTPUTS
EOG1,EIN,VJIA1,VJIG1,VC1A1,E1A1,R1A1,PLOT
XSTPSZ,PLOT
FUNCTIONS
TABLE 1
0,0, 0,1, 5,1
RUN CONTROLS
STOP TIME = 5
MINIMUM STEP SIZE = 5E-6
END
```

The sample and hold model, shown in Fig. 5.21, duplicates the usual situation when one is designing sample and hold circuits. Generally a rise time and fall (or hold) time are specified. The R, C, time constants of the model are picked to duplicate the actual system being analyzed. Assume the rise time is specified as 5 ms and the fall time as 100 seconds. These numbers are consistent with the values of T and t previously given (i.e., 100sec $\gg$ T [period] and 5ms $\ll$ t [sample duration]).

Figure 5.21b illustrates the variable pattern that R1 will follow as indicated by Table 1.

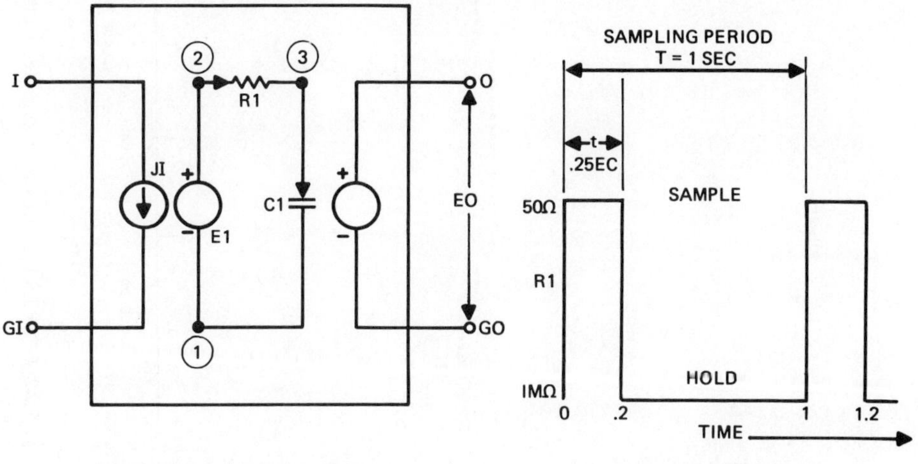

(A) SAMPLE AND HOLD MODEL (B) R1 TIMING DIAGRAM

```
MODEL DESCRIPTION
MODEL SAMP (TEMP) (I-O-GI-GO)
ELEMENTS
JI, I-GI = 0.0
E1, 1-2 = X1(VJI)
R1, 2-3 = TABLE 1 (TIME)
C1, 3-1 = 100E-6
EO, GO-O = X2(VC1)
FUNCTIONS
TABLE 1
0,50, .2,50, .2,1.E6, 1,1.E6, 1,50, 1.2,50, 1.2,1.E6, 2,
1.E6, 2,50, 1.2,50, 1.2,1.E6, 3,1.E6,3,50, 3.2,50, 3.2,
1.E6, 4,1.E6,4,50, 4.2,50, 4.2,1.E6, 5,1.E6
```

Fig. 5.21 Sample and hold model

The output response is shown in Fig. 5.22. The Z transform method of analysis could be used to analytically determine the same results but, obviously, with much additional effort.

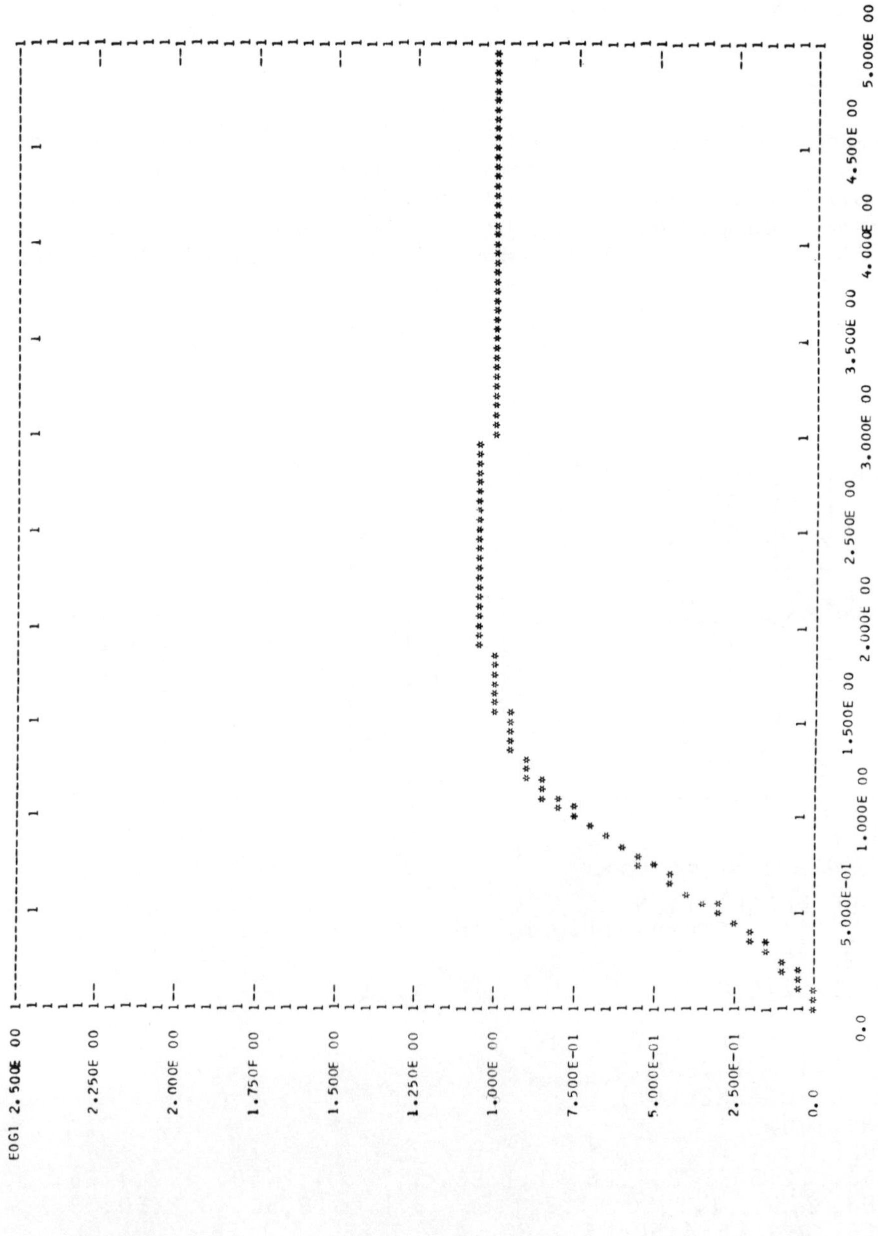

Fig. 5.22 SCEPTRE solution of sampled-data system

5.5 CONCLUSION

It is hoped that the variety of examples illustrated in this chapter will convince the user of the ease with which such problems can be solved. Certainly anyone who has had to labor with the manual methods such as phase-planes, describing functions, and Z transform techniques will appreciate the SCEPTRE approach, especially those who are not programmers and are, therefore, not writing their own special purpose programs for each individual problem.

PROBLEMS

5.1 An automatic depth-control system for a submarine is shown in Fig. p5.1. The depth is measured by a transducer. The gain of the stern plane actuator is $K = 10^{-3}$ when the velocity is 20 ft./sec. The submarine has approximate transfer function

$$G(S) = \frac{(S + 0.1)}{(S^2 + 0.02)},$$

and the feedback transducer is H(S) = 1. Write a SCEPTRE system analysis program including model descriptions, for a desired depth of 20 ft.

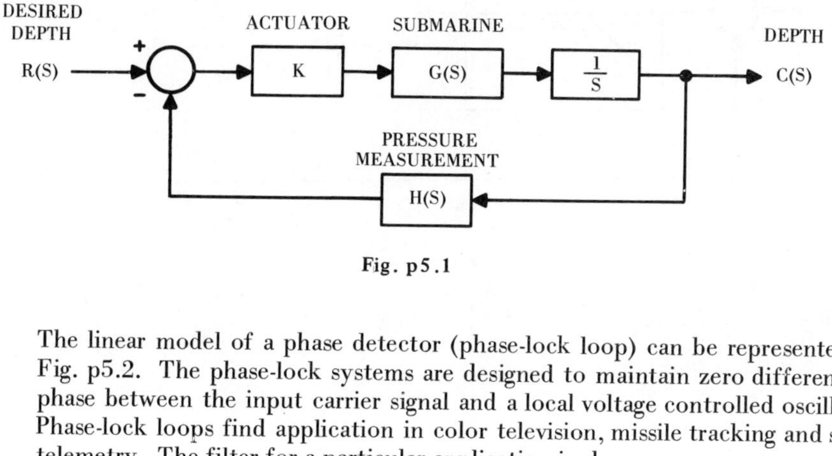

Fig. p5.1

5.2 The linear model of a phase detector (phase-lock loop) can be represented by Fig. p5.2. The phase-lock systems are designed to maintain zero difference in phase between the input carrier signal and a local voltage controlled oscillator. Phase-lock loops find application in color television, missile tracking and space telemetry. The filter for a particular application is chosen as

$$F(S) = \frac{S + 5}{(S + 1)(S + 200)}$$

It is desired to minimize the steady-state error of the system for a ramp change in the phase information signal. Amplifier gain Ka = 10. Write a SCEPTRE system analysis program including model description.

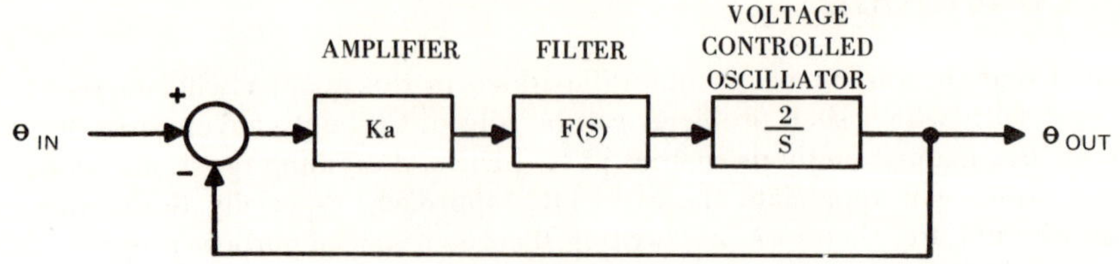

Fig. p5.2

5.3 A block diagram of an instrument servo with dead zone in the error transducer is shown in Fig. p5.3.1. Dead zone input-output characteristic is shown in Fig. 5.3.2.

a. Write a model description of dead zone.

b. Write a SCEPTRE system analysis program to obtain the output phase trajectory for a reference input of $\Theta_R = 10$.

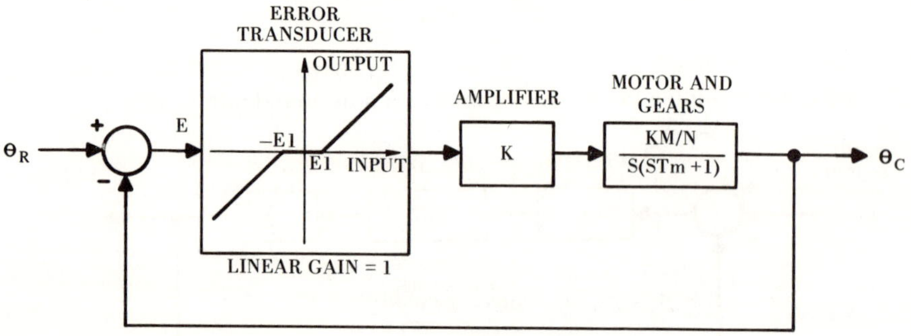

Fig p5.3.1

System constants are

KM/N = 4
Tm = 2
K = 10

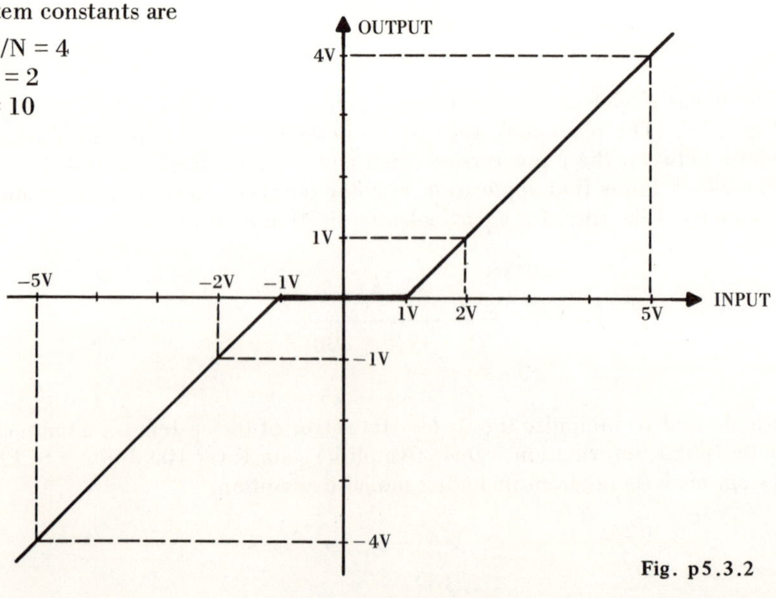

Fig. p5.3.2

5.4 Figure p5.4.1 shows a third order nonlinear control system using on-off controller. On-off controller characteristic is shown in Fig. p5.4.2. Write a SCEPTRE system analysis program for an input of unit step.

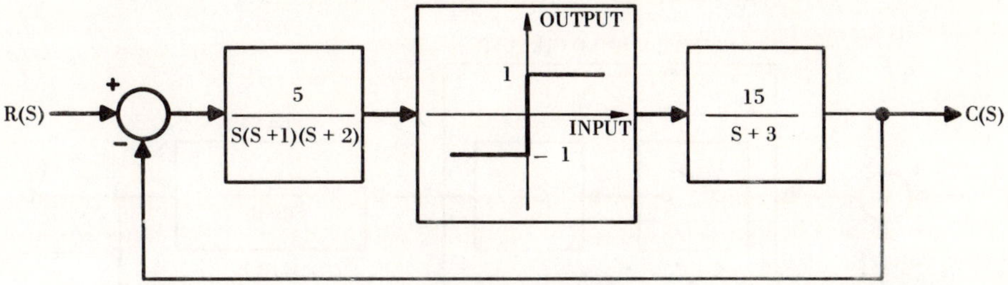

Fig. p5.4.1

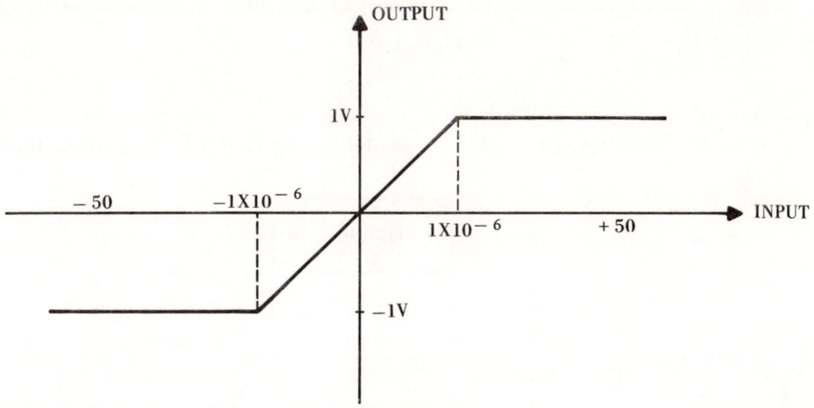

Fig. p5.4.2

5.5 Figure p5.5 shows the sampled data control system. The value of G1(S) is shown and the sampling period, 7, is two seconds with the sampling duration, t, being 0.1 second. Write a SCEPTRE program to plot the output response for 6 seconds, for an input of step function = 2. Assume rise time = 1 millisecond and fall time = 200 seconds.

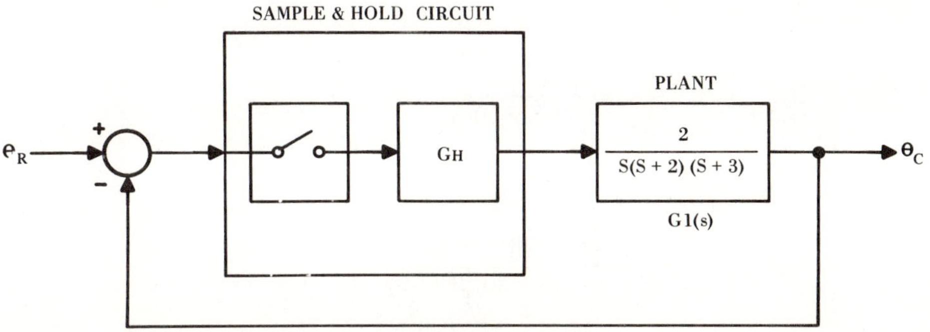

Fig. p5.5

5.6 Figure p5.6 shows the sampled data control system. G1(S) is shown in the figure. Write a SCEPTRE program to plot the output response for 5 seconds for an input step function = 5. Use the sample-hold model of the example in Sec. 5.4.4.

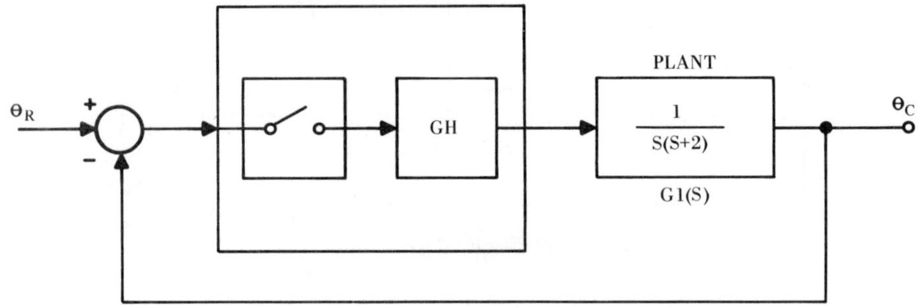

Fig. p5.6

5.7 Figure p5.7 shows a multiple loop feedback control system. Write a SCEPTRE system analysis program, including model description, for an input of unit step.

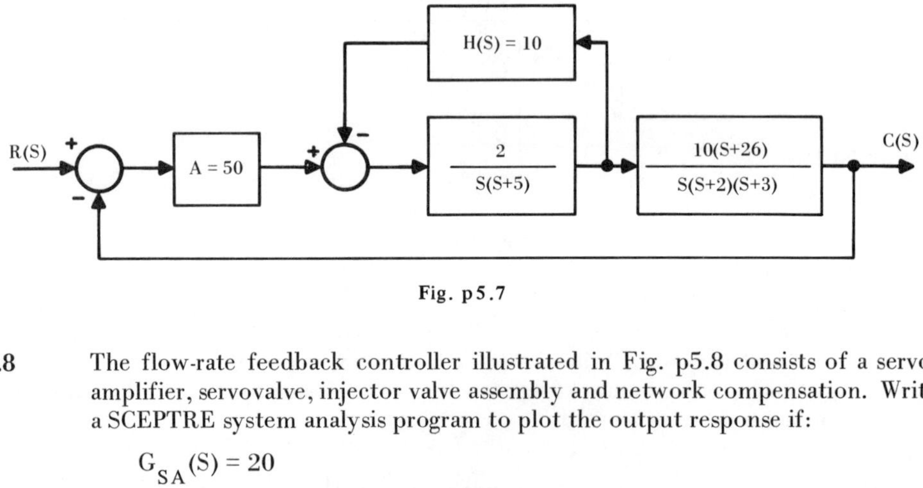

Fig. p5.7

5.8 The flow-rate feedback controller illustrated in Fig. p5.8 consists of a servo-amplifier, servovalve, injector valve assembly and network compensation. Write a SCEPTRE system analysis program to plot the output response if:

$$G_{SA}(S) = 20$$

$$G_{SV}(S) = \frac{200}{S(S^2 + 20\,S + 300)}$$

$$G_{IV}(S) = 1$$

$$G_C(S) = \frac{10(S^2 + 20 + 300)}{(S + 100)(S + 20)(S + 50)}$$

$$E_{IN}(S) = 5$$

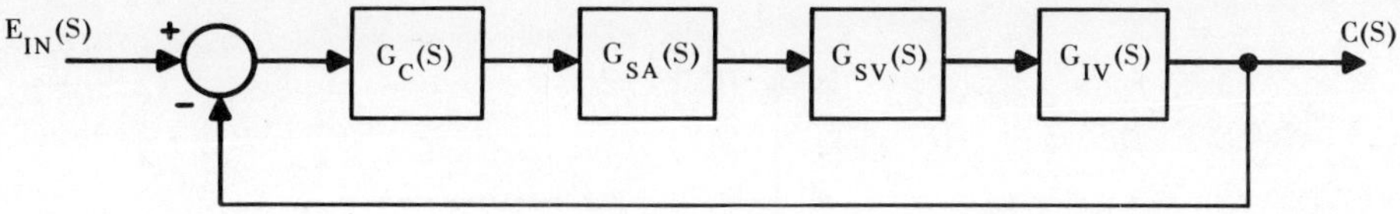

Fig. p5.8

5.9 Figure p5.9.1 shows a second order nonlinear system with saturating amplifier. Saturating amplifier characteristics are shown in Fig. 5.9.2. Write a SCEPTRE system analysis program to plot the phase trajectory for step-function input.

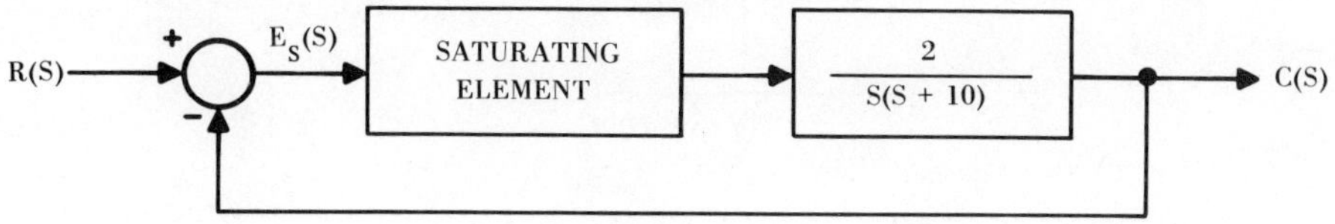

Fig. p5.9.1

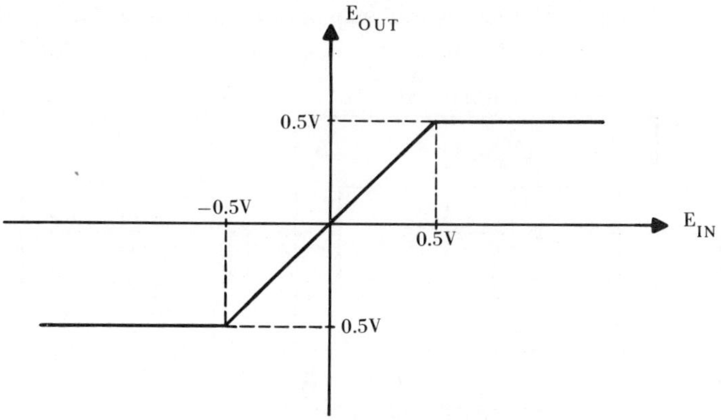

Fig. p5.9.2

5.10 Figure p5.10.1 shows a nonlinear controller with idealized relay. Write a SCEPTRE program to compute the system transient response over the first two seconds. Input is step function of 5. Fig. p5.10.2 shows the characteristic of ideal relay.

Fig. p5.10.1

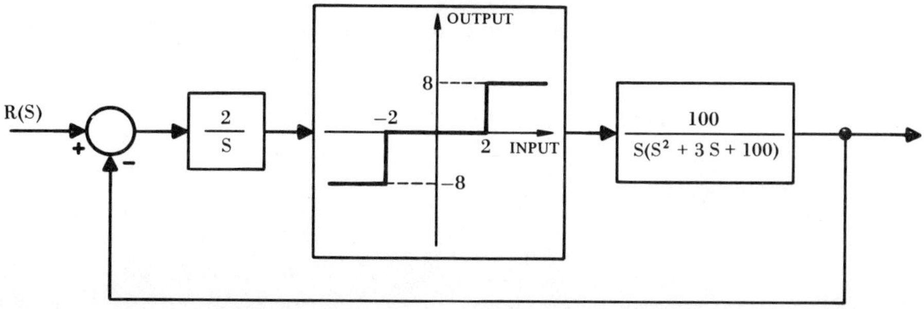

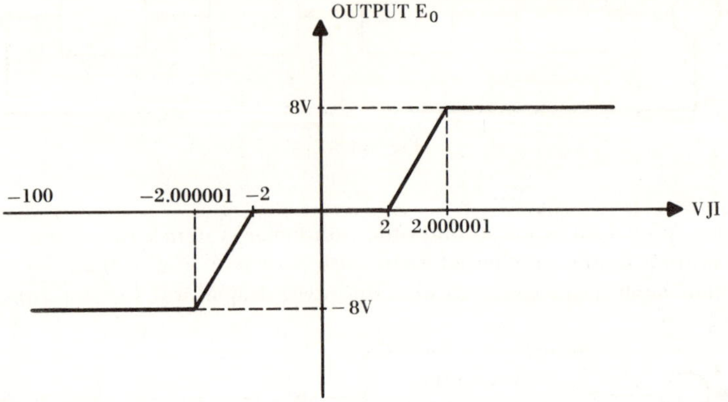

Fig. p5.10.2

5.11 Figure p5.11 shows the sampled data control system. For sampling duration, t, of 0.5 sec. and sampling period, T, of 5 secs. Write a SCEPTRE program to plot the output response for 10 seconds. Rise time is 2 milliseconds and fall time is 120 sec. Assume a unit step input. Also draw the sample and hold model circuit and variable resistor timing diagram. Let C = 2000 μF.

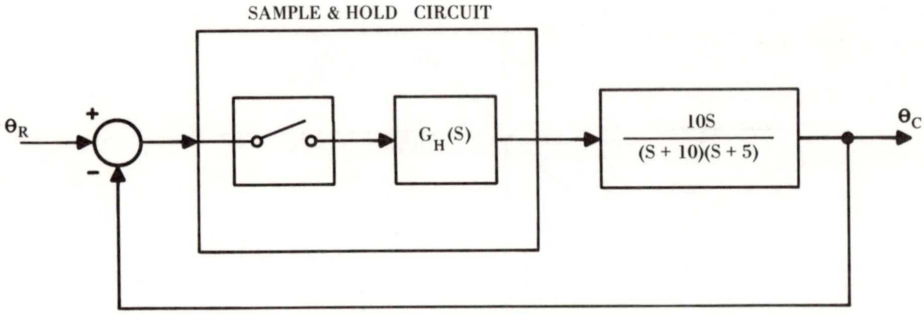

Fig. p5.11

5.12 Figure p5.12 shows the sampled data control system with the plant transfer function ,

$$G(S) = \frac{4(S+2)}{S[(S+2)^2+1]}$$

Let sampling duration t = 0.3 seconds and sampling period T = 3 seconds. For rise time = 1.5 ms and fall time = 150 seconds. Write a SCEPTRE program to plot the output response for 9 seconds. Assume a ramp input. Draw the sample and hold circuit and variable resistor timing diagram. Take C = 500 μF.

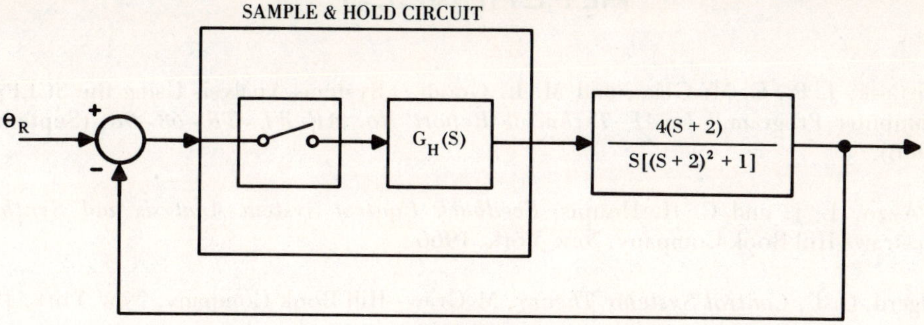

Fig. p5.12

REFERENCES

1. Portasik, J. P., C. M. Glass, and M. L. Goode, "Systems Analysis Using the SCEPTRE Computer Program," *USAF Technical Report No. AF WL–TR–68–98*, (September, 1968).

2. D'Azzo, J. J. and C. H. Houpis, *Feedback Control System Analysis and Synthesis,* McGraw–Hill Book Company, New York, 1966.

3. Elgerd, O. I., *Control Systems Theory*, McGraw–Hill Book Company, New York, 1967.

6

ANALYSIS OF OTHER NON-ELECTRICAL SYSTEMS

6.1 INTRODUCTION

As seen in Chapter 5, the SCEPTRE program is quite versatile. This chapter illustrates two separate methods for applying SCEPTRE to still other non-electrical systems, but in this case, those which do not lend themselves to the block diagram form, as did the systems of Chapter 5.

One method for solving non-electrical problems, with a program which was originally designed for electronic networks, is the rather obvious procedure of converting the non-electrical system to its electrical analogy. Once the conversion is complete, the SCEPTRE program is then written for the electrical equivalent in the normal fashion.

The second method involves deriving the differential equations and then using the DEFINED PARAMETER mode in the manner illustrated in Sec. 5.2.2. Four widely varying examples are now developed in detail, each using both of the methods mentioned as a basis for comparison.

6.2 EXAMPLE: MECHANICAL SYSTEM [1]

Figure 6.1 shows a mechanical system with two masses. The electrical analog of the mechanical system is shown in Fig. 6.2.

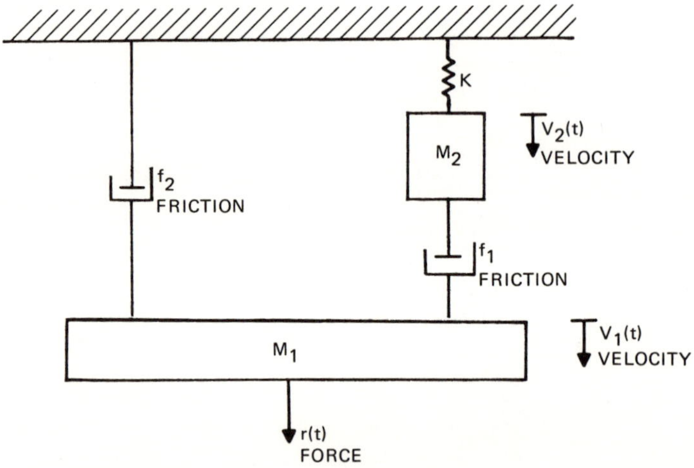

Fig. 6.1 Mechanical system

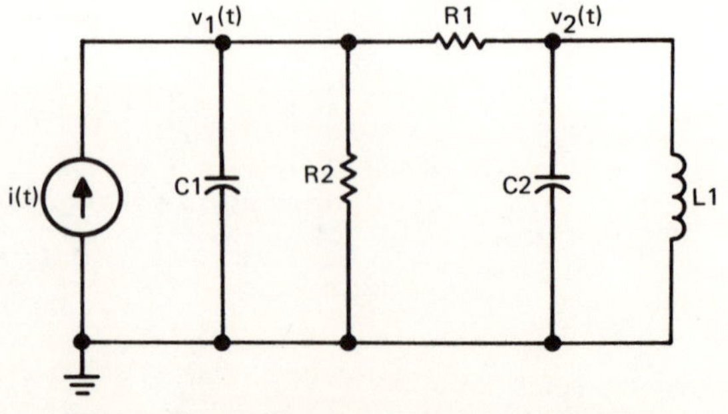

Fig. 6.2 Electrical analog of mechanical system

Table 6.1 shows the analogous quantities and the values taken for this example.

Table 6.1 Mechanical–electrical analogies

Mechanical Quantity	Electrical Analog	Value
MASS M1	C_1	1 lb.
MASS M2	C_2	1 lb.
SPRING CONST. K	1/L	10 lbs/in.
FRICTION COEF. f_1	$1/R_1$	.001
FRICTION COEF. f_2	$1/R_2$	.001
FORCE r(t)	i (t)	5 lbs.
VELOCITY $V_1(t)$, $V_2(t)$	$v_1(t)$, $v_2(t)$	TO BE FOUND

6.2.1 Defined Parameter Solution

Summing the vertical forces in Fig. 6.1 yields,

$$\frac{M_1 dV_1}{dt} + (f_1 + f_2)V_1 - f_1 V_2 = r(t) \tag{6.1}$$

and,

$$\frac{M_2 dV_2}{dt} + f_1 (V_2 - V_1) + \int K V_2 dt = 0. \tag{6.2}$$

Differentiating Eq. (6.2) with respect to time yields,

$$\frac{M_2 d^2 V_2}{dt^2} + f_1 \left(\frac{dV_2}{dt} - \frac{dV_1}{dt}\right) + K V_2 = 0. \tag{6.3}$$

Let,

$$V_1 = X_1,$$

$$V_2 = X_2,$$

and,

$$\frac{dV_2}{dt} = X_3 .$$

Substituting these values in Eqs. (6.1) and (6.3) yields,

$$M_1 \dot{X}_1 + (f_1 + f_2)X_1 - f_1 X_2 = r \tag{6.4}$$

and,

$$M_2 \dot{X}_3 + f_1 (X_3 - \dot{X}_1) + KX_2 = 0. \tag{6.5}$$

Substituting the value of X_1 in the above equation from Eq. (6.4) yields,

$$M_2 \dot{X}_3 + f_1 X_3 - \frac{f_1}{M_1} [r + f_1 X_2 - (f_1 + f_2)X_1] + KX_2 = 0. \tag{6.6}$$

Rearranging Eqs. (6.4) and (6.5), the following three first order differential equations are obtained:

$$\dot{X}_1 = -(\frac{f_1 + f_2}{M_1})X_1 + \frac{f_1}{M_1}X_2 + \frac{r}{M_1} \tag{6.7}$$

$$\dot{X}_2 = X_3 \tag{6.8}$$

$$\dot{X}_3 = -f_1 (\frac{f_1 + f_2}{M_1 M_2}) X_1 + \frac{1}{M_2} [\frac{f_1{}^2}{M_1} - K] X_2 - \frac{f_1}{M_2} X_3 + \frac{f_1 r}{M_1 M_2} \tag{6.9}$$

Substituting in for the values M_1, M_2, f_1, f_2, K and r given in Table 6.1 yields,

$$\dot{X}_1 = -.002X_1 + .001X_2 + 5 \tag{6.10}$$

$$\dot{X}_2 = X_3 \tag{6.11}$$

$$\dot{X}_3 = -2 \times 10^{-6} X_1 - (10 - 10^{-6})X_2 - .001X_3 + .005 \tag{6.12}$$

Following the example given in Sec. 5.2.2, an acceptable SCEPTRE program is as follows:

```
CIRCUIT DESCRIPTION
EXAMPLE 6.2.1
DEFINED PARAMETERS
DPX1 = X1(-.002*PX1 + .001*PX2 + 5.)
DPX2 = X2(PX3)
DPX3 = X3(-2.E-6*PX1-(10.-10.E-6)*PX2-.001*PX3 + .005)
PX1 = 0
PX2 = 0
PX3 = 0
OUTPUTS
PX1 (V1), PLOT
PX2 (V2), PLOT
PX3 (A2), PLOT
XSTPSZ
RUN CONTROLS
STOP TIME = 100
END
```

6.2.2. Electrical Analogy Solution

The problem is now solved with the electrical analogy using the same values of coefficients.

$C_1 = C_2 = 1$	Farad
$L = .1$	Henry
$R_1 = R_2 = 1000$	Ohms
$i(t) = 5$	Amps

Figure 6.3 is the electrical equivalent circuit in SCEPTRE form. The program shows the outputs VC1, VC2, and PA2 correspond to the mechanical quantities $V_1(t)$, $V_2(t)$, and $\frac{dV_2(t)}{dt}$.

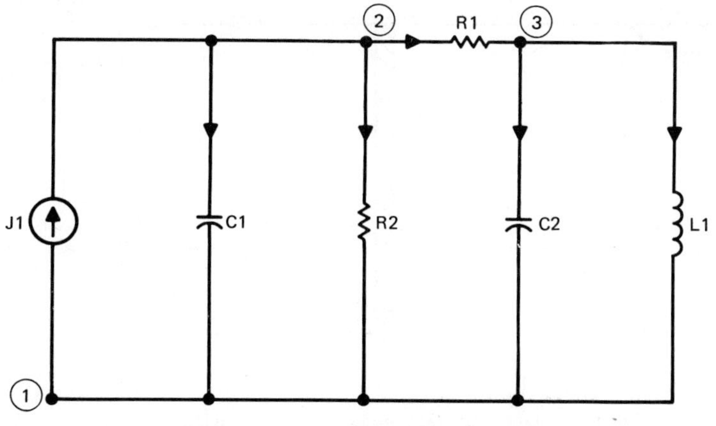

Fig. 6.3 Circuit in SCEPTRE form

The SCEPTRE program is now easily written directly from the electrical equivalent circuit shown in Fig. 6.3.

```
CIRCUIT DESCRIPTION
EXAMPLE 6.2.2
ELEMENTS
J1, 1-2 = 5.0
C1, 2-1 = 1.0
C2, 3-1 = 1.0
R1, 2-3 = 1000
R2, 2-1 = 1000
L1, 3-1 = 0.1
DEFINED PARAMETERS
PA2 = X1(IC2/C2)     [or, PA2 = X1(DC2)]
OUTPUTS
VC1 (V1), PLOT
VC2 (V2), PLOT
PA2 (A2), PLOT
RUN CONTROLS
STOP TIME = 100
END
```

The two preceding versions were run on SCEPTRE and it was observed that the system oscillated with a very small damping factor. This result is not unexpected.

6.3 EXAMPLE: THERMAL SYSTEM [2]

Figure 6.4 shows a heat conduction model for a rocket engine wall. The temperature inside the chamber, Ti, and the temperature outside the chamber To, are known. It is desired to know the temperature, Tb, of the engine wall and temperature, Tc, of the outer insulating layer, as a function of time. Figure 6.5 shows the analogous electrical system.

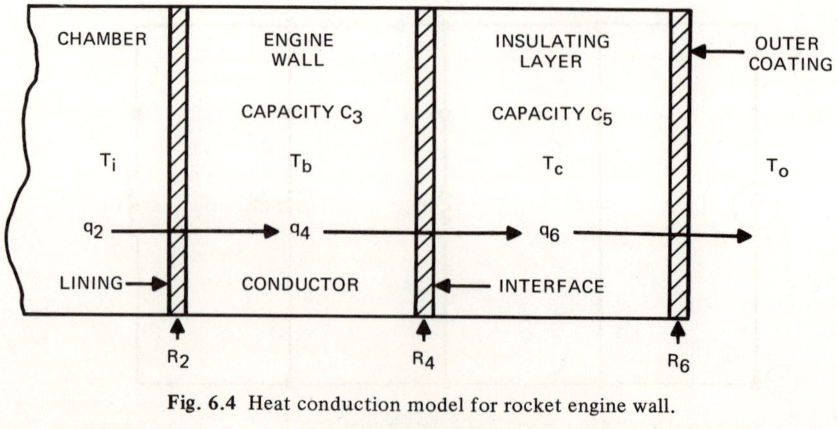

Fig. 6.4 Heat conduction model for rocket engine wall.

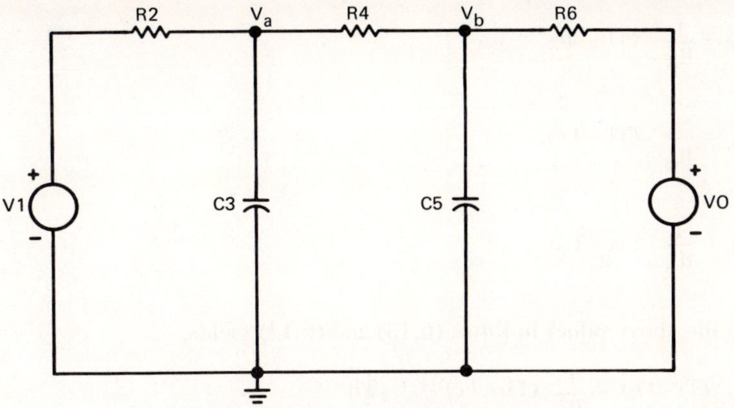

Fig. 6.5 Electrical analog of thermal system

Table 6.2 shows the analogous quantities in thermal and electrical notation and their values taken for this particular problem.

Table 6.2. Thermal–electrical analogies

THERMAL QUANTITY	ELECTRICAL ANALOG	VALUE
Temperature Ti	V_1	3000
Temperature To	V_o	80
Thermal Resistance R_2	R_2	10^6
Thermal Resistance R_4	R_4	10^6
Thermal Resistance R_6	R_6	10^6
Thermal Capacity C_3	C_3	10^{-6}
Thermal Capacity C_5	C_5	10^{-6}

6.3.1. Defined Parameter Solution

From Fig. 6.4. using the first law of thermo dynamics with work equal to zero,

$$q_2 - q_4 = \frac{C_3 dT_b}{dt} \tag{6.13}$$

and,

$$q_4 - q_6 = \frac{C_5 dT_c}{dt} \tag{6.14}$$

where "q" is the heat flow rate (electrical analog is current).

But,

$$q_2 = \frac{1}{R_2} (Ti-Tb), \tag{6.15}$$

$$q_4 = \frac{1}{R_4} (Tb-Tc), \tag{6.16}$$

and,

$$q_6 = \frac{1}{R_6} (Tc-To) . \tag{6.17}$$

Substituting the above values in Equs. (6.13) and (6.14) yields,

$$\frac{1}{R_2} (Ti-Tb) - \frac{1}{R_4} (Tb-Tc) = C_3 \dot{T}b \tag{6.18}$$

and,

$$\frac{1}{R_4} (Tb-Tc) - \frac{1}{R_6} (Tc-To) = C_5 \dot{T}c . \tag{6.19}$$

Let, $Tb = X_1$ and, $Tc = X_2$

Then,

$$\dot{X}_1 = \frac{-1}{C_3}\left(\frac{1}{R_2} + \frac{1}{R_4}\right) X_1 + \frac{1}{R_4 C_3} X_2 + \frac{1}{R_2 C_3} Ti \tag{6.20}$$

and,

$$\dot{X}_2 = \frac{1}{R_4 C_5} X_1 - \frac{1}{C_5}\left(\frac{1}{R_4} + \frac{1}{R_6}\right) X_2 + \frac{1}{C_5 R_6} To . \tag{6.21}$$

Substituting the values of R's and C's in the above equations from Table 6.2 yields,

$$\dot{X}_1 = -2X_1 + X_2 + 3000 \tag{6.22}$$

and,

$$\dot{X}_2 = X_1 - 2X_2 + 80 . \tag{6.23}$$

The SCEPTRE program for these equations is as follows:

```
CIRCUIT DESCRIPTION
EXAMPLE 6.3.1
DEFINED PARAMETERS
DPX1 = X1(-2.*PX1 + PX2 + 3000.)
DPX2 = X2(PX1-2.*PX2 + 80.)
PX1 = 0
PX2 = 0
OUTPUTS
PX1(TB), PLOT
PX2(TC), PLOT
```

```
XSTPSZ
RUN CONTROLS
STOP TIME = 5
END
```

6.3.2 Electrical Analogy Solution

The problem is now solved again using the electrical analogy using the same values of coefficients.

$C_3 = C_5 = 10^{-6}$ Farad

$R_2 = R_4 = R_6 = 10^6$ Ohms

$V_1 = 3000$ Volts

$V_0 = 80$ Volts

Figure 6.6 is the electrical equivalent circuit in SCEPTRE form.

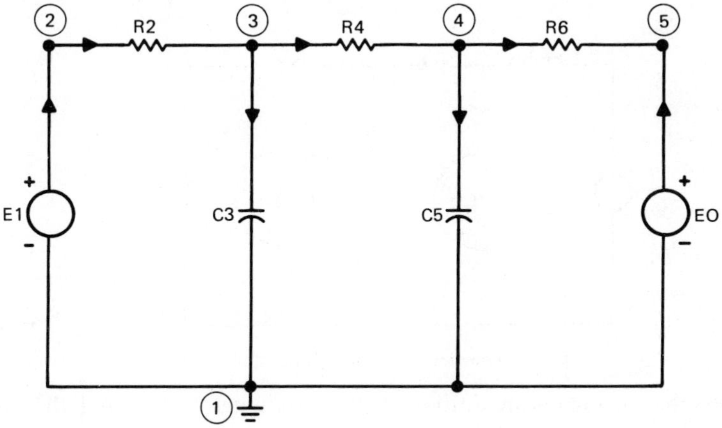

Fig. 6.6 Circuit in SCEPTRE form

The SCEPTRE program format is:

```
CIRCUIT DESCRIPTION
EXAMPLE 6.3.2
E1, 1-2 = 3000
EO, 1-5 = 80
C3, 3-1 = 1.E-6
C5, 4-1 = 1.E-6
R2, 2-3 = 1.E+6
R4, 3-4 = 1.E+6
R6, 4-5 = 1.E+6
OUTPUTS
VC3 (VB), PLOT
```

```
VC5 (VC), PLOT
RUN CONTROLS
STOP TIME = 5
END
```

6.4 EXAMPLE: ROTATIONAL SYSTEM [3]

Figure 6.7 shows a rotational system with two degrees of freedom. Two bearing frictions are shown, and a driving torque is applied to the shaft. The electrical analog of this rotational system is shown in Fig. 6.8. It is desired to find angular velocities ω_1 and ω_2 as a function of time and the torque on the spring, Ts, as a function of time.

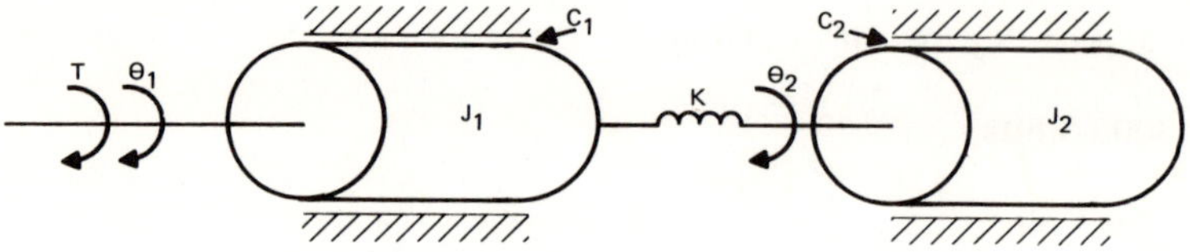

Fig. 6.7 A rotating system with two degrees of freedom

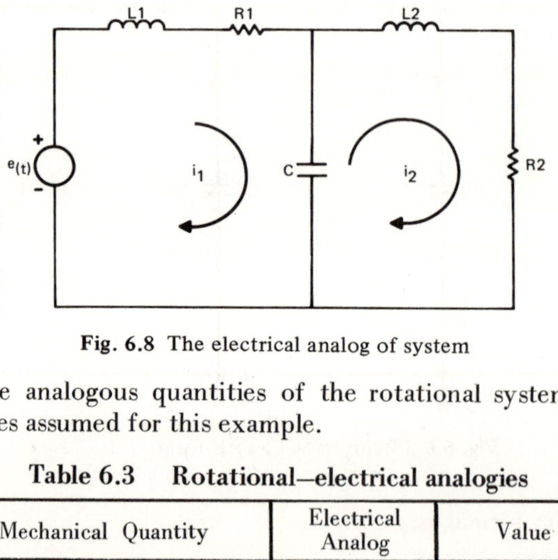

Fig. 6.8 The electrical analog of system

Table 6.3 shows the analogous quantities of the rotational system and the electrical system with the values assumed for this example.

Table 6.3 Rotational–electrical analogies

Mechanical Quantity	Electrical Analog	Value
Polar Moment of Inertia J_1	L_1	10^{-3} slug–ft.^{2}
Polar Moment of Inertia J_2	L_2	10^{-3} slug–ft.^{2}
Angular Velocity ω_1	i_1	To be found
Angular Velocity ω_2	i_2	To be found
Viscous Friction C_1	R_1	.1
Viscous Friction C_2	R_2	.1
Spring Constant K	$1/c$	10
Driving Torque T(t)	e(t)	10 lb–ft.
Torque on Spring Ts(t)	Vc(t)	To be found

6.4.1 Defined Parameter Solution

The basic relationship, comparable to Newton's law, for rotation through a fixed axis is,

$$\frac{Jd\omega}{dt} = \pm \text{ [sum of the torques acting on a body about a fixed axis]}$$

Using this relation with Fig. 6.7 yields,

$$\frac{J_1 d\omega_1}{dt} = T - C_1\omega_1 - K\int \omega_1(t)dt + K\int \omega_2(t)dt \tag{6.24}$$

and,

$$\frac{J_2 d\omega_2}{dt} = -C_2\omega_2 - K\int \omega_2(t)dt + K\int \omega_1(t)dt \tag{6.25}$$

But torque on the spring is,

$$Ts = K\int [\omega_1(t) - \omega_2(t)]dt \tag{6.26}$$

Therefore,

$$\frac{dTs}{dt} = K(\omega_1 - \omega_2) \; . \tag{6.27}$$

Using the relation of Eq. (6.26) and rearranging Eqs. (6.24) and (6.25) yields,

$$\frac{d\omega_1}{dt} = \frac{-C_1}{J_1}\omega_1 - \frac{Ts}{J_1} + \frac{T}{J_1} \; . \tag{6.28}$$

and,

$$\frac{d\omega_2}{dt} = \frac{-C_2}{J_2}\omega_2 + \frac{Ts}{J_2} \tag{6.29}$$

Let, $\omega_1 = X_1$,

$\omega_2 = X_2$

and,

$$Ts = X_3 \tag{6.30}$$

Using above relations in Eqs. (6.27), (6.28), and (6.29), yields,

$$\dot{X}_1 = \frac{-C_1}{J_1}X_1 - \frac{X_3}{J_1} + \frac{T}{J_1} \tag{6.31}$$

$$\dot{X}_2 = \frac{-C_2}{J_2}X_2 + \frac{X_3}{J_2} \tag{6.32}$$

$$\dot{X}_3 = KX_1 - KX_2 \tag{6.33}$$

Substituting the values of the coefficients from Table 6.3 gives the final form,

$$\dot{X}_1 = -\ 100X_1\ -1000X_3 + 10^4 \tag{6.34}$$

$$\dot{X}_2 = -100X_2 + 1000X_3 \tag{6.35}$$

and,

$$\dot{X}_3 = 10X_1 - 10X_2\ . \tag{6.36}$$

The SCEPTRE program is written as before,

```
CIRCUIT DESCRIPTION
EXAMPLE 6.4.1
DEFINED PARAMETERS
DPX1  =  X1(-100.*PX1-1000.*PX3+1.E4)
DPX2  =  X2(-100.*PX2+1000.*PX3)
DPX3  =  X3(10.*PX1-10.*PX2)
PX1  = 0
PX2  = 0
PX3  = 0
OUTPUTS
PX1(W1), PLOT
PX2(W2), PLOT
PX3(TS), PLOT
XSTPSZ
RUN CONTROLS
STOP TIME  =  .1
END
```

6.4.2 Electrical Analogy Solution

In the previous examples, the electrical system is solved directly from the circuit, using the SCEPTRE listing. For this example only, let us also derive the electrical equations from the circuit in the state variable form, to compare them directly with the Rotational system state equations. Obviously this step is not necessary, and is only done here for academic interest.

From Fig. 6.10, writing,

$$e = L_1 \frac{di_1}{dt} + R_1 i_1 + Vc\ , \tag{6.37}$$

$$0 = L_2 \frac{di_2}{dt} + R_2 i_2 - Vc\ , \tag{6.38}$$

and,
$$C \frac{dVc}{dt} = i_1 - i_2\ , \tag{6.39}$$

Let,
$$i_{L1} = X_1\ ,$$

$$i_{L2} = X_2\ , \tag{6.40}$$

and, $V_c = X_3$.

Using the relations of Eq. (6.40) and rearranging Eqs. (6.37), (6.38), and (6.39) yields,

$$\dot{X}_1 = -\frac{R_1}{L_1} X_1 - \frac{X_3}{L_1} + \frac{e}{L_1} \tag{6.41}$$

$$\dot{X}_2 = -\frac{R_2}{L_2} X_2 + \frac{X_3}{L_2} \tag{6.42}$$

$$\dot{X}_3 = \frac{X_1}{C} - \frac{X_2}{C} \tag{6.43}$$

It is observed that Eqs. (6.41), (6.42), and (6.43) of the electrical system are identical to Eqs. (6.31), (6.32), and (6.33), respectively, of the rotational system.

The reader must remember that the state equations compare identically only because the state variables picked for each domain were directly analogous. However, the user need not be concerned with this factor since using the SCEPTRE program negates the need for writing any electrical equations.

Figure 6.9 shows the SCEPTRE form of the analogous electrical system.

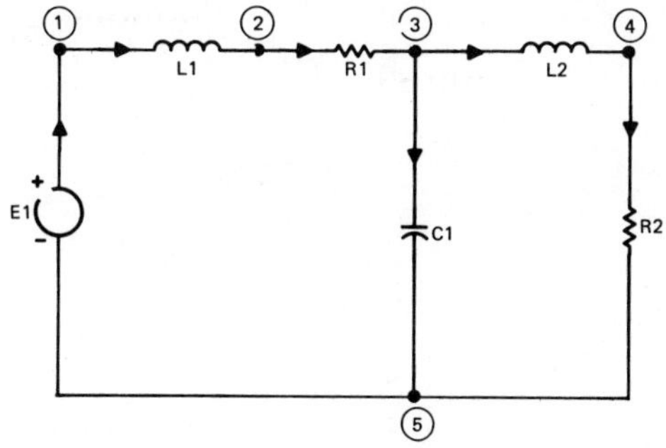

Fig. 6.9 Circuit in SCEPTRE form

From the electrical circuit diagram, the SCEPTRE format is written as follows:

```
CIRCUIT DESCRIPTION
EXAMPLE 6.4.2
ELEMENTS
E1,5-1 = 10
C1, 3-5 = .1
R1, 2-3 = .1
R2, 4-5 = .1
L1, 1-2 = 1.E-3
L2, 3-4 = 1.E-3
OUTPUTS
IR1(I1), PLOT
IR2(I2), PLOT
VC1(VC), PLOT
RUN CONTROLS
STOP TIME = .1
END
```

To illustrate that the results are indeed identical, two curves showing actual results are shown in Figs. 6.10 and 6.11. Figure 6.10 indicates the response of the torque on the spring (Ts) as a function of time. This is one of the outputs requested in EXAMPLE 6.4.1. The response compares identically with Fig. 6.11 which outputs the voltage across the capacitor (Vc) as a function of time.

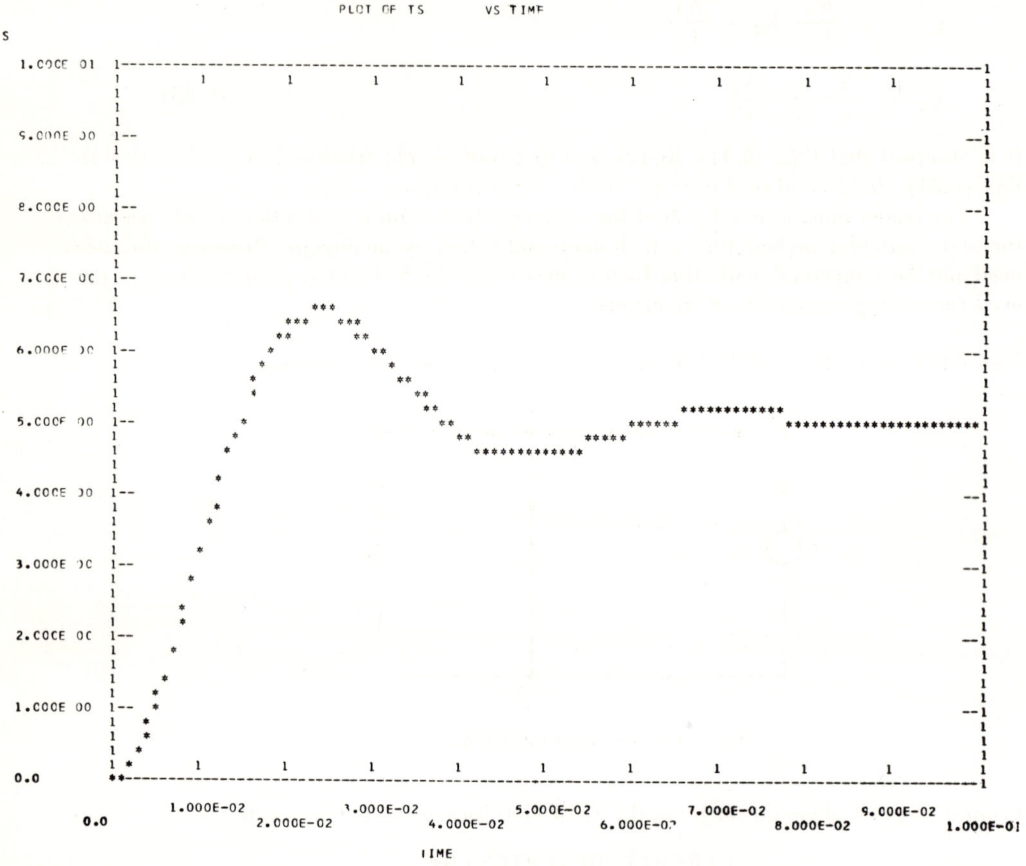

Fig. 6.10

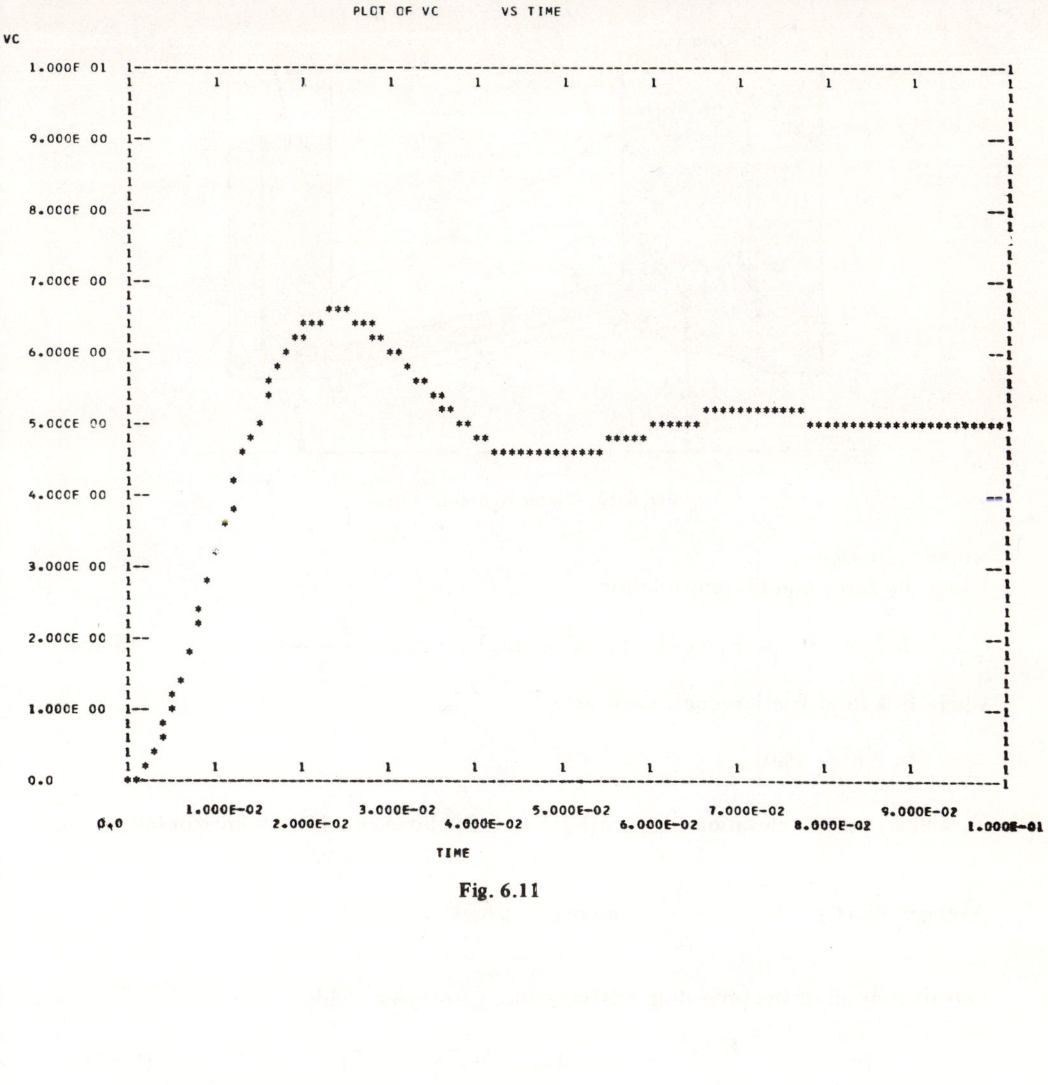

Fig. 6.11

6.5 EXAMPLE: HYDRAULIC SYSTEM [2]

The system shown in Fig. 6.12 represents two large hydraulic reservoirs (part of a residential water system for a town) connected by a long uniform line. It is desired to determine the equations of motion of the system, in order to determine what the time history of the heights of the two reservoirs are for any given time history of the usage rate, $W_2(t)$.

Considering each reservoir as an open system,

For reservoir 1: $\rho A_1 \dot{h}_1 = -W_3$ (6.44)

For reservoir 2: $\rho A_2 \dot{h}_2 = W_3 - W_2$, (6.45)

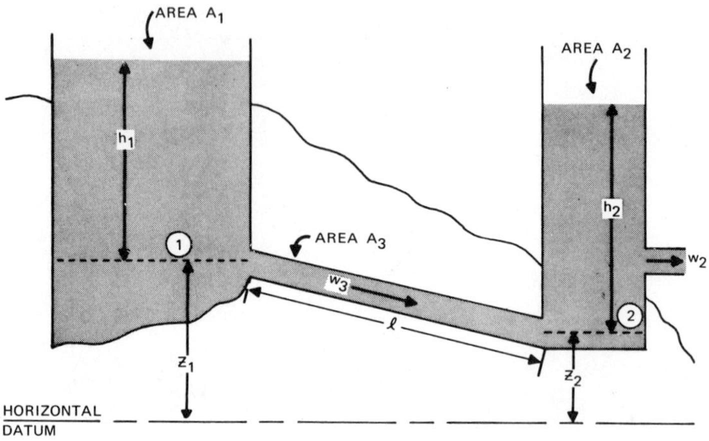

Fig. 6.12 A large hydraulic system

where ρ = Density.
Using the force equilibrium relations,

$$\Sigma f = 0 = P_1 A_3 - P_2 A_3 - bV_3 - m_3 \dot{V}_3 + m_3 g \left(\frac{Z_1 - Z_2}{\ell}\right) \quad , \quad (6.46)$$

where b is fluid friction coefficient and,

$$P_1 = \rho g h_1, \qquad P_2 = \rho g h_2 \quad .$$

Z_1 and Z_2 are the elevations of points (1) and (2) above an arbitrary horizontal datum.

Average velocity $V_3 = \dfrac{W_3}{\rho A_3}$ and $m_3 = \rho A_3 \ell$.

Substituting all of the preceding relations and rearranging yields,

$$-(h_1 - h_2) + \frac{\ell}{g\rho A_3}\dot{W}_3 + \frac{b}{g(\rho A_3)^2} W_3 = Z_1 - Z_2 \ . \quad (6.47)$$

Inertial force $\ell\dot{\omega}_3$ is negligible, so solving for W_3 from Eq. (6.47) using Eq. (6.44) and (6.45) yields,

$$\rho A_1 \dot{h}_1 + K[(h_1 + Z_1) - (h_2 + Z_2)] = 0 \quad (6.48)$$

$$\rho A_2 \dot{h}_2 - K[(h_1 + Z_1) - (h_2 + Z_2)] = -W_2 \quad (6.49)$$

where,

$$K = \frac{g(\rho A_3)^2}{b} \quad .$$

These equations are of the form

$$\frac{C_1 dV_1}{dt} + \frac{V_1 - V_2}{R} + i_1 = 0 \qquad (6.50)$$

and,

$$\frac{C_2 dV_2}{dt} - \frac{(V_1 - V_2)}{R} - i_1 = -i_2 \qquad (6.51)$$

Figure 6.13 shows the analogous electrical system. Also, Table 6.4 shows the analogous quantities with the values taken for this example.

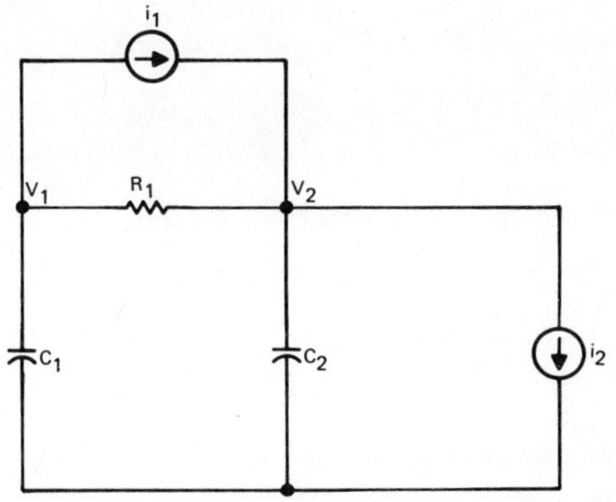

Fig. 6.13 The electrical analogy of the hydraulic system

Hydraulic Quantity	Electrical Analog	Electrical Values
ρA_1	C_1	2×10^6 Farads
ρA_2	C_2	2×10^5 Farads
$1/K$	R_1	$.78125 \times 10^{-9}\ \Omega$
W_2	I_2	10^4 amps
$K(Z_1 - Z_2)$	I_1	256×10^7

where: $\rho = 2$ lb. Sec^2/Ft^4

$b = .1$ (friction coefficient)

$A_1 = 10^6$ SqFt.

$A_2 = 10^5$ SqFt.

$A_3 = 10^3$ SqFt.

$$K = \frac{\rho^2 g A_3}{b} = 128 \times 10^7$$

$W_2 = 10^4$ Ft/Sec

$\dot{W}_3 \simeq 0$ (since area is so large)

Table 6.4. Hydraulic—electrical analogies

6.5.1. Defined Parameter Solution

Rearranging Eqs. (6.48) and (6.49) yields,

$$\dot{h}_1 = -\frac{K}{\rho A_1} h_1 + \frac{K}{\rho A_1} h_2 + \frac{K}{\rho A_1} [Z_1 - Z_2] \tag{6.52}$$

and,

$$\dot{h}_2 = \frac{K}{\rho A_2} h_1 - \frac{K}{\rho A_2} h_2 + \frac{K}{\rho A_2} (Z_1 - Z_2) - \frac{W_2}{\rho A_2} . \tag{6.53}$$

Let,

$Z_1 = 5$ and $Z_2 = 3$,

where Z_1 and Z_2 are the levels above the arbitrary datum,

and let,

$h_{1(o)} = 10 = h_{2(o)}$,

the initial conditions for the heights shown in Fig. 6.12.

Substituting above values and those from Table 6.4 yields,

$$\dot{h}_1 = -640\, h_1 + 640\, h_2 - 1280 \tag{6.54}$$

and,

$$\dot{h}_2 = 6400\, h_1 - 6400\, h_2 + 12800 - .05 \tag{6.55}$$

The SCEPTRE program for the solution of Eqs. (6.54) and (6.55) is written as follows:

```
CIRCUIT DESCRIPTION
EXAMPLE 6.5.1
DEFINED PARAMETERS
DPH1 = X1(-640.*PH1+640.*PH2-1280.)
DPH2 = X2(+6400.*PH1-6400.*PH2+12800.-.05)
PH1 = 10
PH2 = 10
OUTPUTS
PH1 (H1), PLOT
PH2 (H2), PLOT
XSTPSZ
RUN CONTROLS
STOP TIME = 10
END
```

6.5.2. Electrical Analogy Solution

Figure 6.14 shows the equivalent circuit in SCEPTRE form and the program follows readily.

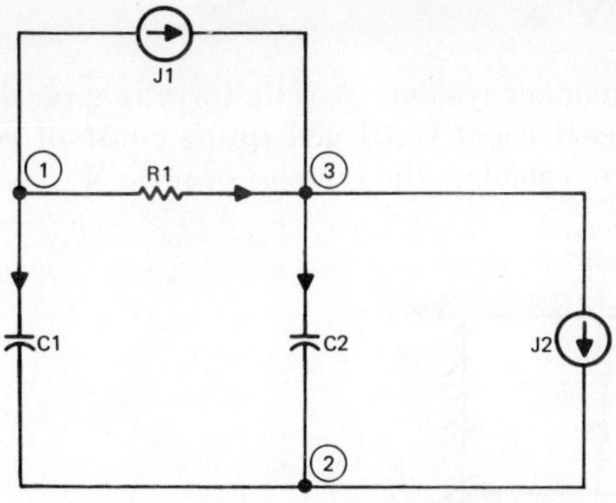

Fig. 6.14 Circuit in SCEPTRE form

```
CIRCUIT DESCRIPTION
EXAMPLE 6.5.2
ELEMENTS
J1, 1-3 = 256.E7
J2, 3-2 = 10000
C1, 1-2 = 2.E6
C2, 3-2 = 2.E5
R1, 1-3 = .78125E-9
OUTPUTS
VC1(H1), PLOT
VC2(H2), PLOT
INITIAL CONDITIONS
VC1 = 10
VC2 = 10
RUN CONTROLS
STOP TIME = 10
END
```

The outputs are to be relabeled to indicate H1 and H2. This provides the user with plots that are directly related to the problem at hand. Notice that absolute units (amperes, volts, seconds, farads, etc.) are used in this problem as indeed they are for all problems in this Chapter. This is a natural consequence of the fact that the physical systems involved all have response times on the order of seconds and, therefore, it is most convenient to use seconds as the basic unit of time.

From an academic point of view, the reader should realize that this particular problem has no steady state solution. Unlike most transient problems that feature a constant forcing function, the state variables of the differential equations (H1 and H2) and those of the electrical analog (VC1 and VC2) continue to vary significantly as long as the problem is allowed to run. If the physical constraints of the system are applied to the solution, however, the problem as it is represented here should be terminated if and when $H1 \leq 0$ (See Problem 6.7).

6.6 SUMMARY

The four preceding examples show how SCEPTRE is used to solve non-electrical problems. Which of the two methods is more convenient depends upon the situation. Sometimes the differential equations are readily available and the DEFINED PARAMETER mode would be more useful; however, in other cases the equations are quite involved, but the electrical analogy is readily attainable.* User experience is, at any rate, the most valuable tool in determining the approach to be followed. It would seem fair to say, however, that both of the foregoing methods are more convenient than having to develop a "new" Fortran program for each and every problem that comes along.

*Many books and handbooks contain charts indicating the electrical analogues for various non-electrical components. Such books include:

Dynamics of Physical Systems, R. H. Cannon, Jr., McGraw-Hill, 1967.

Electromechanical System Theory, H. E. Koenig and W. A. Blackwell, McGraw-Hill, 1961.

Introduction to System Dynamics, J. L. Shearer, A. T. Murphy, H. H. Richardson, Addison-Wesley, 1967.

PROBLEMS

6.1 Figure p6.1 shows a simple spring-mass-damper system. A 5 lb. force is applied to a mass weighing 10 lbs. Friction coefficient is .01 and spring constant is 50 lbs./in.. Write a SCEPTRE program to calculate the velocity of mass M.

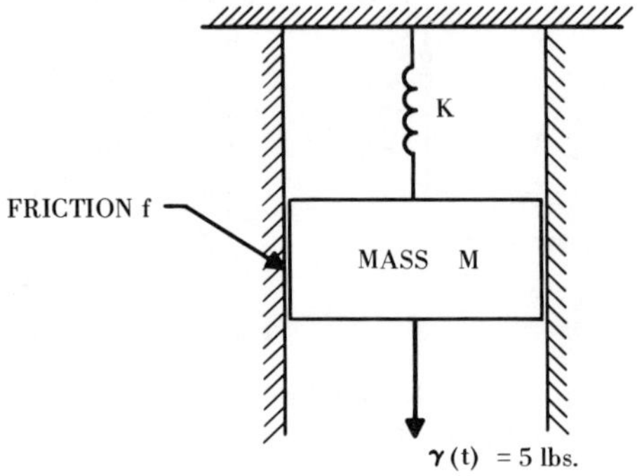

Fig. p6.1

6.2 Figure p6.2 shows a rotational system. Two bearing frictions are shown and a driving torque is applied to the shaft. Following are the mechanical quantities and their values.

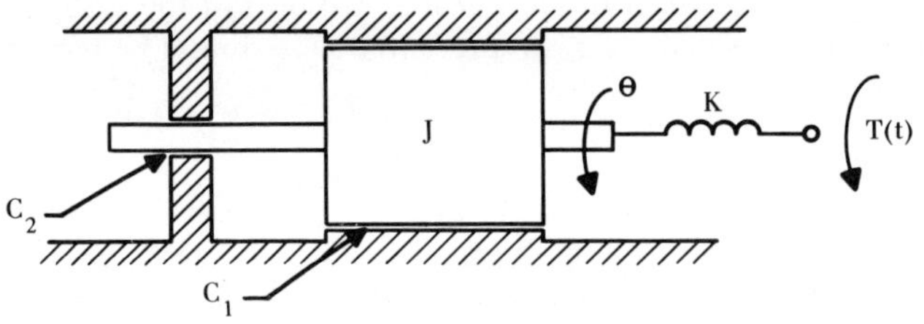

Fig. p6.2

J = Polar moment of inertia = 10^{-3} slug-ft ^{2}.

$C_1 = C_2$ = Viscous friction = 0.2

K = Spring Constant = 10

$T(t)$ = Driving torque = 20 lb.-ft.

Draw an analogous electrical circuit and write a SCEPTRE program to find the angular velocity ω.

6.3 Figure p6.3 shows a complex mechanical system with 3 masses.

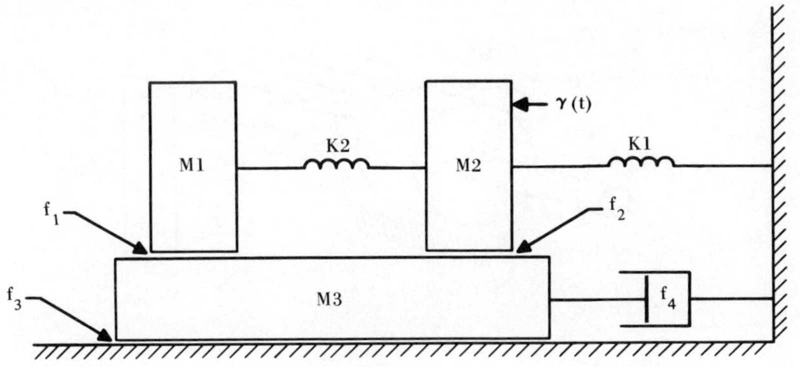

Fig. p6.3

a. Draw an analogous electrical circuit.
b. Make a table of mechanical-electrical analogies.
c. Using electrical circuit write a SCEPTRE program to find the velocities V_1, V_2, and V_3 of masses M1, M2 and M3.

Mechanical quantities are as follows:

M1 = .5 lb.
M2 = 2 lb.
M3 = 1 lb.
Spring constant K1 = 10 lbs./1n.
Spring constant K2 = 5 lbs./1n.
Friction coefficient $f_1 = f_3 = 0.001$
Friction coefficient $f_2 = .005$
Friction coefficient $f_4 = .002$
Force $\gamma(t) = 10$ lbs.

6.4 The Fig. p6.4 shows a simple rotating mechanical system. Mechanical quantities and their values are shown below. Draw an analogous electrical circuit and write a SCEPTRE program to find the angular velocity.

M = Torque applied = 20 ft.-lbs.
J = 5 slug-ft.^{2}
K = Rotational spring constant = 10 ft.-lb./γad.
b = Friction coefficient = .005.
$\Omega = \frac{d\Theta}{dt}$ = Angular velocity.

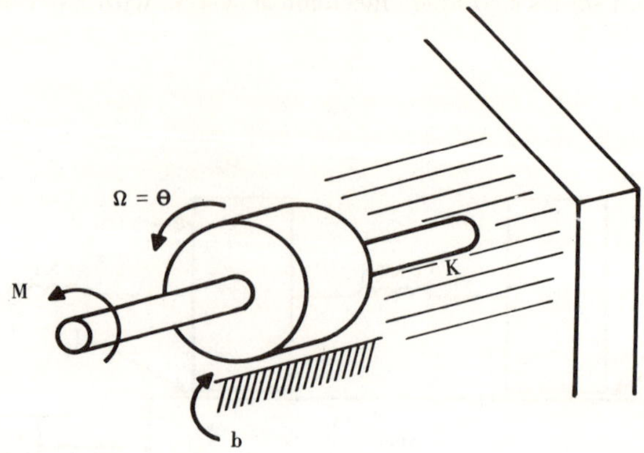

Fig. p6.4

6.5 Figure p6.5 shows an ideal torsionally vibrating mass system. Torque T(1) = 10 lb-ft is applied. Write a SCEPTRE program to find the angular velocity **ω** of the mass for the given mechanical quantities.

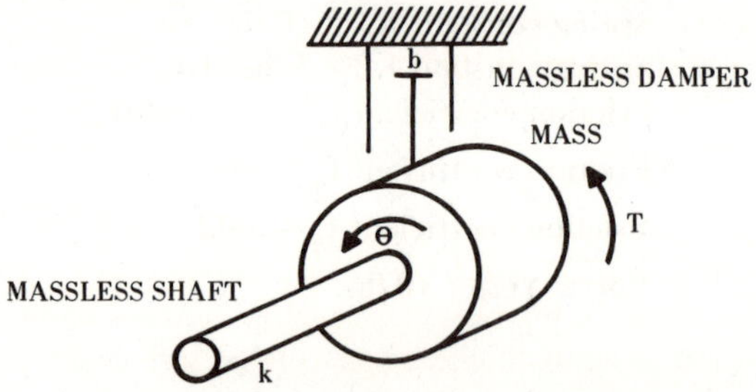

Fig. p6.5

J = Polar moment of inertia = 10^{-4} slug-ft^2
b = Viscous damping coefficient = 0.05.
k = Torsional flexibility of shaft = 10
$\omega = \frac{d\Theta}{dt}$ = Angular velocity of mass.

6.6 Figure p6.6 shows the model of a torsional system with dynamic absorbers. A torque of 15 lb-ft is applied by the engine. Mechanical quantities are

J1 = Polar moment of intertia = 10^{-3} slug-ft^2.
J2 = Polar moment of intertia = 10^{-4} slug-ft^2.
b1 = b2 = Viscous damping coefficient = 0.1.
k = Torsional flexibility of shaft = 20.

a. Draw an electrical analog of the system in SCEPTRE form.
b. Write a SCEPTRE program to find the angular velocities ω 1 and ω 2.

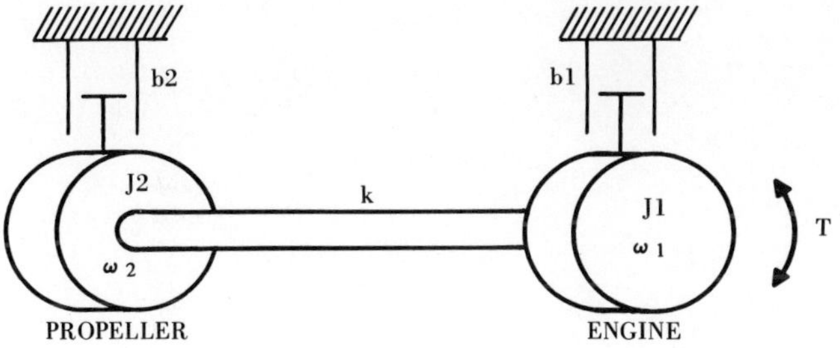

Fig. p6.6

6.7
a. Make the additions necessary to the listing in Sec. 6.5.1 that will terminate the run if H1 = 0.
b. Do the same thing to the listing in Sec. 6.5.2.

6.8 What changes must be made to the listing of Sec. 6.2.2 if it is decided to use the system of units usually used for electronic transient problems (time in nanoseconds, resistance in kilohms, capacitance in pirofarads, inductance in microhenries, current in milliamperes, and voltage in volts)?

REFERENCES

1. Dorf, R. C., *Modern Control Systems*, Addison–Wesley Publishing Company, 1967.
2. Cannon, R. H., Jr., *Dynamics of Physical Systems*, McGraw–Hill Book Company, New York, 1967.
3. Pfeiffer, P. E., *Linear Systems Analysis*, McGraw–Hill Book Company, New York, 1961.

APPENDIX A

MESH DIRECTIONS

The statement $Q = [-F^T\ U]$ and $T = [U\ F]$ is the same as stating $Q\ T^T = 0$ and $T\ Q^T = 0$. The obvious importance of this statement dictates a demand for its proof. Since $T\ Q^T = Q\ T^T$, it is sufficient to consider $Q\ T^T$ only. The (m, n)–th element of the product is the inner product of the m–th row of Q and the n–th row of T. Non-zero terms can occur only if an element is in both fundamental cut set m and fundamental circuit n. Since each row of Q contains exactly one tree branch and each row of T contains exactly one link, there must be exactly two such elements. Further, one of the elements must contribute a +1 and the other a −1. The following illustrations make this clear by showing all possible orientations of the mesh directions and the element reference directions. The sum must, therefore, be zero which completes the proof.

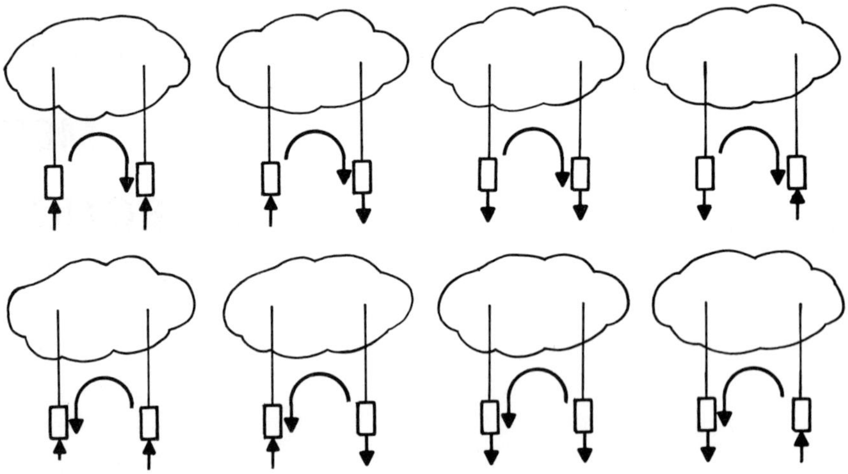

Fig. A.1 Possible Mesh and Element Reference Directions

APPENDIX B

MATRIX CONSIDERATIONS

The partition of the F matrix leads to the general submatrix F_{rs} where r is one of the link classifications (1, 2, 3, 8) and s is one of the tree branch classifications (4, 5, 6, 7). The general submatrix F_{rs} exists only if at least one element of each class is present in the network, and if at least one class r element closes a loop which contains at least one class s element. The condition $F_{15} \neq 0$ would then require at least one capacitor link (class 1) which closes a loop which contains at least one resistor tree branch (class 5). The condition $F_{16} \neq 0$ would require at least one capacitor link (class 1) which closes a loop which contains at least one inductor tree branch (class 6). The condition $F_{26} \neq 0$ would require at least one resistor link (class 2) which closes a loop which contains at least one inductor tree branch (class 6). All of these conditions are prohibited by the preference order given for the formation of trees in Sec. 1.1; i.e., capacitors have priority over both resistors and inductors, and resistors have priority over inductors.

APPENDIX C

MATRIX DERIVATIONS

The utility of the derivations in Sec. 1.6 depends to a large degree on the non-singularity of the matrices given in Eqs. (1.24), (1.29), (1.36), and (1.43). It is the purpose here to show that, subject to certain constraints, these matrices are non-singular and therefore their inverses exist.

Eq. 1.24 is repeated here for convenience,

$$M_R = R_{22} + F_{25} R_{55} F_{25}^T , \tag{1.24}$$

where R_{22} and R_{55} are diagonal matrices containing only positive, real, and finite quantities. It will be sufficient to prove that (1.24) is always positive definite, and therefore non-singular.

1. Any matrix A is positive definite, if and only if, $X^T A X > 0$ where X is any non-zero vector. Any diagonal matrix composed entirely of positive elements (R_{55} for example) must be positive definite since, if, $X = \mathrm{col}(X_1, X_2, \ldots X_n)$, and $R_{55} = \mathrm{diag}(R_1, R_2, \ldots R_n)$, then, $X^T R_{55} X = X_1^2 R_1 + X_2^2 R_2 + \ldots + X_n^2 R_n > 0$

   Therefore, R_{55} is a positive definite matrix. In the same way $X^T R_{22} X > 0$ and R_{22} is also positive definite.

2. Define $Y = F_{25}^T X \Rightarrow Y^T = X^T F_{25}$

   Then $Y^T R_{55} Y = Y_1^2 R_1 + Y_2^2 R_2 + \ldots + Y_n^2 R_n \geq 0$

   So $Y^T R_{55} Y = X^T F_{25} R_{55} F_{25}^T X \geq 0$

   Therefore, $F_{25} R_{55} F_{25}^T$ is at least positive semidefinite

3. Since $M_R = R_{22} + F_{25} R_{55} F_{25}^T$, $X^T M_R X = X^T [R_{22} + F_{25} R_{55} F_{25}^T] X =$

   $X^T R_{22} X + X^T F_{25} R_{55} F_{25}^T X \geq X^T R_{22} X > 0$

   Therefore M_R is positive definite and its inverse exists.

   The same proof applies to M_G(1.29) and M_C(1.36), but M_L(1.43) is another matter. Eq. (1.43) is repeated for convenience.

$$M_L = L_{33} + L_{36} F_{36}^T + F_{36} L_{63} + F_{36} L_{66} F_{36}^T \tag{1.43}$$

Since neither L_{33} nor L_{66} are necessarily diagonal, the above proof does not apply. In fact, it can easily be shown that M_L is singular when unity coupling is used between inductors.

APPENDIX D

DIODE REPRESENTATION IN THE INITIAL-CONDITIONS PROGRAM

The current in a diode or transistor junction at a point on its voltage/current (V/I) characteristic may be separated into two components as $J = G_J V_J + Q$. Consider the typical diode curve below.

Angle α_1 = angle α_2 is

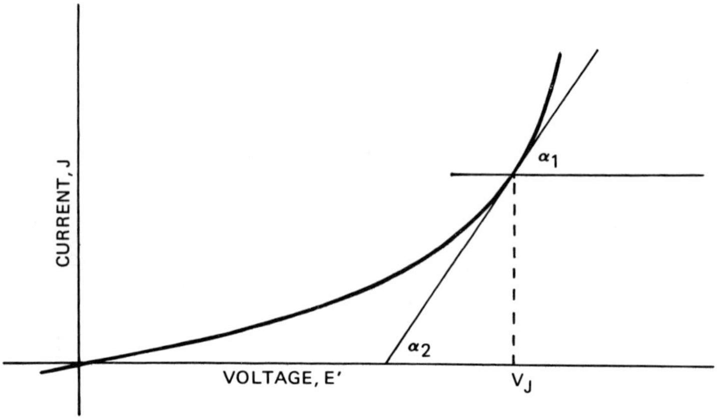

Fig. D. 1

enclosed by the slope of the diode characteristic at any point and the horizontal at that point. E′ is an offset voltage that marks the intersection of a continuation of the slope G_J and the line $J = 0$.

From the figure:

$$\tan \alpha_1 = \tan \alpha_2 = G_J = \frac{J}{V_J - E'}$$

or $$J = G_J V_J - G_J E'$$

or $$J = G_J V_J + Q \text{ if } Q \text{ is defined as } - G_J E'$$

APPENDIX E

TEST
JIGS
PHOTOGRAPHS

Figures E.1 through E.4 illustrate the types of test jigs which are used for the measurements discussed in Chapter Three.

Fig. E.1 Transistor socket

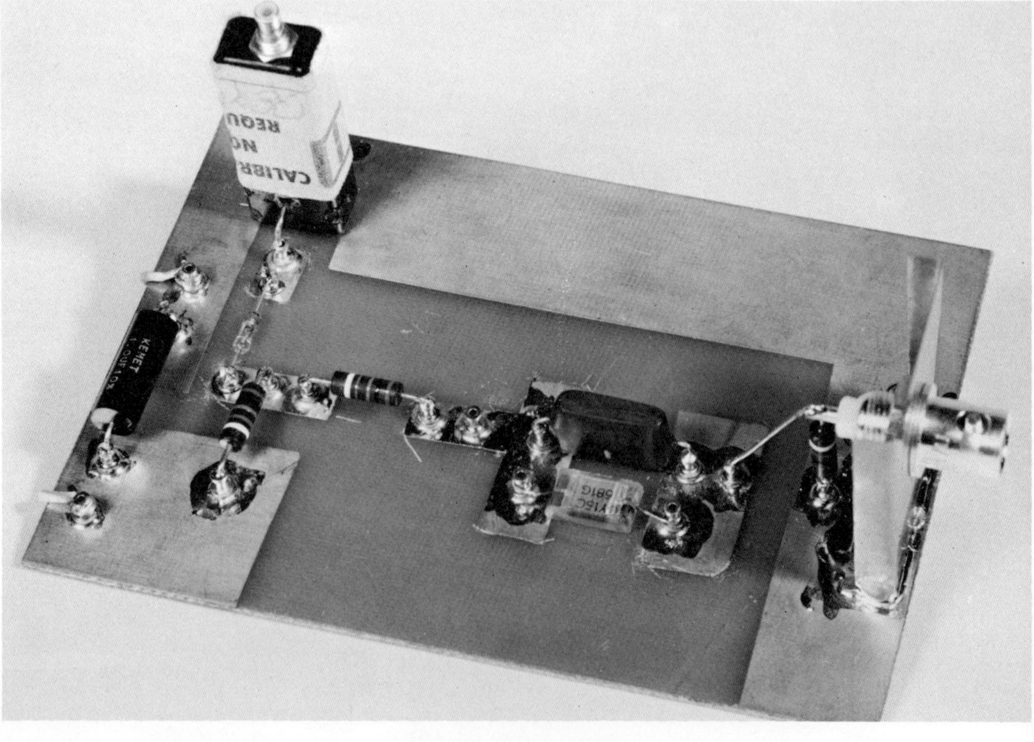

Fig. E.2 Diode test jig

Fig. E.3 Transistor test jig

Fig. E.4 JFET test jig

APPENDIX F

TABULAR DATA CORRECTIONS

Throughout Chapter 3 it is necessary to add initial and final points to many of the Tables. The reason for this is quite basic. At the very low end and at the extremely high end of the various measurement spectrums, it becomes impossible to make accurate measurements. In order for SCEPTRE to calculate realistic values in these regions, it is necessary to supply approximate initial and final values that do not cause SCEPTRE to extrapolate to unrealistic extremes. Assume Fig. F.1a is a graph constructed from the actual test data. The data points X and Y are added to "flatten out" the curve, so that SCEPTRE does not incorrectly compute the values shown by X′ and Y′ in Fig. F.1b, using linear extrapolation.

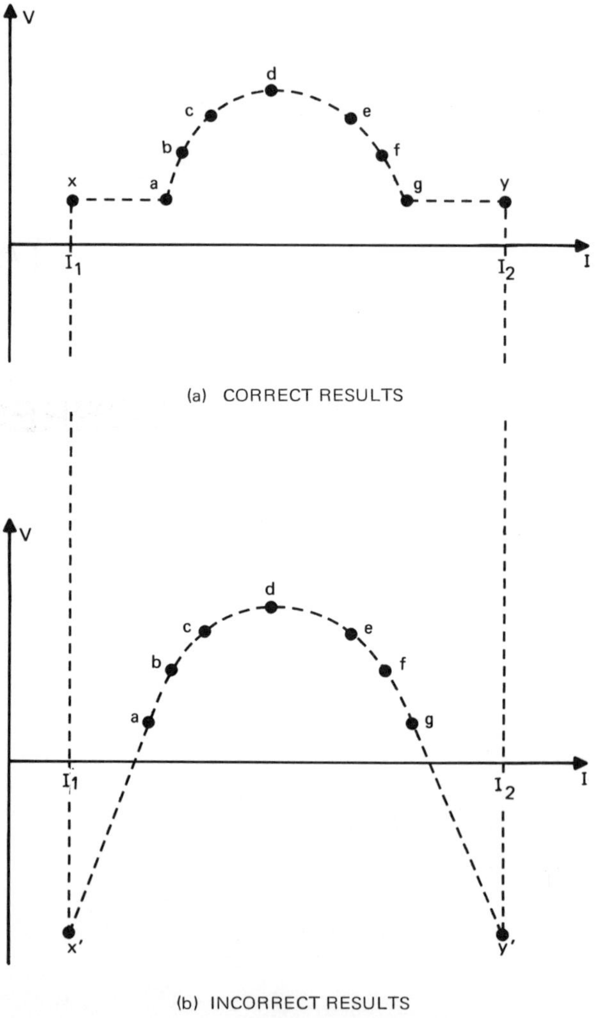

Fig. F.1 Tabular data corrections

APPENDIX G

SCEPTRE AVAILABILITY

Tapes containing the SCEPTRE program are available from:

Air Force Weapons Laboratory (WLET)

Attention: SCEPTRE Project Officer

Kirtland Air Force Base

New Mexico 87117

Tapes are available for the IBM 7094, IBM S/360, CDC 6600 and the Univac 1108. The user should send a 2400 foot, 1/2 inch, 7 track, identified tape along with a formal request for the program.

Technical advice is available from the SCEPTRE Project Officer at Kirtland or in some cases he may refer the inquirer to a specific engineer at the IBM facilities in Owego, New York for detailed answers.

While many Universities and Companies have their own computers, the rented remote terminal is becoming increasingly popular. Two organizations which have the SCEPTRE program available and which lease terminals world-wide are the Control Data Corporation and the University Computing Company. The following paragraphs describe their particular SCEPTRE versions.

CONTROL DATA 6600 SCEPTRE

The Control Data 6600 version of SCEPTRE at the Boston Data Center was originally adapted to the CDC 6600 under the Chippewa operating system by the Air Force Weapons Laboratory. Further conversion was performed by in-house engineers. It is presently operational under the SCOPE operating system at all domestic CDC Data Centers and is subject to a continuing process of development, debugging and test by a registered professional electrical engineer who is also available for user consultation. A monthly Bulletin is sent to all users with the latest informal data on revisions and suspected problem areas. All major features of SCEPTRE are operational. Subprograms are presently handled with system control cards rather than by a subprogram Heading. A completely stable code is not expected since improvements are made with a continuing development and support program. All revisions are available to assure that input decks will remain as constant as the operating system allows.

Major strengths of this version include the CYBERNET service which allows reproducible results under the same object code at seven domestic locations, the speed of the CDC 6600 with its inherent parallelism and high speed instruction stack, and the choice of tape or disk for all temporary and permanent files. An optional optimizing compiler may be used for compiling the second phase of SCEPTRE in order to further reduce running time. Core availability has proven sufficient for the largest allowable circuits and user subprograms. Word lengths of 60-bits allow extremely high dynamic ranges without loss of precision.

UNIVERSITY COMPUTING COMPANY 1108 SCEPTRE

The UCC version of SCEPTRE is based on the original IBM 7090 version distributed by the Air Force Weapons Laboratory. Users of SCEPTRE will see no difference in the methods for coding circuits or otherwise directing the analysis of their circuits. Because of the much higher running speed of the 1108, the computer save interval and the restart feature along with the reoutput feature of SCEPTRE have been disabled for the 1108 version. The computer time limit feature has been implemented on the 1108 version. Basic differences in the implementation of the 1108 version as contrasted to the 7090 version are due to the availability of high speed mass random access storage on the 1108. The present UCC version of SCEPTRE requires no tapes. Scratch files and other storage files use the high speed 432 drums and the Fastrand mass storage. Use of the 432 high speed drums for the overlays used in phase 1 of SCEPTRE result in greatly enhanced running speeds over those that could be achieved by using magnetic tape as storage for the overlay segment.

UCC has adapted the 1108 operating system to increase its reliability and throughout the system makes heavy usage of high speed 432 drums and the Fastrand mass storage device. The Fortran V compiler has also been refined and has had many errors removed so that its speed and reliability are considerably enhanced. The 1108 operating system allows users to access the 1108 by means of two types of remote terminals. These are the remote batch terminal system which includes the Cope terminal as well as other devices such as the Univac 1004, IBM 1130, Honeywell 200, GE 400, and other systems having card readers and line printers which have communications interfaces and can function as remote terminals to the UCC 1108 system. The second type of terminal is the Keyboard terminal serviced through the UCC FASBAC system. Through the FASBAC system, users of teletype and similar terminals can have access to the full power of the UCC 1108 system. All software and applications programs which are available on the 1108 are available through these two remote terminals systems. Hence, users of SCEPTRE can have access to this program through Keyboard terminals or remote batch terminals.

APPENDIX H

FCAL SUBPROGRAM

The subprogram FCAL* may be used in SCEPTRE to calculate the junction capacitances. Its use provides protection against run failures or excessive running time due to unrealistic computations. A typical expression for the combined transition and diffusion capacitance is,

$$CE = \frac{COE}{[\phi_E - VCE]^{n_E}} + \Theta_E T_E [JE + IES]$$

Although ϕ_E is carefully chosen, it is possible in a particular solution step that VCE is greater than ϕ_E. In SCEPTRE this would terminate the run since a negative antilog is not possible. With FCAL, the expression (ϕ_E − VCE) cannot go smaller than $.05\phi_E$. The subprogram listing is:

```
FUNCTION CCAL (COE, ØE, VCE, NE, THT, XJE, XIES)
IF (VCE.LE. (.95*ØE)) GO TO 10
CT = COE / (.05*ØE)**NE
GO TO 20
10 CT = COE/(ØE - VCE)**NE
20 CCAL = CT +THT * (XJE + XIES)
IF (CCAL . GE ..5) RETURN
CCAL = .5
RETURN
END
```

The subprogram is entered into the SCEPTRE listing in the normal fashion such as,

```
CE, 1-3 = FCAL (10., .8, VCE, .33, 37., JE, 1.9E-10)
```

*This information obtained from a Honeywell, Inc. interoffice memo dated 12 March 1967, written by D. R. Tierney.

APPENDIX I

TABULATED MODEL DATA

```
MODEL   1N100      (PERM) (A-K)
SUPPLIED BY  DAVID N. HARSTAD, SANDIA CORPORATION
SANDIA BASE,ALBUQUERQUE,NEW MEXICO 87115
EBERS-MOLL DIODE MODEL
UNITS-MILLIAMPS-VOLTS-KOHMS-PF-MICROHENRY-NANOSEC
ELEMENTS
RB,A-1=   0.170E-01
RC,1-K=   0.100E 04
CT, 1-K = Q1  (  0.354E 00 , 0.500E 00 ,VCT, 0.500E 00,  0.234E 02 ,
JD,  0.202 )
JD, 1-K = DIODE EQUATION (    0.250E-02 ,  0.147E 02 )
FUNCTIONS
Q1(A,B,C,D,E,F,G)=((A/ABS(B-C)**D)+E*(F+G))
```

```
MODEL 1N191  (PERM) (A-C)
SUPPLIED BY ALLAN G LINDGREN, CONTROL DATA CORP, ST PAUL,MINN 55112
UNITS-MILLIAMPS-VOLTS-KOHMS-PF-MICROHENRY-NANOSEC
EBERS-MOLL DIODE MODEL
ELEMENTS
RB,A-AIN= 5.000E-02
RS,AIN-C= 4.000E+02
CD,AIN-C= Q1(3.536E-01, 5.000E-01,VCD,5.000E-01, 6.191E+2, JD,1.250E-12)
JD,AIN-C=DIODE EQUATION( 1.25CE-12, 3.890E+01)
FUNCTIONS
EQUATION 1(A,B,C,C,E,F,G) = (A/(B-C)**D+E*(F+G))
```

```
MODEL UT262 (PERM) (C-A)
SUPPLIED BY CONVAIR DIVISION, GENERAL DYNAMICS, SAN DIEGO CALIF.92112
UNITS  OHMS,VOLTS,AMPS,FARADS,HENRIES,SECONDS
ELEMENTS
RB,A-1=.10
RS,1-C=3.9E9
JX,C-1=0.
CD,1-C=EQUATION1(24.E-12,1.0,VCD,.41,4.E-5,JD,1.52E-9)
JD,1-C=DIODE EQUATION(1.52E-9,24.0)
FUNCTIONS
EQUATION1(A,B,C,D,E,F,G)=(A/(B-C)**D+E*(F+G))
```

```
MODEL   1N279      (PERM) (A-K)
SUPPLIED BY  DAVID N. HARSTAD, SANDIA CORPORATION
SANDIA BASE,ALBUQUERQUE,NEW MEXICO 87115
EBERS-MOLL DIODE MODEL
UNITS-MILLIAMPS-VOLTS-KOHMS-PF-MICROHENRY-NANOSEC
ELEMENTS
CT, 1-K = Q1  (  0.245E 02 , 0.500E 00 ,VCT, 0.500E 00,  0.737E 03 ,
JD,  0.102)
RB,A-1=   0.600E-02
RC,1-K=   0.380E 04
JD, 1-K = DIODE EQUATION (    0.124E-02 ,  0.231E 02 )
FUNCTIONS
Q1(A,B,C,D,E,F,G)=((A/ABS(B-C)**D)+E*(F+G))
```

```
MODEL CT484 (PERM) (C-A)
SUPPLIED BY CONVAIR DIVISION, GENERAL DYNAMICS, SAN DIEGO CALIF.92112
UNITS  OHMS,VOLTS,AMPS,FARADS,HENRIES,SECONDS
ELEMENTS
RB,A-1=.25
RS,1-C=3.E8
JX,C-1=0.
CD,1-C=EQUATION1(190.E-12,.75,VCD,.50,6.55E-7,JD,1.72E-11)
JD,1-C=DIODE EQUATION(1.72E-11,29.4)
FUNCTIONS
EQUATION1(A,B,C,D,E,F,G)=(A/(B-C)**D+E*(F+G))
```

```
MODEL   SD500      (PERM) (A-K)
SUPPLIED BY  DAVID N. HARSTAD, SANDIA CORPORATION
SANDIA BASE,ALBUQUERQUE,NEW MEXICO 87115
EBERS-MOLL DIODE MODEL
UNITS-MILLIAMPS-VOLTS-KOHMS-PF-MICROHENRY-NANOSEC
ELEMENTS
RB,A-1=   0.100E-02
CT, 1-K = Q1  (  0.324E 04 , 0.500E 00 ,VCT, 0.500E 00,  0.619E 06 ,
JD,  0.310)
RC,1-K=   0.570E 03
JD, 1-K = DIODE EQUATION (    0.397E-10 ,  0.389E 02 )
FUNCTIONS
Q1(A,B,C,D,E,F,G)=((A/ABS(B-C)**D)+E*(F+G))
```

```
MODEL FD624 (PERM) (A-C)
MODEL SUPPLIED BY ALLAN G LINDGREN, CONTROL DATA CORP, ST PAUL,MINN 55112
UNITS-MILLIAMPS-VOLTS-KOHMS-PF-MICROHENRY-NANOSEC
EBERS-MOLL DIODE MODEL
ELEMENTS
CD,A-1=20.
JD,A-1=DIODE EQUATION (.1E-3,20.)
RS,A-1=1.6E3
RB,1-C=2.E-3
```

```
MODEL   1N645      (PERM) (A-K)
SUPPLIED BY  DAVID N. HARSTAD, SANDIA CORPORATION
SANDIA BASE,ALBUQUERQUE,NEW MEXICO 87115
EBERS-MOLL DIODE MODEL
UNITS-MILLIAMPS-VOLTS-KOHMS-PF-MICROHENRY-NANOSEC
ELEMENTS
CT, 1-K = Q1  (  0.108E 02 , (.864E 00 ,VCT, 0.577E 00,  0.125E 06 ,
JD,  0.206)
RB,A-1=   0.532E-03
RC,1-K=   0.123E 09
JD, 1-K = DIODE EQUATION (    0.251E-06 ,  0.261E 02 )
FUNCTIONS
Q1(A,B,C,D,E,F,G)=((A/ABS(B-C)**D)+E*(F+G))
```

```
MODEL 1N647 (PERM) (C-A)
SUPPLIED BY CONVAIR DIVISION, GENERAL DYNAMICS, SAN DIEGO CALIF.92112
UNITS  OHMS,VOLTS,AMPS,FARADS,HENRIES,SECONDS
ELEMENTS
RB,A-1=.1
RS,1-C=4.2E10
JX,C-1=0.
CD,1-C=EQUATION1(7.7E-12,1.0,VCD,.51,1.9E-5,JD,1.6E-9)
JD,1-C=DIODE EQUATION(1.6E-9,22.0)
FUNCTIONS
EQUATION1(A,B,C,D,E,F,G)=(A/(B-C)**D+E*(F+G))
```

```
MODEL   1N649      (PERM) (A-K)
SUPPLIED BY  DAVID N. HARSTAD, SANDIA CORPORATION
SANDIA BASE,ALBUQUERQUE,NEW MEXICO 87115
EBERS-MOLL DIODE MODEL
UNITS-MILLIAMPS-VOLTS-KOHMS-PF-MICROHENRY-NANOSEC
ELEMENTS
CT, 1-K = G1  (  0.108E 02 , 0.864E 00 ,VCT, 0.577E 00,  0.        ,
JD,  0.206)
RB,A-1=    0.532E-03
RC,1-K=    0.123E 09
JD, 1-K = DIODE EQUATION (     0.251E-06 ,  0.261E 02 )
FUNCTIONS
G1(A,B,C,D,E,F,G)=((A/ABS(B-C)**D)+E*(F+G))
```

```
MODEL 1N659 (PERM) (C-A)
SUPPLIED BY CONVAIR DIVISION, GENERAL DYNAMICS, SAN DIEGO CALIF.92112
UNITS  OHMS,VOLTS,AMPS,FARADS,HENRIES,SECONDS
ELEMENTS
RB,A-1=45.0
RS,1-C=5.6E8
JX,C-1=0.
CD,1-C=EQUATION1(1.9E-11,.80,VCD,.40,3.E-6,JD,1.4E-11)
JD,1-C=DIODE EQUATION(1.4E-11,33.7)
FUNCTIONS
EQUATION1(A,B,C,D,E,F,G)=(A/(B-C)**D+E*(F+G))
```

```
MODEL 1N695 (PERM) (C-A)
SUPPLIED BY CONVAIR DIVISION, GENERAL DYNAMICS, SAN DIEGO CALIF.92112
UNITS  OHMS,VOLTS,AMPS,FARADS,HENRIES,SECONDS
ELEMENTS
RB,A-1=1.0
RS,1-C=3.2E5
JX,C-1=0.
CD,1-C=EQUATION1(.7E-12,.60,VCD,.17,3.6E-7,JD,7.8E-7)
JD,1-C=DIODE EQUATION(7.8E-7,25.0)
FUNCTIONS
EQUATION1(A,B,C,D,E,F,G)=(A/(B-C)**D+E*(F+G))
```

```
MODEL 1N746A (PERM) (C-A)
SUPPLIED BY CONVAIR DIVISION, GENERAL DYNAMICS, SAN DIEGO CALIF.92112
UNITS  OHMS,VOLTS,AMPS,FARADS,HENRIES,SECONDS
ELEMENTS
RB,1-A=28.
RS,C-1=1.E5
JX,C-1=0.
CD,1-C=EQUATION1(4.63E-10,.75,VCD,.50,4.09E-6,JD,2.69E-10)
JD,1-C=DIODE EQUATION(2.69E-10,35.3)
JZ,C-1=DIODE TABLE1
FUNCTIONS
EQUATION1(A,B,C,D,E,F,G)=(A/(B-C)**D+E*(F+G))
DIODE TABLE1
-3.0,-10., -1.,-6., -.8,-.5, -.6,-.05, -.5,-.002, -.48,-1.E-9, 0.,1.E-9,
1.,10.E-6, 2.7,10.E-6, 2.72,1.E-3, 2.74,20.E-3, 2.76,50.E-3
```

```
MODEL 1N748A (PERM) (C-A)
SUPPLIED BY CONVAIR DIVISION, GENERAL DYNAMICS, SAN DIEGO CALIF.92112
UNITS  OHMS,VOLTS,AMPS,FARADS,HENRIES,SECONDS
ELEMENTS
RB,1-A=23.
RS,C-1=1.E5
JX,C-1=0.
CD,1-C=EQUATION1(4.14E-10,.75,VCD,.50,3.88E-6,JD,1.14E-10)
JD,1-C=DIODE EQUATION(1.14E-10,33.6)
JZ,C-1=DIODE TABLE1
FUNCTIONS
EQUATION1(A,B,C,D,E,F,G)=(A/(B-C)**D+E*(F+G))
DIODE TABLE1
-3.0,-10., -1.,-6., -.8,-.5, -.6,-.05, -.5,-.002, -.48,-1.E-9, 0.,1.E-9,
1.,10.E-6, 3.3,10.E-6, 3.35,1.E-3, 3.44,20.E-3, 3.46,50.E-3
```

```
MODEL 1N750A (PERM) (C-A)
SUPPLIED BY CONVAIR DIVISION, GENERAL DYNAMICS, SAN DIEGO CALIF.92112
UNITS  OHMS,VOLTS,AMPS,FARADS,HENRIES,SECONDS
ELEMENTS
RB,1-A=19.
RS,C-1=5.E5
JX,C-1=0.
CD,1-C=EQUATION1(3.68E-10,.75,VCD,.50,3.45E-6,JD,4.84E-11)
JD,1-C=DIODE EQUATION(4.84E-11,32.7)
JZ,C-1=DIODE TABLE1
FUNCTIONS
EQUATION1(A,B,C,D,E,F,G)=(A/(B-C)**D+E*(F+G))
DIODE TABLE1
-3.0,-10., -1.,-6., -.8,-.5, -.6,-.05, -.5,-.002, -.48,-1.E-9, 0.,1.E-9,
1.,2.0E-6, 4.0,2.0E-6, 4.23,1.E-3, 4.32,20.E-3, 4.35,50.E-3
```

```
MODEL 1N752A (PERM) (C-A)
SUPPLIED BY CONVAIR DIVISION, GENERAL DYNAMICS, SAN DIEGO CALIF.92112
UNITS  OHMS,VOLTS,AMPS,FARADS,HENRIES,SECONDS
ELEMENTS
RB,1-A=11.
RS,C-1=1.E6
JX,C-1=0.
CD,1-C=EQUATION1(3.31E-10,.75,VCD,.50,3.1E-6,JD,1.25E-11)
JD,1-C=DIODE EQUATION(1.25E-11,30.8)
JZ,C-1=DIODE TABLE1
FUNCTIONS
EQUATION1(A,B,C,D,E,F,G)=(A/(B-C)**D+E*(F+G))
DIODE TABLE1
-3.0,-10., -1.,-6., -.8,-.5, -.6,-.05, -.5,-.002, -.48,-1.E-9, 0.,1.E-9,
1.,1.0E-6, 4.9,1.0E-6, 5.28,1.E-3, 5.34,20.E-3, 5.41,50.E-3
```

```
MODEL 1N754A (PERM) (C-A)
SUPPLIED BY CONVAIR DIVISION, GENERAL DYNAMICS, SAN DIEGO CALIF.92112
UNITS  OHMS,VOLTS,AMPS,FARADS,HENRIES,SECONDS
ELEMENTS
RB,1-A=5.0
RS,C-1=1.E7
JX,C-1=0.
CD,1-C=EQUATION1(2.85E-10,.75,VCD,.50,2.67E-6,JD,4.6E-12)
JD,1-C=DIODE EQUATION(4.6E-12,29.7)
JZ,C-1=DIODE TABLE1
FUNCTIONS
EQUATION1(A,B,C,D,E,F,G)=(A/(B-C)**D+E*(F+G))
DIODE TABLE1
-3.0,-10., -1.,-6., -.8,-.5, -.6,-.05, -.5,-.002, -.48,-1.E-9, 0.,1.E-9,
1.,1.0E-7, 6.2,1.0E-7, 6.60,1.E-3, 6.70,20.E-3, 6.73,50.E-3
```

```
MODEL 1N756A (PERM) (C-A)
SUPPLIED BY CONVAIR DIVISION, GENERAL DYNAMICS, SAN DIEGO CALIF.92112
UNITS  OHMS,VOLTS,AMPS,FARADS,HENRIES,SECONDS
ELEMENTS
RB,1-A=8.
RS,C-1=1.E7
JX,C-1=0.
CD,1-C=EQUATION1(2.38E-10,.75,VCD,.50,2.23E-6,JD,7.24E-12)
JD,1-C=DIODE EQUATION(7.24E-12,30.3)
JZ,C-1=DIODE TABLE1
FUNCTIONS
EQUATION1(A,B,C,D,E,F,G)=(A/(B-C)**D+E*(F+G))
DIODE TABLE1
-3.0,-10., -1.,-6., -.8,-.5, -.6,-.05, -.5,-.002, -.48,-1.E-9, 0.,1.E-9,
1.,1.0E-7, 7.6,1.0E-7, 7.93,1.E-3, 8.04,20.E-3, 8.07,50.E-3
```

```
MODEL 1N758A (PERM) (C-A)
SUPPLIED BY CONVAIR DIVISION, GENERAL DYNAMICS, SAN DIEGO CALIF.92112
UNITS  OHMS,VOLTS,AMPS,FARADS,HENRIES,SECONDS
ELEMENTS
RB,1-A=17.
RS,C-1=1.E7
JX,C-1=0.
CD,1-C=EQUATION1(1.99E-10,.75,VCD,.50,1.864E-6,JD,3.4E-11)
JD,1-C=DIODE EQUATION(3.4E-11,32.2)
JZ,C-1=DIODE TABLE1
FUNCTIONS
EQUATION1(A,B,C,D,E,F,G)=(A/(B-C)**D+E*(F+G))
DIODE TABLE1
-3.0,-10., -1.,-6., -.8,-.5, -.6,-.05, -.5,-.002, -.48,-1.E-9, 0.,1.E-9,
1.,1.0E-7, 9.4,1.0E-7, 9.52,1.E-3, 9.66,20.E-3, 9.68,50.E-3
```

```
MODEL 1N827A (PERM) (C-A)
SUPPLIED BY CONVAIR DIVISION, GENERAL DYNAMICS, SAN DIEGO CALIF.92112
UNITS  OHMS,VOLTS,AMPS,FARADS,HENRIES,SECONDS
ELEMENTS
RB,1-A=10.
RS,C-1=10.E6
JX,C-1=0.
CD,1-C=EQUATION1(2.19E-10,.75,VCD,.50,2.05E-6,JD,1.04E-11)
JD,1-C=DIODE EQUATION(1.04E-11,30.6)
JZ,C-1=DIODE TABLE1
FUNCTIONS
EQUATION1(A,B,C,D,E,F,G)=(A/(B-C)**D+E*(F+G))
DIODE TABLE1
-3.0,-10., -1.,-6., -.8,-.5, -.6,-.05, -.5,-.002, -.48,-1.E-9, 0.,1.E-9,
1.,1.0E-9, 5.6,1.0E-9, 6.04,1.E-3, 6.13,7.5E-3, 6.16,15.E-3, 6.19,50.E-3
```

```
MODEL 1N914 (PERM) (C-A)
SUPPLIED BY CONVAIR DIVISION, GENERAL DYNAMICS, SAN DIEGO CALIF.92112
UNITS  OHMS,VOLTS,AMPS,FARADS,HENRIES,SECONDS
ELEMENTS
RB,A-1=2.0
RS,1-C=1.1E6
JX,C-1=0.
CD,1-C=EQUATION1(2.4E-11,.90,VCD,.50,1.81E-5,JD,1.9E-9)
JD,1-C=DIODE EQUATION(2.9E-9,21.5)
FUNCTIONS
EQUATION1(A,B,C,D,E,F,G)=(A/(B-C)**D+E*(F+G))
```

```
MODEL 1N961 (PERM) (A-K)
SUPPLIED BY  DAVID N. HARSTAD, SANDIA CORPORATION
SANDIA BASE,ALBUQUERQUE,NEW MEXICO 87115
DANIEL  ZENER MODEL UNIT-OHMS-ETC
ELEMENTS
RC , 1-K = 7.51E6
CT, 1-K =      G1(217.2,.9516,VCT,.4547)
RB, A-1 = .01
J1,1-K = DIODE EQUATION(7.145E-10,31.)
J2, K-1 = Q2 (586., -9.86,-1.304E-3,VCT )
J3, K-1 = Q3(0.474E+01,-0.986E+1,0.373E+1,VCT)
FUNCTIONS
Q1(A,B,C,D)    =   ((A/ABS(B-C)**D))
Q2 (B,C,D,E) =(EXP (B *(C-D*E -E)) )
Q3 (F,C,G,E) =(EXP(F*C - G*E) )
```

```
MODEL 1N963B (PERM) (C-A)
SUPPLIED BY CONVAIR DIVISION, GENERAL DYNAMICS, SAN DIEGO CALIF.92112
UNITS  OHMS,VOLTS,AMPS,FARADS,HENRIES,SECONDS
ELEMENTS
RB,1-A=11.5
RS,C-1=1.82E6
JX,C-1=0.
CD,1-C=EQUATION1(3.32E-10,.75,VCD,.50,4.34E-6,JD,4.25E-12)
JD,1-C=DIODE EQUATION(4.25E-12,33.3)
JZ,C-1=DIODE TABLE1
FUNCTIONS
EQUATION1(A,B,C,D,E,F,G)=(A/(E-C)**D+E*(F+G))
DIODE TABLE1
-3.0,-10., -1.,-6., -.8,-.5, -.6,-.05, -.5,-.002, -.48,-1.E-9, 0.,1.E-9,
1.,5.0E-6, 9.1,5.0E-6, 11.79,1.E-3, 11.88,10.5E-3, 11.91,30.E-3
```

```
MODEL 1N965B (PERM) (C-A)
SUPPLIED BY CONVAIR DIVISION, GENERAL DYNAMICS, SAN DIEGO CALIF.92112
UNITS  OHMS,VOLTS,AMPS,FARADS,HENRIES,SECONDS
ELEMENTS
RB,1-A=16.
RS,C-1=2.28E6
JX,C-1=0.
CD,1-C=EQUATION1(2.90E-12,.75,VCD,.50,3.81E-6,JD,1.37E-11)
JD,1-C=DIODE EQUATION(1.37E-11,34.1)
JZ,C-1=DIODE TABLE1
FUNCTIONS
EQUATION1(A,B,C,D,E,F,G)=(A/(E-C)**D+E*(F+G))
DIODE TABLE1
-3.0,-10., -1.,-6., -.8,-.5, -.6,-.05, -.5,-.002, -.48,-1.E-9, 0.,1.E-9,
1.,5.0E-6, 11.4,5.E-6, 14.78,1.E-3, 14.86,8.5E-3, 14.97,30.E-3
```

```
MODEL 1N967B (PERM) (C-A)
SUPPLIED BY CONVAIR DIVISION, GENERAL DYNAMICS, SAN DIEGO CALIF.92112
UNITS  OHMS,VOLTS,AMPS,FARADS,HENRIES,SECONDS
ELEMENTS
RB,1-A=21.
RS,C-1=2.75E6
JX,C-1=0.
CD,1-C=EQUATION1(2.90E-10,.75,VCD,.50,3.81E-6,JD,4.35E-11)
JD,1-C=DIODE EQUATION(4.35E-11,35.7)
JZ,C-1=DIODE TABLE1
FUNCTIONS
EQUATION1(A,B,C,D,E,F,G)=(A/(B-C)**D+E*(F+G))
DIODE TABLE1
-3.0,-10., -1.,-6., -.8,-.5, -.6,-.05, -.5,-.002, -.48,-1.E-9, 0.,1.E-9,
1.,5.0E-6, 13.7,5.E-6, 17.77,1.E-3, 17.85,7.E-3, 17.88,20.E-3
```

```
MODEL 1N969B (PERM) (C-A)
SUPPLIED BY CONVAIR DIVISION, GENERAL DYNAMICS, SAN DIEGO CALIF.92112
UNITS  OHMS,VOLTS,AMPS,FARADS,HENRIES,SECONDS
ELEMENTS
RB,1-A=29.
RS,C-1=3.34E6
JX,C-1=0.
CD,1-C=EQUATION1(2.65E-10,.75,VCD,.50,3.48E-6,JD,2.53E-10)
JD,1-C=DIODE EQUATION(2.53E-10,38.1)
JZ,C-1=DIODE TABLE1
FUNCTIONS
EQUATION1(A,B,C,D,E,F,G)=(A/(B-C)**D+E*(F+G))
DIODE TABLE1
-3.0,-10., -1.,-6., -.8,-.5, -.6,-.05, -.5,-.002, -.48,-1.E-9, 0.,1.E-9,
1.,5.0E-7, 16.7,5.E-7, 21.82,1.E-3, 21.84,5.6E-3, 21.87,20.E-3
```

```
MODEL 1N971B (PERM) (C-A)
SUPPLIED BY CONVAIR DIVISION, GENERAL DYNAMICS, SAN DIEGO CALIF.92112
UNITS  OHMS,VOLTS,AMPS,FARADS,HENRIES,SECONDS
ELEMENTS
RB,1-A=41.0
RS,C-1=4.14E6
JX,C-1=0.
CD,1-C=EQUATION1(1.99E-10,.75,VCD,.50,2.61E-6,JD,5.6E-9)
JD,1-C=DIODE EQUATION(5.6E-9,42.3)
JZ,C-1=DIODE TABLE1
FUNCTIONS
EQUATION1(A,B,C,D,E,F,G)=(A/(B-C)**D+E*(F+G))
DIODE TABLE1
-3.0,-10., -1.,-6., -.8,-.5, -.6,-.05, -.5,-.002, -.48,-1.E-9, 0.,1.E-9,
1.,5.0E-7, 20.6,5.E-7, 26.74,.5E-3, 26.81,4.6E-3, 26.85,15.E-3
```

```
MODEL 1N973B (PERM) (C-A)
SUPPLIED BY CONVAIR DIVISION, GENERAL DYNAMICS, SAN DIEGO CALIF.92112
UNITS  OHMS,VOLTS,AMPS,FARADS,HENRIES,SECONDS
ELEMENTS
RB,1-A=58.
RS,C-1=5.02E6
JX,C-1=0.
CD,1-C=EQUATION1(1.85E-10,.75,VCD,.50,2.44E-6,JD,2.41E-10)
JD,1-C=DIODE EQUATION(2.41E-10,38.2)
JZ,C-1=DIODE TABLE1
FUNCTIONS
EQUATION1(A,B,C,D,E,F,G)=(A/(B-C)**D+E*(F+G))
DIODE TABLE1
-3.0,-10., -1.,-6., -.8,-.5, -.6,-.05, -.5,-.002, -.48,-1.E-9, 0.,1.E-9,
1.,5.0E-7, 25.1,5.E-7, 32.67,.1E-3, 32.78,3.8E-3, 32.82,10.E-3
```

```
MODEL   FDM1000   (PERM) (A-K)
SUPPLIED BY  DAVID N. HARSTAD, SANDIA CORPORATION
SANDIA BASE,ALBUQUERQUE,NEW MEXICO 87115
EBERS-MOLL DIODE MODEL
UNITS-MILLIAMPS-VOLTS-KOHMS-PF-MICROHENRY-NANOSEC
ELEMENTS
CT, 1-K = Q1  (  0.977E 00 , 0.100E 01 ,VCT, 0.274E 00,  0.994E 02 ,
JD,  0.405)
RB,A-1=   0.450E-02
RC,1-K=   0.974E 06
JD, 1-K = DIODE EQUATION (    0.404E-05 ,  0.197E 02 )
FUNCTIONS
Q1(A,B,C,D,E,F,G)=((A/ABS(B-C)**D)+E*(F+G))
```

```
MODEL H2969 (PERM) (A-C)
SUPPLIED BY ALLAN G LINDGREN, CONTROL DATA CORP,
UNITS-MILLIAMPS-VOLTS-KOHMS-PF-MICROHENRY-NANOSEC
EBERS-MOLL DIODE MODEL
ELEMENTS
RB,A-AIN=2.E-3
RS,AIN-C=1.6E6
CD,AIN-C=Q1(4.,.8,VCD,.08,58.2,JD,.1)
JD,AIN-C= DIODE EQUATION(.1,9.7)
FUNCTIONS
EQUATION 1(A,B,C,D,E,F,G) = (A/(B-C)**D+E*(F+G))
```

```
MODEL 1N3016B (PERM) (C-A)
SUPPLIED BY CONVAIR DIVISION, GENERAL DYNAMICS, SAN DIEGO CALIF.92112
UNITS  OHMS,VOLTS,AMPS,FARADS,HENRIES,SECONDS
ELEMENTS
RB,1-A=3.5
RS,C-1=69.5E3
JX,C-1=0.
CD,1-C=EQUATION1(5.95E-10,.75,VCD,.50,7.85E-6,JD,2.53E-10)
JD,1-C=DIODE EQUATION(2.53E-10,27.8)
JZ,C-1=DIODE TABLE1
FUNCTIONS
EQUATION1(A,B,C,D,E,F,G)=(A/(B-C)**D+E*(F+G))
DIODE TABLE1
-3.0,-10., -1.,-6., -.8,-.5, -.6,-.05, -.5,-.002, -.48,-1.E-9, 0.,1.E-9,
1.,1.5E-4, 5.2,1.5E-4, 6.58,10.E-3, 6.67,37.E-3, 6.70,.10
```

```
MODEL 1N3018B (PERM) (C-A)
SUPPLIED BY CONVAIR DIVISION, GENERAL DYNAMICS, SAN DIEGO CALIF.92112
UNITS  OHMS,VOLTS,AMPS,FARADS,HENRIES,SECONDS
ELEMENTS
RB,1-A=4.5
RS,C-1=2.48E5
JX,C-1=0.
CD,1-C=EQUATION1(5.95E-12,.75,VCD,.50,7.85E-6,JD,2.81E-10)
JD,1-C=DIODE EQUATION(2.81E-10,27.8)
JZ,C-1=DIODE TABLE1
FUNCTIONS
EQUATION1(A,B,C,D,E,F,G)=(A/(B-C)**D+E*(F+G))
DIODE TABLE1
-3.0,-10., -1.,-6., -.8,-.5, -.6,-.05, -.5,-.002, -.48,-1.E-9, 0.,1.E-9,
1.,50.E-6, 6.2,50.E-6, 7.97,10.E-3, 8.06,31.E-3, 8.10,.10
```

```
MODEL 1N3020B (PERM) (C-A)
SUPPLIED BY CONVAIR DIVISION, GENERAL DYNAMICS, SAN DIEGO CALIF.92112
UNITS  OHMS,VOLTS,AMPS,FARADS,HENRIES,SECONDS
ELEMENTS
RB,1-A=7.0
RS,C-1=7.6E6
JX,C-1=0.
CD,1-C=EQUATION1(4.96E-10,.75,VCD,.50,6.52E-6,JD,4.55E-10)
JD,1-C=DIODE EQUATION(4.55E-10,28.2)
JZ,C-1=DIODE TABLE1
FUNCTIONS
EQUATION1(A,B,C,D,E,F,G)=(A/(B-C)**D+E*(F+G))
DIODE TABLE1
-3.0,-10., -1.,-6., -.8,-.5, -.6,-.05, -.5,-.002, -.48,-1.E-9, 0.,1.E-9,
1.,10.E-6, 7.6,10.E-6, 9.73,5.0E-3, 9.83,25.E-3,9.86,75.E-3
```

```
MODEL 1N3022B (PERM) (C-A)
SUPPLIED BY CONVAIR DIVISION, GENERAL DYNAMICS, SAN DIEGO CALIF.92112
UNITS  OHMS,VOLTS,AMPS,FARADS,HENRIES,SECONDS
ELEMENTS
RB,1-A=9.0
RS,C-1=1.82E6
JX,C-1=0.
CD,1-C=EQUATION1(4.96E-12,.75,VCD,.50,6.52E-6,JD,1.77E-10)
JD,1-C=DIODE EQUATION(1.77E-10,28.6)
JZ,C-1=DIODE TABLE1
FUNCTIONS
EQUATION1(A,B,C,D,E,F,G)=(A/(E-C)**D+E*(F+G))
DIODE TABLE1
-3.0,-10., -1.,-6., -.8,-.5, -.6,-.05, -.5,-.002, -.48,-1.E-9, 0.,1.E-9,
1.,5.E-6, 9.1,5.E-6, 11.71,1.E-3, 11.81,21.E-3, 11.84,50.E-3
```

```
MODEL 1N3024B (PERM) (C-A)
SUPPLIED BY CONVAIR DIVISION, GENERAL DYNAMICS, SAN DIEGO CALIF.92112
UNITS  OHMS,VOLTS,AMPS,FARADS,HENRIES,SECONDS
ELEMENTS
RB,1-A=14.0
RS,C-1=2.28E6
JX,C-1=0.
CD,1-C=EQUATION1(4.77E-10,.75,VCD,.50,6.28E-6,JD,2.84E-10)
JD,1-C=DIODE EQUATION(2.84E-10,29.7)
JZ,C-1=DIODE TABLE1
FUNCTIONS
EQUATION1(A,B,C,D,E,F,G)=(A/(E-C)**D+E*(F+G))
DIODE TABLE1
-3.0,-10., -1.,-6., -.8,-.5, -.6,-.05, -.5,-.002, -.48,-1.E-9, 0.,1.E-9,
1.,5.E-6, 11.4,5.E-6, 14.66,1.E-3, 14.76,17.E-3, 14.79,50.E-3
```

```
MODEL 1N3026B (PERM) (C-A)
SUPPLIED BY CONVAIR DIVISION, GENERAL DYNAMICS, SAN DIEGO CALIF.92112
UNITS  OHMS,VOLTS,AMPS,FARADS,HENRIES,SECONDS
ELEMENTS
RB,1-A=20.
RS,C-1=2.74E6
JX,C-1=0.
CD,1-C=EQUATION1(4.50E-10,.75,VCD,.50,5.92E-6,JD,4.46E-10)
JD,1-C=DIODE EQUATION(4.46E-10,30.6)
JZ,C-1=DIODE TABLE1
FUNCTIONS
EQUATION1(A,B,C,D,E,F,G)=(A/(E-C)**D+E*(F+G))
DIODE TABLE1
-3.0,-10., -1.,-6., -.8,-.5, -.6,-.05, -.5,-.002, -.48,-1.E-9, 0.,1.E-9,
1.,5.E-6, 13.7,5.E-6, 17.64,1.E-3, 17.72,14.0E-3, 17.77,50.E-3
```

```
MODEL 1N3028B (PERM) (C-A)
SUPPLIED BY CONVAIR DIVISION, GENERAL DYNAMICS, SAN DIEGO CALIF.92112
UNITS  OHMS,VOLTS,AMPS,FARADS,HENRIES,SECONDS
ELEMENTS
RB,1-A=23.
RS,C-1=3.34E6
JX,C-1=0.
CD,1-C=EQUATION1(4.29E-10,.75,VCD,.50,5.57E-6,JD,4.55E-10)
JD,1-C=DIODE EQUATION(4.55E-10,31.5)
JZ,C-1=DIODE TABLE1
FUNCTIONS
EQUATION1(A,B,C,D,E,F,G)=(A/(B-C)**D+E*(F+G))
DIODE TABLE1
-3.0,-10., -1.,-6., -.8,-.5, -.6,-.05, -.5,-.002, -.48,-1.E-9, 0.,1.E-9,
1.,5.E-6,16.7,5.E-6, 21.66,1.E-3, 21.74,11.5E-3, 21.78,50.E-3
```

```
MODEL   1N3070     (PERM) (A-K)
SUPPLIED BY  DAVID N. HARSTAD, SANDIA CORPORATION
SANDIA BASE,ALBUQUERQUE,NEW MEXICO 87115
EBERS-MOLL DIODE MODEL
UNITS-MILLIAMPS-VOLTS-KOHMS-PF-MICROHENRY-NANOSEC
ELEMENTS
CT, 1-K = G1  (  0.176E 01 , 0.100E 01 ,VCT, 0.166E 00,  0.628E 03 ,
JD,  0.405)
RB,A-1=   0.165E-02
RC,1-K=   0.864E 07
JD, 1-K = DIODE EQUATION (    0.423E-05 ,  0.218E 02 )
FUNCTIONS
G1(A,B,C,D,E,F,G)=((A/ABS(B-C)**D)+E*(F+G))
```

```
MODEL 1N3600 (PERM) (C-A)
SUPPLIED BY CONVAIR DIVISION, GENERAL DYNAMICS, SAN DIEGO CALIF.92112
UNITS  OHMS,VOLTS,AMPS,FARADS,HENRIES,SECONDS
ELEMENTS
RB,A-1=.60
RS,1-C=1.36E9
JX,C-1=0.
CD,1-C=EQUATION1(.90E-12,.85,VCD,.05,3.98E-9,JD,5.06E-9)
JD,1-C=DIODE EQUATION(5.06E-9,20.9)
FUNCTIONS
EQUATION1(A,B,C,D,E,F,G)=(A/(B-C)**D+E*(F+G))
```

```
MODEL 1N3669 (PERM) (C-A)
SUPPLIED BY CONVAIR DIVISION, GENERAL DYNAMICS, SAN DIEGO CALIF.92112
UNITS  OHMS,VOLTS,AMPS,FARADS,HENRIES,SECONDS
ELEMENTS
RB,A-1=2.0
RS,1-C=1.1E6
JX,C-1=0.
CD,1-C=EQUATION1(23.E-12,.80,VCD,.46,1.3E-5,JD,1.3E-10)
JD,1-C=DIODE EQUATION(1.3E-10,28.0)
FUNCTIONS
EQUATION1(A,B,C,D,E,F,G)=(A/(B-C)**D+E*(F+G))
```

```
MODEL   1N4003     (PERM) (A-K)
SUPPLIED BY  DAVID N. HARSTAD, SANDIA CORPORATION
SANDIA BASE,ALBUQUERQUE,NEW MEXICO 87115
EBERS-MOLL DIODE MODEL
UNITS-MILLIAMPS-VOLTS-KOHMS-PF-MICROHENRY-NANOSEC
ELEMENTS
RB,A-1=    0.500E-03
RC,1-K=    0.200E 08
CT, 1-K = Q1  (  0.230E 03 , 0.100E 01 ,VCT, 0.480E 00,  0.150E 06,
JD, 0.420E-05 )
JD, 1-K = DIODE EQUATION (    0.420E-05 ,  0.220E 02 )
FUNCTIONS
Q1(A,B,C,D,E,F,G)=((A/ABS(B-C)**D)+E*(F+G))
```

```
MODEL 1N4572 (PERM) (C-A)
SUPPLIED BY CONVAIR DIVISION, GENERAL DYNAMICS, SAN DIEGO CALIF.92112
UNITS  OHMS,VOLTS,AMPS,FARADS,HENRIES,SECONDS
ELEMENTS
RB,A-1=10.0
RS,1-C=24.E8
JX,C-1=0.
CD,1-C=EQUATION1(5.6E-12,.85,VCD,.615,1.35E-5,JD,7.42E-12)
JD,1-C=DIODE EQUATION(7.42E-12,28.5)
FUNCTIONS
EQUATION1(A,B,C,D,E,F,G)=(A/(B-C)**D+E*(F+G))
```

```
MODEL   FD6666     (PERM) (A-K)
SUPPLIED BY  DAVID N. HARSTAD, SANDIA CORPORATION
SANDIA BASE,ALBUQUERQUE,NEW MEXICO 87115
EBERS-MOLL DIODE MODEL
UNITS-MILLIAMPS-VOLTS-KOHMS-PF-MICROHENRY-NANOSEC
ELEMENTS
CT, 1-K = G1  (  0.105E 01 , 0.100E 01 ,VCT, 0.102E 00,  0.900E 02 ,
JD, 0.505)
RB,A-1=    0.433E-03
RC,1-K=    0.120E 07
JD, 1-K = DIODE EQUATION (    0.536E-05 ,  0.208E 02 )
FUNCTIONS
G1(A,B,C,D,E,F,G)=((A/ABS(B-C)**D)+E*(F+G))
```

```
MODEL   2N174      (PERM) (E-B-C)
SUPPLIED BY  DAVID N. HARSTAD, SANDIA CORPORATION
SANDIA BASE,ALBUQUERQUE,NEW MEXICO 87115
A MODIFIED EBERS-MOLL EQUIVALENT CIRCUIT      PNP
UNITS-MILLIAMPS-VOLTS-KOHMS-PF-MICROHENRY-NANOSEC
ELEMENTS
CET,E-1 = G1   (   0.141E 04,  0.100E 01,VCET, 0.500E 00, 0.511E 05 ,
J1,  0.545E 00 )
CCT,2-1 = G1   (   0.141E 04,  0.100E 01,VCCT, 0.500E 00, 0.516E 06 ,
J2,  0.551E 00 )
J1, E-1 = DIODE EQUATION (  0.545E 00 ,  0.321E 02 )
J2, 2-1 = DIODE EQUATION (  0.551E 00 ,  0.324E 02 )
RBB, 1-B =  0.300E-02
RE,  E-1 =  0.400E 03
RC,  2-1 =  0.160E 03
RCC, C-2 =  0.200E-04
JA, 1-E = 0.833E 00 * J2
JB, 1-2 = 0.982E 00 * J1
FUNCTIONS
G1(A,B,C,D,E,F,G)=((A/ABS(B-C)**D)+E*(F+G))
```

```
MODEL   TC229      (PERM) (E-B-C)
SUPPLIED BY  DAVID N. HARSTAD, SANDIA CORPORATION
SANDIA BASE,ALBUQUERQUE,NEW MEXICO 87115
EBERS MOLL TRANSISTOR EQUIVALENT CIRCUIT  NPN
UNITS-MILLIAMPS-VOLTS-KOHMS-PF-MICROHENRY-NANOSEC
ELEMENTS
CET,1-E = Q1   (    0.438E 01,  0.106E 01,VCET, 0.310E 00, 0.252E 03 ,
J1,  0.102E-08 )
CCT,1-2 = Q1   (    0.265E 01,  0.110E 01,VCCT, 0.794E-01, 0.385E 04 ,
J2,  0.533E-06 )
J1, 1-E = DIODE EQUATION (  0.102E-08 ,  0.317E 02 )
J2, 1-2 = DIODE EQUATION (  0.533E-06 ,  0.242E 02 )
RBB, B-1 =  0.161E-02
RE,  1-E =  0.288E 07
RC,  1-2 =  0.106E 08
RCC, 2-C =  0.100E-03
JA, E-1 = 0.667E 00 * J2
JB, 2-1 = 0.988E 00 * J1
FUNCTIONS
Q1(A,B,C,D,E,F,G)=((A/ABS(B-C)**D)+E*(F+G))
```

```
MODEL 2N329A (PERM) (E-C-B)
SUPPLIED BY CONVAIR DIVISION, GENERAL DYNAMICS, SAN DIEGO CALIF.92112
UNITS  OHMS,VOLTS,AMPS,FARADS,HENRIES,SECONDS
ELEMENTS
RB,1-B=350.
RC,C-2=32.
RE,E-3=2.5
RCC,2-1=5.E7
REE,3-1=5.E7
CE,3-1=EQUATION1(28.E-12,.80,VCE,.47,39.,8.8E-8,JE,2.16E-14)
CC,2-1=EQUATION1(96.E-12,.80,VCC,.50,29.,1.4E-9,JC,6.22E-11)
JE,3-1=DIODE EQUATION(2.16E-14,39.)
JC,2-1=DIODE EQUATION(6.22E-11,29.)
JI,1-3=.866*JC
JN,1-2=.972*JE
JP1,1-3=0.
JP2,1-2=0.
FUNCTIONS
EQUATION1(A,B,C,D,E,F,G,H)=(A/(B-C)**D+E*F*(G+H))
```

```
MODEL 2N336 (PERM) (E-C-B)
SUPPLIED BY CONVAIR DIVISION, GENERAL DYNAMICS, SAN DIEGO CALIF.92112
UNITS  OHMS,VOLTS,AMPS,FARADS,HENRIES,SECONDS
ELEMENTS
RB,B-1=1000.
RC,2-C=60.
RE,3-E=2.5
RCC,1-2=5.E7
REE,1-3=5.E7
CE,1-3=EQUATION1(57.E-12,1.0,VCE,.43,36.7,5.2E-9,JE,2.91E-12)
CC,1-2=EQUATION1(10.E-12,.80,VCC,.32,28.2,4.6E-7,JC,1.46E-11)
JE,1-3=DIODE EQUATION(2.91E-12,36.7)
JC,1-2=DIODE EQUATION(1.46E-11,28.2)
JI,3-1=.584*JC
JN,2-1=.991*JE
JP1,3-1=0.
JP2,2-1=0.
FUNCTIONS
EQUATION1(A,B,C,D,E,F,G,H)=(A/(B-C)**D+E*F*(G+H))
```

```
MODEL   2N356       (PERM) (E-B-C)
SUPPLIED BY  DAVID N. HARSTAD, SANDIA CORPORATION
SANDIA BASE,ALBUQUERQUE,NEW MEXICO 87115
EBERS MOLL TRANSISTOR EQUIVALENT CIRCUIT  NPN
UNITS-MILLIAMPS-VOLTS-KOHMS-PF-MICROHENRY-NANOSEC
ELEMENTS
CET,1-E = Q1   (    0.235E 02,  0.500E 00,VCET, 0.500E 00, 0.206E 04 ,
J1,  0.183E-04 )
CCT,1-2 = Q1   (    0.328E 02,  0.500E 00,VCCT, 0.500E 00, 0.774E 04 ,
J2,  0.183E-04 )
J1, 1-E = DIODE EQUATION (  0.183E-04 ,  0.389E 02 )
J2, 1-2 = DIODE EQUATION (  0.183E-04 ,  0.389E 02 )
RBB, B-1 =  0.200E-01
RE,  1-E =  0.800E 03
RC,  1-2 =  0.800E 03
RCC, 2-C =  0.100E-02
JA, E-1 = 0.833E 00 * J2
JB, 2-1 = 0.937E 00 * J1
FUNCTIONS
Q1(A,B,C,D,E,F,G)=((A/ABS(B-C)**D)+E*(F+G))
```

```
MODEL   2N384      (PERM) (E-B-C)
SUPPLIED BY  DAVID N. HARSTAD, SANDIA CORPORATION
SANDIA BASE,ALBUQUERQUE,NEW MEXICO 87115
A MODIFIED EBERS-MOLL EQUIVALENT CIRCUIT          PNP
UNITS-MILLIAMPS-VOLTS-KOHMS-PF-MICROHENRY-NANOSEC
ELEMENTS
CET,E-1 = Q1    (    0.109E 02,   0.500E 00,VCET, 0.333E 00, 0.619E 02 ,
J1,  0.330E-04 )
CCT,2-1 = Q1    (    0.896E 01,   0.500E 00,VCCT, 0.333E 00, 0.310E 03 ,
J2,  0.330E-04 )
J1, E-1 = DIODE EQUATION (  0.330E-04 ,  0.389E 02 )
J2, 2-1 = DIODE EQUATION (  0.330E-04 ,  0.389E 02 )
RBB, 1-E =  0.100E 00
RE,  E-1 =  0.400E 02
RC,  2-1 =  0.100E 04
RCC, C-2 =  0.200E-01
JA, 1-E = 0.900E 00 * J2
JB, 1-2 = 0.909E 00 * J1
FUNCTIONS
Q1(A,B,C,D,E,F,G)=((A/ABS(B-C)**D)+E*(F+G))
```

```
MODEL   2N393      (PERM) (E-B-C)
SUPPLIED BY  DAVID N. HARSTAD, SANDIA CORPORATION
SANDIA BASE,ALBUQUERQUE,NEW MEXICO 87115
A MODIFIED EBERS-MOLL EQUIVALENT CIRCUIT          PNP
UNITS-MILLIAMPS-VOLTS-KOHMS-PF-MICROHENRY-NANOSEC
ELEMENTS
CET,E-1 = Q1    (    0.707E 01,   0.500E 00,VCET, 0.500E 00, 0.124E 03 ,
J1,  0.572E-06 )
CCT,2-1 = Q1    (    0.112E 02,   0.500E 00,VCCT, 0.500E 00, 0.619E 04 ,
J2,  0.572E-06 )
J2, 2-1 = DIODE EQUATION (  0.572E-06 ,  0.389E 02 )
RBB, 1-E =  0.115E 00
RE,  E-1 =  0.500E 03
RC,  2-1 =  0.100E 04
RCC, C-2 =  0.100E-01
JA, 1-E = 0.900E 00 * J2
JB, 1-2 = 0.909E 00 * J1
FUNCTIONS
Q1(A,B,C,D,E,F,G)=((A/ABS(B-C)**D)+E*(F+G))
```

```
MODEL   2N404      (PERM) (E-B-C)
SUPPLIED BY  DAVID N. HARSTAD, SANDIA CORPORATION
SANDIA BASE,ALBUQUERQUE,NEW MEXICO 87115
A MODIFIED EBERS-MOLL EQUIVALENT CIRCUIT       PNP
UNITS-MILLIAMPS-VOLTS-KOHMS-PF-MICROHENRY-NANOSEC
ELEMENTS
CET,E-1 = Q1    (    0.156E 02,   0.118E 01,VCET, 0.590E 00, 0.286E 03 ,
J1,  0.203E-01 )
CCT,2-1 = Q1    (    0.366E 02,   0.313E 01,VCCT, 0.826E 00, 0.348E 04 ,
J2,  0.214E-01 )
J1, E-1 = DIODE EQUATION (  0.203E-01 ,   0.233E 02 )
J2, 2-1 = DIODE EQUATION (  0.214E-01 ,   0.219E 02 )
RBB, 1-B =  0.190E-02
RE,  E-1 =  0.647E 04
RC,  2-1 =  0.746E 04
RCC, C-2 =  0.100E-03
JA, 1-E = 0.900E 00 * J2
JB, 1-2 = 0.984E 00 * J1
FUNCTIONS
Q1(A,B,C,D,E,F,G)=((A/ABS(B-C)**D)+E*(F+G))
```

```
MODEL   MPS404A    (PERM) (E-B-C)
SUPPLIED BY  DAVID N. HARSTAD, SANDIA CORPORATION
SANDIA BASE,ALBUQUERQUE,NEW MEXICO 87115
A MODIFIED EBERS-MOLL EQUIVALENT CIRCUIT       PNP
UNITS-MILLIAMPS-VOLTS-KOHMS-PF-MICROHENRY-NANOSEC
ELEMENTS
CET,E-1 = Q1    (    0.755E 01,   0.500E 00,VCET, 0.333E 00, 0.322E 04 ,
J1,  0.273E-06 )
CCT,2-1 = Q1    (    0.630E 01,   0.500E 00,VCCT, 0.333E 00, 0.330E 05 ,
J2,  0.188E-06 )
J1, E-1 = DIODE EQUATION (  0.273E-06 ,   0.242E 02 )
J2, 2-1 = DIODE EQUATION (  0.188E-06 ,   0.249E 02 )
RBB, 1-B =  0.822E-03
RE,  E-1 =  0.100E 08
RC,  2-1 =  0.100E 08
RCC, C-2 =  0.100E-03
JA, 1-E = 0.950E 00 * J2
JB, 1-2 = 0.993E 00 * J1
FUNCTIONS
Q1(A,B,C,D,E,F,G)=((A/ABS(B-C)**D)+E*(F+G))
```

```
MODEL   2N457      (PERM) (E-B-C)
SUPPLIED BY  DAVID N. HARSTAD, SANDIA CORPORATION
SANDIA BASE,ALBUQUERQUE,NEW MEXICO 87115
A MODIFIED EBERS-MOLL EQUIVALENT CIRCUIT        PNP
UNITS-MILLIAMPS-VOLTS-KOHMS-PF-MICROHENRY-NANOSEC
ELEMENTS
CET,E-1 = G1   (   0.122E 03,  0.500E 00,VCET, 0.500E 00, 0.144E 05 ,
J1,  0.258E-04 )
CCT,2-1 = G1   (   0.122E 03,  0.500E 00,VCCT, 0.500E 00, 0.619E 05 ,
J2,  0.258E-04 )
J1, E-1 = DIODE EQUATION (  0.258E-04 ,  0.389E 02 )
J2, 2-1 = DIODE EQUATION (  0.258E-04 ,  0.389E 02 )
RBB, 1-B =  0.100E-01
RE,  E-1 =  0.600E 02
RC,  2-1 =  0.600E 02
RCC, C-2 =  0.100E-03
JA, 1-E = 0.833E 00 * J2
JB, 1-2 = 0.968E 00 * J1
FUNCTIONS
G1(A,B,C,D,E,F,G)=((A/ABS(B-C)**D)+E*(F+G))
```

```
MODEL   GE489A  (PERM) (B1-E-B2)
SUPPLIED BY  DAVID N. HARSTAD, SANDIA CORPORATION
SANDIA BASE,ALBUQUERQUE,NEW MEXICO 87115
UNIJUNCTION TRANSISTOR MODEL-M E DANIEL
UNITS  MILLIAMPS-VOLTS-KOHMS-PF-MICROHENRIES-NANOSECONDS
ELEMENTS
R1,I-B1=G3(.6,.0019,VJ1,10.,6.52,31.5,1.,PIE,PIE1,.5758)
JB,B2-I = G2(.1752,PIE)
C1,E-I = 1000
RC,E-I = 6.E5
JE,E-I = DIODE EQUATION(2.99E-7,22.4)
R2,B2-I = G2(1.,.6,.0019,VJ1,10.,6.52,31.5)
FUNCTIONS
EQUATION 1(A,B,C,D,E,F,G)=((A-B+C*(D-E))*(F+G*(D-E)))
EQUATION 2(A,B)=(A*B)
EQUATION 3(A,B,C,D,E,F,G,C,P,C,)=(((A-B*(C-D))*(E+F*(C-D)))/((G+(C/F))
**G))
```

```
MODEL     GE493          (PERM)(B1-E-B2)
SUPPLIED BY  DAVID N. HARSTAD, SANDIA CORPORATION
SANDIA BASE,ALBUQUERQUE,NEW MEXICO 87115
UNIJUNCTION TRANSISTOR MODEL-M E DANIEL
UNITS  MILLIAMPS-VOLTS-KOHMS-PF-MICROHENRIES-NANOSECONDS
ELEMENTS
R1,I-B1 = Q3(.72,.0024,VJ1,10.,5.69,27.50,1.,PIE,PIE1,.6134)
JB, B2-I  = Q2(.1645,PIE)
C1, E-I =1000
RC, E-I =  6.E5
JE, E-I =  DIODE EQUATION(9.056E-6,14.4)
R2, B2-I = Q1(1.,.72,.0024,VJ1,10.,5.69,27.5)
FUNCTIONS
EQUATION 1(A,B,C,D,E,F,G)=((A-B+C*(D-E))*(F+G*(D-E)))
EQUATION 2(A,B)=(A*B)
EQUATION 3(A,B,C,D,E,F,G,O,P,Q,)=(((A-B*(C-D))*(E+F*(C-D)))/((G+(O/P))
**Q))
```

```
MODEL    MJE521     (PERM) (E-B-C)
SUPPLIED BY  DAVID N. HARSTAD, SANDIA CORPORATION
SANDIA BASE,ALBUQUERQUE,NEW MEXICO 87115
EBERS MOLL TRANSISTOR EQUIVALENT CIRCUIT  NPN
UNITS-MILLIAMPS-VOLTS-KOHMS-PF-MICROHENRY-NANOSEC
ELEMENTS
CET,1-E = Q1   (   0.447E-01,  0.400E 01,VCET, 0.500E 00, 0.103E 04 ,
J1,  0.586E-07 )
CCT,1-2 = Q1   (   0.230E 03,  0.600E 00,VCCT, 0.300E 00, 0.200E 06 ,
J2,  0.195E-06 )
J1, 1-E = DIODE EQUATION (  0.586E-07 ,  0.324E 02 )
J2, 1-2 = DIODE EQUATION (  0.195E-06 ,  0.251E 02 )
RBB, B-1 =  0.200E-03
RE,  1-E =  0.600E 08
RC,  1-2 =  0.100E-05
RCC, 2-C =  0.100E-04
JA, E-1 = 0.500E 00 * J2
JB, 2-1 = 0.976E 00 * J1
FUNCTIONS
Q1(A,B,C,D,E,F,G)=((A/ABS(B-C)**D)+E*(F+G))
```

```
MODEL   2N597     (PERM) (E-B-C)
SUPPLIED BY  DAVID N. HARSTAD, SANDIA CORPORATION
SANDIA BASE,ALBUQUERQUE,NEW MEXICO 87115
A MODIFIED EBERS-MOLL EQUIVALENT CIRCUIT       PNP
UNITS-MILLIAMPS-VOLTS-KOHMS-PF-MICROHENRY-NANOSEC
ELEMENTS
CET,E-1 = G1   (    0.208E 02,   0.500E 00,VCET, 0.500E 00, 0.208E 04 ,
J1,  0.522E-05 )
CCT,2-1 = G1   (    0.704E 02,   0.500E 00,VCCT, 0.500E 00, 0.124E 05 ,
J2,  0.522E-05 )
J1, E-1 = DIODE EQUATION (  0.522E-05 ,   0.389E 02 )
J2, 2-1 = DIODE EQUATION (  0.522E-05 ,   0.389E 02 )
RBB, 1-B =  0.600E-02
RE,  E-1 =  0.300E 07
RC,  2-1 =  0.300E 03
RCC, C-2 =  0.250E 01
JA, 1-E = 0.900E 00 * J2
JB, 1-2 = 0.988E 00 * J1
FUNCTIONS
G1(A,B,C,D,E,F,G)=((A/ABS(B-C)**D)+E*(F+G))
```

```
MODEL 2N657 (PERM) (E-C-B)
SUPPLIED BY CONVAIR DIVISION, GENERAL DYNAMICS, SAN DIEGO CALIF.92112
UNITS  OHMS,VOLTS,AMPS,FARADS,HENRIES,SECONDS
ELEMENTS
RB,B-1=18.0
RC,2-C=6.1
RE,3-E=4.5
RCC,1-2=5.E7
REE,1-3=5.E7
CE,1-3=EQUATION1(80.E-12,.90,VCE,.41,20.7,1.8E-8,JE,1.E-7)
CC,1-2=EQUATION1(47.E-12,.90,VCC,.395,20.7,4.44E-7,JC,1.96E-7)
JE,1-3=DIODE EQUATION(1.E-7,20.7)
JC,1-2=DIODE EQUATION(1.96E-7,20.7)
JI,3-1=.50*JC
JN,2-1=.981*JE
JP1,3-1=0.
JP2,2-1=0.
FUNCTIONS
EQUATION1(A,B,C,D,E,F,G,H)=(A/(B-C)**D+E*F*(G+H))
```

```
MODEL    2N705        (PERM) (E-B-C)
SUPPLIED BY  DAVID N. HARSTAD, SANDIA CORPORATION
SANDIA BASE,ALBUQUERQUE,NEW MEXICO 87115
A MODIFIED EBERS-MOLL EQUIVALENT CIRCUIT       PNP
UNITS-MILLIAMPS-VOLTS-KOHMS-PF-MICROHENRY-NANOSEC
ELEMENTS
CET,E-1 = G1   (    0.350E 01,  0.300E 00,VCET, 0.700E 00, 0.225E 02 ,
J1,  0.286E-01 )
CCT,2-1 = G1   (    0.119E 02,  0.300E 00,VCCT, 0.250E 00, 0.128E 04 ,
J2,  0.286E-01 )
J1, E-1 = DIODE EQUATION (  0.286E-01 ,  0.212E 02 )
J2, 2-1 = DIODE EQUATION (  0.286E-01 ,  0.212E 02 )
RBB, 1-B =  0.500E-01
RE,  E-1 =  0.100E 04
RC,  2-1 =  0.100E 04
RCC, C-2 =  0.120E-01
JA, 1-E = 0.667E 00 * J2
JB, 1-2 = 0.976E 00 * J1
FUNCTIONS
G1(A,B,C,D,E,F,G)=((A/ABS(B-C)**D)+E*(F+G))
```

```
MODEL    2N706A     (PERM) (E-B-C)
SUPPLIED BY  DAVID N. HARSTAD, SANDIA CORPORATION
SANDIA BASE,ALBUQUERQUE,NEW MEXICO 87115
EBERS MOLL TRANSISTOR EQUIVALENT CIRCUIT  NPN
UNITS-MILLIAMPS-VOLTS-KOHMS-PF-MICROHENRY-NANOSEC
ELEMENTS
CET,1-E = G1   (    0.814E 01,  0.500E 00,VCET, 0.333E 00, 0.251E 02 ,
J1,  0.471E-09 )
CCT,1-2 = G1   (    0.696E 01,  0.500E 00,VCCT, 0.333E 00, 0.378E 03 ,
J2,  0.779E-06 )
J1, 1-E = DIODE EQUATION (  0.471E-09 ,  0.316E 02 )
J2, 1-2 = DIODE EQUATION (  0.779E-06 ,  0.238E 02 )
RBB, B-1 =  0.130E-02
RE,  1-E =  0.105E 08
RC,  1-2 =  0.359E 07
RCC, 2-C =  0.100E-03
JA, E-1 = 0.833E 00 * J2
JB, 2-1 = 0.           * J1
FUNCTIONS
G1(A,B,C,D,E,F,G)=((A/ABS(B-C)**D)+E*(F+G))
```

```
MODEL 2N718 (PERM) (E-C-E)
SUPPLIED BY CONVAIR DIVISION, GENERAL DYNAMICS, SAN DIEGO CALIF.92112
UNITS  OHMS,VOLTS,AMPS,FARADS,HENRIES,SECONDS
ELEMENTS
RB,B-1=30.
RC,2-C=7.4
RE,E-3=.10
RCC,1-2=5.E7
REE,3-1=5.E7
CE,1-3=EQUATION1(37.E-12,.90,VCE,.41,40.,7.7E-10,JE,3.3E-14)
CC,1-2=EQUATION1(25.E-12,.80,VCC,.38,38.,2.5E-6,JC,2.E-13)
JE,1-3=DIODE EQUATION(3.3E-14,40.)
JC,1-2=DIODE EQUATION(2.E-13,38.)
JI,3-1=.50*JC
JN,2-1=.991*JE
JP1,3-1=0.
JP2,2-1=0.
FUNCTIONS
EQUATION1(A,B,C,D,E,F,G,H)=(A/(B-C)**D+E*F*(G+H))
```

```
MODEL 2N720A (PERM) (E-C-E)
SUPPLIED BY CONVAIR DIVISION, GENERAL DYNAMICS, SAN DIEGO CALIF.92112
UNITS  OHMS,VOLTS,AMPS,FARADS,HENRIES,SECONDS
ELEMENTS
RB,B-1=50.
RC,2-C=8.7
RE,E-3=1.0
RCC,1-2=5.E7
REE,3-1=5.E7
CE,1-3=EQUATION1(78.E-12,1.0,VCE,.41,40.,1.5E-9,JE,6.6E-14)
CC,1-2=EQUATION1(32.E-12,.90,VCC,.39,39.,4.2E-6,JC,1.86E-13)
JE,1-3=DIODE EQUATION(6.6E-14,40.)
JC,1-2=DIODE EQUATION(1.86E-13,39.)
JI,3-1=.50*JC
JN,2-1=.991*JE
JP1,3-1=0.
JP2,2-1=0.
FUNCTIONS
EQUATION1(A,B,C,D,E,F,G,H)=(A/(B-C)**D+E*F*(G+H))
```

```
MODEL  2N743   (PERM) (E-B-C)
TAKEN FROM AFWL-TR-69-44
UNITS-VOLTS-KOHMS-PFARADS-MICROHENRIES-MILLIAMPS-NANOSEC
NPN   EBERS-MOLL EQUIVALENT CIRCUIT
ELEMENTS
RC,C-3=.0083
RB,B-1=.2
CC,3-1=EQUATION 1(5.3,.8,VCC,.25,1642.,JC,1.49E-11)
CE,E-1=EQUATION 1(4.6,.9,VCE,.39,15.88,JE,3.26E-11)
JE,1-E=DIODE EQUATION(3.26E-11,36.)
JC,1-3=DIODE EQUATION(1.49E-11,38.)
JI,E-1=.80*JC
JN,3-1=.98*JE
JP1,3-1=0.
JP2,E-1=0.
FUNCTIONS
EQUATION 1 (A,B,C,D,E,F,G)=(A/(B-C)**D+E*(F+G))
```

```
MODEL 2N915 (PERM) (B-C-E)
SUPPLIED BY CONVAIR DIVISION, GENERAL DYNAMICS, SAN DIEGO CALIF.92112
UNITS  OHMS,VOLTS,AMPS,FARADS,HENRIES,SECONDS
ELEMENTS
RB,B-1=15.0
RC,2-C=17.0
RE,E-3=1.0
RCC,1-2=5.E7
REE,3-1=5.E7
CE,1-3=EQUATION1(7.1E-12,.90,VCE,.31,40.,5.2E-10,JE,6.65E-15)
CC,1-2=EQUATION1(4.6E-12,.80,ECC,.41,27.,1.8E-6,JC,1.86E-13)
JE,1-3=DIODE EQUATION(6.65E-15,40.)
JC,1-2=DIODE EQUATION(1.86E-13,27.)
JI,3-1=.50*JC
JN,2-1=.992*JE
JP1,3-1=0.
JP2,2-1=0.
FUNCTIONS
EQUATION1(A,B,C,D,E,F,G,H)=(A/(B-C)**D+E*F*(G+H))
```

```
MODEL 2N910 (PERM) (B-C-E)
SUPPLIED BY CONVAIR DIVISION, GENERAL DYNAMICS, SAN DIEGO CALIF.92112
UNITS  OHMS,VOLTS,AMPS,FARADS,HENRIES,SECONDS
ELEMENTS
RB,B-1=13.0
RC,2-C=8.0
RE,E-3=.3
RCC,1-2=5.E7
REE,3-1=5.E7
CE,1-3=EQUATION1(50.E-12,.90,VCE,.20,37.5,2.71E-9,JE,4.07E-14)
CC,1-2=EQUATION1(22.E-12,.90,VCC,.31,37.5,9.25E-8,JC,4.87E-14)
JE,1-3=DIODE EQUATION(4.07E-14,37.5)
JC,1-2=DIODE EQUATION(4.87E-14,37.5)
J1,3-1=.285*JC
JN,2-1=.985*JE
JP1,3-1=0.
JP2,2-1=0.
FUNCTIONS
EQUATION1(A,B,C,D,E,F,G,H)=(A/(B-C)**D+E*F*(G+H))
```

```
MODEL 2N834     (PERM) (E-B-C)
TAKEN FROM AFWL-TR-69-44
UNITS-VOLTS-KOHMS-PFARADS-MICROHENRIES-MILLIAMPS-NANOSEC
NPN   EBERS-MOLL EQUIVALENT CIRCUIT
ELEMENTS
RC,C-3=.011
RB,B-1=.01
CC,3-1=EQUATION 1(7.3,.9,VCC,.31,1868.4,JC,1.14E-11)
CE,E-1=EQUATION 1(9.0,.8,VCE,.50,18.56,JE,1.32E-12)
JE,1-E=DIODE EQUATION(1.32E-12,41.)
JC,1-3=DIODE EQUATION(1.14E-11,38.)
JI,E-1=.878*JC
JN,3-1=.9843*JE
JP1,3-1=0.
JP2,E-1=0.
FUNCTIONS
EQUATION 1 (A,B,C,D,E,F,G)=(A/(B-C)**D+E*(F+G))
```

```
MODEL 2N918 (PERM) (B-C-E)
SUPPLIED BY CONVAIR DIVISION, GENERAL DYNAMICS, SAN DIEGO CALIF.92112
UNITS  OHMS,VOLTS,AMPS,FARADS,HENRIES,SECONDS
ELEMENTS
RB,B-1=12.0
RC,2-C=11.5
RE,E-3=8.5
RCC,1-2=5.E7
REE,3-1=5.E7
CE,1-3=EQUATION1(2.E-12,.50,VCE,.15,38.,2.18E-10,JE,2.58E-15)
CC,1-2=EQUATION1(1.7E-12,.80,VCC,.12,36.3,1.27E-8,JC,2.12E-14)
JE,1-3=DIODE EQUATION(2.58E-15,38.)
JC,1-2=DIODE EQUATION(2.12E-14,36.3)
JI,3-1=.492*JC
JN,2-1=.977*JE
JP1,3-1=0.
JP2,2-1=0.
FUNCTIONS
EQUATION1(A,B,C,D,E,F,G,H)=(A/(B-C)**D+E*F*(G+H))
```

```
MODEL   2N964      (PERM) (E-B-C)
SUPPLIED BY  DAVID N. HARSTAD, SANDIA CORPORATION
SANDIA BASE,ALBUQUERQUE,NEW MEXICO 87115
A MODIFIED EBERS-MOLL EQUIVALENT CIRCUIT        PNP
UNITS-MILLIAMPS-VOLTS-KOHMS-PF-MICROHENRY-NANOSEC
ELEMENTS
CET,E-1 = G1   (    0.229E 01,  0.500E 00,VCET, 0.333E 00, 0.970E 01 ,
J1,  0.941E-03 )
CCT,2-1 = G1   (    0.481E 01,  0.500E 00,VCCT, 0.333E 00, 0.920E 02 ,
J2,  0.227E-02 )
J1, E-1 = DIODE EQUATION (  0.941E-03 ,  0.280E 02 )
J2, 2-1 = DIODE EQUATION (  0.227E-02 ,  0.289E 02 )
RBB, 1-B =  0.149E-02
RE,  E-1 =  0.183E 04
RC,  2-1 =  0.729E 03
RCC, C-2 =  0.400E-04
JA, 1-E = 0.909E 00 * J2
JB, 1-2 = 0.980E 00 * J1
FUNCTIONS
G1(A,B,C,D,E,F,G)=((A/ABS(B-C)**D)+E*(F+G))
```

```
MODEL   2N995      (PERM) (E-B-C)
SUPPLIED BY  DAVID N. HARSTAD, SANDIA CORPORATION
SANDIA BASE,ALBUQUERQUE,NEW MEXICO 87115
A MODIFIED EBERS-MOLL EQUIVALENT CIRCUIT          PNP
UNITS-MILLIAMPS-VOLTS-KOHMS-PF-MICROHENRY-NANOSEC
ELEMENTS
CET,E-1 = Q1    (    0.127E 02,  0.108E 01,VCET, 0.320E 00, 0.423E 02 ,
J1,  0.274E-08 )
CCT,2-1 = Q1    (    0.152E 02,  0.101E 01,VCCT, 0.176E 00, 0.415E 03 ,
J2,  0.616E-08 )
J1, E-1 = DIODE EQUATION (  0.274E-08 ,  0.266E 02 )
J2, 2-1 = DIODE EQUATION (  0.616E-08 ,  0.261E 02 )
RBB, 1-B =  0.250E-01
RE,  E-1 =  0.130E 09
RC,  2-1 =  0.130E 09
RCC, C-2 =  0.200E-02
JA, 1-B = 0.667E 00 * J2
JB, 1-2 = 0.986E 00 * J1
FUNCTIONS
Q1(A,B,C,D,E,F,G)=((A/ABS(B-C)**D)+E*(F+G))
```

```
MODEL 2N1016B  (PERM) (E-B-C)
TAKEN FROM AFWL-TR-69-44
UNITS-VOLTS-KOHMS-PFARADS-MICROHENRIES-MILLIAMPS-NANOSEC
NPN   EBERS-MOLL EQUIVALENT CIRCUIT
ELEMENTS
RC,C-3=5.E-4
RB,B-1=.03
JC,3-1=EQUATION 1(1600.,.7,VCC,.5,81195.,JC,2.20E-7)
JE,B-1=EQUATION 1(350.,.8,VCE,.47,13696.,JE,6.44E-8)
JE,1-B=DIODE EQUATION(6.44E-10,40.)
JC,1-3=DIODE EQUATION(2.20E-10,38.)
JI,B-1=.7368*JC
JN,3-1=.9783*JE
JP1,3-1=0.
JP2,B-1=0.
FUNCTIONS
EQUATION 1 (A,B,C,D,E,F,G)=(A/(B-C)**D+E*(F+G))
```

```
MODEL    2N1039      (PERM) (E-B-C)
SUPPLIED BY  DAVID N. HARSTAD, SANDIA CORPORATION
SANDIA BASE,ALBUQUERQUE,NEW MEXICO 87115
A MODIFIED EBERS-MOLL EQUIVALENT CIRCUIT        PNP
UNITS-MILLIAMPS-VOLTS-KOHMS-PF-MICROHENRY-NANOSEC
ELEMENTS
CET,E-1 = Q1    (    0.122E 04,  0.500E 00,VCET, 0.500E 00, 0.774E 06 ,
J1,  0.321E-04 )
CCT,2-1 = Q1    (    0.255E 03,  0.500E 00,VCCT, 0.500E 00, 0.619E 07 ,
J2,  0.321E-04 )
J1, E-1 = DIODE EQUATION (  0.321E-04 ,  0.389E 02 )
J2, 2-1 = DIODE EQUATION (  0.321E-04 ,  0.389E 02 )
RBB, 1-B =  0.100E-01
RE,  E-1 =  0.270E 02
RC,  2-1 =  0.240E 03
RCC, C-2 =  0.290E-03
JA, 1-E = 0.833E 00 * J2
JB, 1-2 = 0.976E 00 * J1
FUNCTIONS
Q1(A,B,C,D,E,F,G)=((A/ABS(B-C)**D)+E*(F+G))
```

```
MODEL    2N1131      (PERM) (E-B-C)
SUPPLIED BY  DAVID N. HARSTAD, SANDIA CORPORATION
SANDIA BASE,ALBUQUERQUE,NEW MEXICO 87115
A MODIFIED EBERS-MOLL EQUIVALENT CIRCUIT        PNP
UNITS-MILLIAMPS-VOLTS-KOHMS-PF-MICROHENRY-NANOSEC
ELEMENTS
CET,E-1 = Q1    (    0.838E 02,  0.163E 01,VCET, 0.496E 00, 0.105E 03 ,
J1,  0.356E-08 )
CCT,2-1 = Q1    (    0.727E 02,  0.936E 00,VCCT, 0.361E 00, 0.475E 03 ,
J2,  0.144E-05 )
J1, E-1 = DIODE EQUATION (  0.356E-08 ,  0.330E 02 )
J2, 2-1 = DIODE EQUATION (  0.144E-05 ,  0.239E 02 )
RBB, 1-B =  0.182E-02
RE,  E-1 =  0.199E 07
RC,  2-1 =  0.189E 07
RCC, C-2 =  0.100E-03
JA, 1-E = 0.667E 00 * J2
JB, 1-2 = 0.952E 00 * J1
FUNCTIONS
Q1(A,B,C,D,E,F,G)=((A/ABS(B-C)**D)+E*(F+G))
```

```
MODEL 2N1184   (PERM) (E-B-C)
TAKEN FROM AFWL-TR-69-44
UNITS-VOLTS-KOHMS-PFARADS-MICROHENRIES-MILLIAMPS-NANOSEC
PNP   EBERS-MOLL EQUIVALENT CIRCUIT
ELEMENTS
RC,C-2=1.2E-3
RB,B-1=.02
CC,2-1=EQUATION 1 (240.,.4,VCC,.48,5997.6,JC,1.13E-2)
CE,E-1=EQUATION 1 (200.,.4,VCE,.50,3612.8,JE,8.12E-3)
JE,E-1=DIODE EQUATION(8.12E-3,47.)
JC,2-1=DIODE EQUATION(1.13E-2,40.)
JI,1-E=.8820*JC
JN,1-2=.9855*JE
JP1,1-2=0.
JP2,1-E=0.
FUNCTIONS
EQUATION 1 (A,B,C,D,E,F,G)=(A/(B-C)**D+E*(F+G))
```

```
MODEL   2N1228    (PERM) (E-B-C)
SUPPLIED BY  DAVID N. HARSTAD, SANDIA CORPORATION
SANDIA BASE,ALBUQUERQUE,NEW MEXICO 87115
A MODIFIED EBERS-MOLL EQUIVALENT CIRCUIT     PNP
UNITS-MILLIAMPS-VOLTS-KOHMS-PF-MICROHENRY-NANOSEC
ELEMENTS
CET,E-1 = G1   (    0.178E 03,  0.108E 01,VCET, 0.320E 00, 0.305E 04 ,
J1,  0.175E-06 )
CCT,2-1 = G1   (    0.137E 03,  0.101E 01,VCCT, 0.176E 00, 0.332E 05 ,
J2,  0.533E-07 )
J1, E-1 = DIODE EQUATION (  0.175E-06 ,  0.230E 02 )
J2, 2-1 = DIODE EQUATION (  0.533E-07 ,  0.250E 02 )
RBB, 1-B =  0.200E-02
RE,  E-1 =  0.135E 08
RC,  2-1 =  0.809E 08
RCC, C-2 =  0.100E-03
JA, 1-E = 0.667E 00 * J2
JB, 1-2 = 0.941E 00 * J1
FUNCTIONS
G1(A,B,C,D,E,F,G)=((A/ABS(B-C)**D)+E*(F+G))
```

```
MODEL   2N1301     (PERM) (E-B-C)
SUPPLIED BY  DAVID N. HARSTAD, SANDIA CORPORATION
SANDIA BASE,ALBUQUERQUE,NEW MEXICO 87115
A MODIFIED EBERS-MOLL EQUIVALENT CIRCUIT       PNP
UNITS-MILLIAMPS-VOLTS-KOHMS-PF-MICROHENRY-NANOSEC
ELEMENTS
CET,E-1 = Q1   (   0.136E 02,  0.500E 00,VCET, 0.333E 00, 0.177E 03 ,
J1,  0.572E-06 )
CCT,2-1 = Q1   (   0.224E 02,  0.500E 00,VCCT, 0.333E 00, 0.619E 03 ,
J2,  0.572E-06 )
J1, E-1 = DIODE EQUATION (  0.572E-06 ,  0.389E 02 )
J2, 2-1 = DIODE EQUATION (  0.572E-06 ,  0.389E 02 )
RBB, 1-B =  0.500E-01
RE,  E-1 =  0.400E 02
RC,  2-1 =  0.650E 03
RCC, C-2 =  0.500E-02
JA, 1-E = 0.833E 00 * J2
JB, 1-2 = 0.986E 00 * J1
FUNCTIONS
Q1(A,B,C,D,E,F,G)=((A/ABS(B-C)**D)+E*(F+G))
```

```
MODEL   2N1306     (PERM) (E-B-C)
SUPPLIED BY  DAVID N. HARSTAD, SANDIA CORPORATION
SANDIA BASE,ALBUQUERQUE,NEW MEXICO 87115
EBERS MOLL TRANSISTOR EQUIVALENT CIRCUIT  NPN
UNITS-MILLIAMPS-VOLTS-KOHMS-PF-MICROHENRY-NANOSEC
ELEMENTS
CET,1-E = Q1   (   0.210E 02,  0.140E 01,VCET, 0.400E 00, 0.340E 03 ,
J1,  0.196E-02 )
CCT,1-2 = Q1   (   0.428E 02,  0.140E 01,VCCT, 0.407E 00, 0.206E 04 ,
J2,  0.200E-02 )
J1, 1-E = DIODE EQUATION (  0.196E-02 ,  0.257E 02 )
J2, 1-2 = DIODE EQUATION (  0.200E-02 ,  0.259E 02 )
RBB, B-1 =  0.400E-01
RE,  1-E =  0.150E 05
RC,  1-2 =  0.400E 04
RCC, 2-C =  0.100E-02
JA, E-1 = 0.833E 00 * J2
JB, 2-1 = 0.990E 00 * J1
FUNCTIONS
Q1(A,B,C,D,E,F,G)=((A/ABS(B-C)**D)+E*(F+G))
```

```
MODEL   2N1308     (PERM) (B-E-C)
SUPPLIED BY  DAVID N. HARSTAD, SANDIA CORPORATION
SANDIA BASE,ALBUQUERQUE,NEW MEXICO 87115
EBERS MOLL TRANSISTOR EQUIVALENT CIRCUIT  NPN
UNITS-MILLIAMPS-VOLTS-KOHMS-PF-MICROHENRY-NANOSEC
ELEMENTS
JET,1-E = G1   (    0.210E 02,  0.140E 01,VCET, 0.400E 00, 0.264E 03 ,
J1,  0.199E-02 )
JCT,1-2 = G1   (    0.428E 02,  0.140E 01,VCCT, 0.407E 00, 0.206E 04 ,
J2,  0.203E-02 )
J1, 1-E = DIODE EQUATION (  0.199E-02 ,  0.257E 02 )
J2, 1-2 = DIODE EQUATION (  0.203E-02 ,  0.259E 02 )
RBB, B-1 =  0.400E-01
RE,  1-E =  0.150E 05
RC,  1-2 =  0.400E 04
RCC, 2-C =  0.100E-02
JA, E-1 = 0.833E 00 * J2
JB, 2-1 = 0.993E 00 * J1
FUNCTIONS
G1(A,B,C,D,E,F,G)=((A/ABS(B-C)**D)+E*(F+G))
```

```
MODEL   2N1483     (PERM) (B-E-C)
SUPPLIED BY  DAVID N. HARSTAD, SANDIA CORPORATION
SANDIA BASE,ALBUQUERQUE,NEW MEXICO 87115
EBERS MOLL TRANSISTOR EQUIVALENT CIRCUIT  NPN
UNITS-MILLIAMPS-VOLTS-KOHMS-PF-MICROHENRY-NANOSEC
ELEMENTS
JET,1-E = G1   (    0.309E 03,  0.500E 00,VCET, 0.333E 00, 0.496E 04 ,
J1,  0.338E-10 )
JCT,1-2 = G1   (    0.600E 03,  0.500E 00,VCCT, 0.333E 00, 0.619E 05 ,
J2,  0.338E-10 )
J1, 1-E = DIODE EQUATION (  0.338E-10 ,  0.389E 02 )
J2, 1-2 = DIODE EQUATION (  0.338E-10 ,  0.389E 02 )
RBB, B-1 =  0.120E-01
RE,  1-E =  0.800E 03
RC,  1-2 =  0.200E 04
RCC, 2-C =  0.267E-02
JA, E-1 = 0.833E 00 * J2
JB, 2-1 = 0.952E 00 * J1
FUNCTIONS
G1(A,B,C,D,E,F,G)=((A/ABS(B-C)**D)+E*(F+G))
```

```
MODEL   2N1499A    (PERM) (E-B-C)
SUPPLIED BY  DAVID N. HARSTAD, SANDIA CORPORATION
SANDIA BASE,ALBUQUERQUE,NEW MEXICO 87115
A MODIFIED EBERS-MOLL EQUIVALENT CIRCUIT      PNP
UNITS-MILLIAMPS-VOLTS-KOHMS-PF-MICROHENRY-NANOSEC
ELEMENTS
GET,E-1 = Q1    (    0.572E 01,  0.500E 00,VCET, 0.333E 00, 0.233E 03 ,
J1,  0.282E-02 )
GCT,2-1 = Q1    (    0.560E 01,  0.500E 00,VCCT, 0.333E 00, 0.449E 03 ,
J2,  0.282E-02 )
J1, E-1 = DIODE EQUATION (  0.282E-02 ,  0.293E 02 )
J2, 2-1 = DIODE EQUATION (  0.282E-02 ,  0.282E 02 )
RBB, 1-B =  0.500E-01
RE,  E-1 =  0.500E 03
RC,  2-1 =  0.160E 04
RCC, C-2 =  0.100E-01
JA, 1-E = 0.667E 00 * J2
JB, 1-2 = 0.968E 00 * J1
FUNCTIONS
Q1(A,B,C,D,E,F,G)=((A/ABS(B-C)**D)+E*(F+G))
```

```
MODEL   2N1613     (PERM) (E-B-C)
SUPPLIED BY  DAVID N. HARSTAD, SANDIA CORPORATION
SANDIA BASE,ALBUQUERQUE,NEW MEXICO 87115
EBERS MOLL TRANSISTOR EQUIVALENT CIRCUIT  NPN
UNITS-MILLIAMPS-VOLTS-KOHMS-PF-MICROHENRY-NANOSEC
ELEMENTS
GET,1-E = Q1    (    0.546E 02,  0.713E 00,VCET, 0.323E 00, 0.565E 02 ,
J1,  0.219E-06 )
GCT,1-2 = Q1    (    0.269E 04,  0.168E 01,VCCT, 0.446E 01, 0.258E 03 ,
J2,  0.226E-06 )
J1, 1-E = DIODE EQUATION (  0.219E-06 ,  0.320E 02 )
J2, 1-2 = DIODE EQUATION (  0.226E-06 ,  0.324E 02 )
RBB, B-1 =  0.100E-01
RE,  1-E =  0.500E 06
RC,  1-2 =  0.600E 07
RCC, 2-C =  0.100E-02
JA, E-1 = 0.833E 00 * J2
JB, 2-1 = 0.988E 00 * J1
FUNCTIONS
Q1(A,B,C,D,E,F,G)=((A/ABS(B-C)**D)+E*(F+G))
```

```
MODEL    2N1724      (PERM) (E-B-C)
SUPPLIED BY  DAVID N. HARSTAD, SANDIA CORPORATION
SANDIA BASE,ALBUQUERQUE,NEW MEXICO 87115
EBERS MOLL TRANSISTOR EQUIVALENT CIRCUIT  NPN
UNITS-MILLIAMPS-VOLTS-KOHMS-PF-MICROHENRY-NANOSEC
ELEMENTS
CET,1-E = G1    (    0.232E 04,  0.921E 00,VCET, 0.399E 00, 0.350E 03 ,
J1,   0.210E-08 )
CCT,1-2 = G1    (    0.968E 03,  0.921E 00,VCCT, 0.399E 00, 0.351E 04 ,
J2,   0.208E-08 )
J1, 1-E = DIODE EQUATION (  0.210E-08 ,  0.277E 02 )
J2, 1-2 = DIODE EQUATION (  0.208E-08 ,  0.278E 02 )
RBB, B-1 = .1
RE,  1-E =  0.390E 08
RC,  1-2 =  0.947E 09
RCC, 2-C = .1
JA, E-1 = 0.600E 00 * J2
JB, 2-1 = 0.976E 00 * J1
FUNCTIONS
G1(A,B,C,D,E,F,G)=((A/ABS(B-C)**D)+E*(F+G))
```

```
MODEL    PT1867      (PERM) (E-B-C)
SUPPLIED BY  DAVID N. HARSTAD, SANDIA CORPORATION
SANDIA BASE,ALBUQUERQUE,NEW MEXICO 87115
EBERS MOLL TRANSISTOR EQUIVALENT CIRCUIT  NPN
UNITS-MILLIAMPS-VOLTS-KOHMS-PF-MICROHENRY-NANOSEC
ELEMENTS
CET,1-E = G1    (    0.245E 02,  0.500E 00,VCET, 0.500E 00, 0.619E 04 ,
J1,   0.280E-07 )
CCT,1-2 = G1    (    0.583E 02,  0.500E 00,VCCT, 0.500E 00, 0.619E 05 ,
J2,   0.280E-07 )
J1, 1-E = DIODE EQUATION (  0.280E-07 ,  0.389E 02 )
J2, 1-2 = DIODE EQUATION (  0.280E-07 ,  0.389E 02 )
RBB, B-1 =  0.500E-01
RE,  1-E =  0.100E 05
RC,  1-2 =  0.600E 05
RCC, 2-C =  0.500E-02
JA, E-1 = 0.900E 00 * J2
JB, 2-1 = 0.952E 00 * J1
FUNCTIONS
G1(A,B,C,D,E,F,G)=((A/ABS(B-C)**D)+E*(F+G))
```

```
MODEL 2N1900    (PERM) (E-B-C)
TAKEN FROM AFWL-TR-69-44
UNITS-VOLTS-KOHMS-PFARADS-MICROHENRIES-MILLIAMPS-NANOSEC
NPN   EBERS-MOLL EQUIVALENT CIRCUIT
ELEMENTS
RC,C-3=6.E-4
RB,B-1=5.E-3
CC,3-1=EQUATION 1(1800.,.7,VCC,.47,40610.,JC,2.316E-8)
CE,E-1=EQUATION 1(7000.,.8,VCE,.38,63.79,JE,1.6E-8)
JE,1-E=DIODE EQUATION(1.6E-8,37.8)
JC,1-3=DIODE EQUATION(2.316E-8,37.8)
JI,E-1=.5782*JC
JN,3-1=.9927*JE
JP1,3-1=0.
JP2,E-1=0.
FUNCTIONS
EQUATION 1 (A,B,C,D,E,F,G)=(A/(B-C)**D+E*(F+G))
```

```
MODEL 2N2060 (PERM) (B-C-E)
SUPPLIED BY CONVAIR DIVISION, GENERAL DYNAMICS, SAN DIEGO CALIF.92112
UNITS  OHMS,VOLTS,AMPS,FARADS,HENRIES,SECONDS
ELEMENTS
RB,B-1=94.0
RC,C-2=2.0
RE,E-3=6.0
RCC,1-2=5.E7
REE,3-1=5.E7
CE,1-3=EQUATION1(71.3E-12,.975,VCE,.445,35.8,1.77E-9,JE,4.61E-15)
CC,1-2=EQUATION1(34.5E-12,.605,VCC,.445,35.8,2.77E-8,JC,2.24E-14)
JE,1-3=DIODE EQUATION(4.61E-15,35.8)
JC,1-2=DIODE EQUATION(2.24E-14,35.8)
JI,3-1=.286*JC
JN,2-1=.992*JE
JP1,3-1=0.
JP2,2-1=0.
FUNCTIONS
EQUATION1(A,B,C,D,E,F,G,H)=(A/(B-C)**D+E*F*(G+H))
```

```
MODEL   2N2102    (PERM) (E-B-C)
SUPPLIED BY  DAVID N. FARSTAD, SANDIA CORPORATION
SANDIA BASE,ALBUQUERQUE,NEW MEXICO 87115
EBERS MOLL TRANSISTOR EQUIVALENT CIRCUIT  NPN
UNITS-MILLIAMPS-VOLTS-KOHMS-PF-MICROHENRY-NANOSEC
ELEMENTS
CET,1-E = G1   (   0.298E 02,  0.125E 01,VCET, 0.446E 00, 0.250E 05 ,
J1,  0.537E-06 )
CCT,1-2 = G1   (   0.        ,  0.256E 01,VCCT, 0.373E 00, 0.145E 05 ,
J2,  0.841E-06 )
J1, 1-E = DIODE EQUATION (  0.537E-06 ,  0.246E 02 )
J2, 1-2 = DIODE EQUATION (  0.841E-06 ,  0.242E 02 )
RBB, B-1 =  0.120E-03
RE,  1-E =  0.600E 11
RC,  1-2 =  0.400E 11
RCC, 2-C =  0.400E-04
JA, E-1 = 0.714E 00 * J2
JB, 2-1 = 0.972E 00 * J1
FUNCTIONS
G1 (A,B,C,D,E,F,G) =((A/(B-C)**D)+ E * (F + G))
```

```
MODEL 2N2187   (PERM) (E-B-C)
TAKEN FROM AFWL-TR-69-44
UNITS-VOLTS-KOHMS-PFARADS-MICROHENRIES-MILLIAMPS-NANOSEC
PNP   EBERS-MOLL EQUIVALENT CIRCUIT
ELEMENTS
RC,C-2=.015
RB,B-1=.1
CC,2-1=EQUATION 1 (7.6,.7,VCC,.41,246.5,JC,5.67E-10)
CE,E-1=EQUATION 1 (5.4,.6,VCE,.36,75.91,JE,5.67E-11)
JE,E-1=DIODE EQUATION(5.67E-11,38.9)
JC,2-1=DIODE EQUATION(5.67E-10,36.1)
JI,1-E=.3793*JC
JN,1-2=.9706*JE
JP1,1-2=0.
JP2,1-E=0.
FUNCTIONS
EQUATION 1 (A,B,C,D,E,F,G)=(A/(B-C)**D+E*(F+G))
```

```
MODEL 2N2192    (PERM) (E-B-C)
TAKEN FROM AFWL-TR-69-44
UNITS-VOLTS-KOHMS-PFARADS-MICROHENRIES-MILLIAMPS-NANOSEC
NPN    EBERS-MOLL EQUIVALENT CIRCUIT
ELEMENTS
RC,C-3=3.5E-3
RB,B-1=.04
CC,3-1=EQUATION 1(21.,.8,VCC,.41,3990.,JC,9.49E-10)
CE,E-1=EQUATION 1(39.,.9,VCE,.45,34.26,JE,1.31E-10)
JE,1-E=DIODE EQUATION(1.31E-10,42.)
JC,1-3=DIODE EQUATION(9.49E-10,38.)
JI,E-1=.875*JC
JN,3-1=.9949*JE
JP1,3-1=0.
JP2,E-1=0.
FUNCTIONS
EQUATION 1 (A,B,C,D,E,F,G)=(A/(B-C)**D+E*(F+G))
```

```
MODEL 2N2223    (PERM) (E-B-C)
TAKEN FROM AFWL-TR-69-44
UNITS-VOLTS-KOHMS-PFARADS-MICROHENRIES-MILLIAMPS-NANOSEC
NPN    EBERS-MOLL EQUIVALENT CIRCUIT
ELEMENTS
RC,C-3=.009
RB,B-1=.032
CC,3-1=EQUATION 1(33.,.9,VCC,.41,22010.,JC,1.78E-9)
CE,E-1=EQUATION 1(59.,.8,VCE,.41,60.63,JE,1.08E-10)
JE,1-E=DIODE EQUATION(1.08E-10,38.4)
JC,1-3=DIODE EQUATION(1.78E-9,35.7)
JI,E-1=.1667*JC
JN,3-1=.9868*JE
JP1,3-1=0.
JP2,E-1=0.
FUNCTIONS
EQUATION 1 (A,B,C,D,E,F,G)=(A/(B-C)**D+E*(F+G))
```

```
MODEL    2N2256     (PERM) (E-B-C)
SUPPLIED BY  DAVID N. HARSTAD, SANDIA CORPORATION
SANDIA BASE,ALBUQUERQUE,NEW MEXICO 87115
A MODIFIED EBERS-MOLL EQUIVALENT CIRCUIT         PNP
UNITS-MILLIAMPS-VOLTS-KOHMS-PF-MICROHENRY-NANOSEC
ELEMENTS
CET,E-1 = Q1    (    0.394E 02,  0.800E 00,VCET, 0.300E 00, 0.113E 03 ,
J1,  0.981E-08 )
CCT,2-1 = Q1    (    0.124E 02,  0.200E 00,VCCT, 0.200E 00, 0.737E 04 ,
J2,  0.981E-07 )
J1, E-1 = DIODE EQUATION (  0.981E-08 ,  0.354E 02 )
J2, 2-1 = DIODE EQUATION (  0.981E-07 ,  0.278E 02 )
RBB, 1-B =  0.150E-02
RE,  E-1 =  0.600E 07
RC,  2-1 =  0.500E 07
RCC, C-2 =  0.100E-03
JA, 1-E = 0.500E 00 * J2
JB, 1-2 = 0.980E 00 * J1
FUNCTIONS
Q1(A,B,C,D,E,F,G)=((A/ABS(B-C)**D)+E*(F+G))
```

```
MODEL    2N2369     (PERM) (E-B-C)
SUPPLIED BY  DAVID N. HARSTAD, SANDIA CORPORATION
SANDIA BASE,ALBUQUERQUE,NEW MEXICO 87115
EBERS MOLL TRANSISTOR EQUIVALENT CIRCUIT  NPN
UNITS-MILLIAMPS-VOLTS-KOHMS-PF-MICROHENRY-NANOSEC
ELEMENTS
J1, 1-E = DIODE EQUATION (  0.230E-10 ,  0.373E 02 )
J2, 1-2 = DIODE EQUATION (  0.104E-05 ,  0.235E 02 )
CCT,1-2 = Q1    (    0.293E 01,  0.110E 01,VCCT, 0.794E-01, 0.747E 02 ,
J2,  0.104E-05 )
CET,1-E = Q1    (    0.331E 01,  0.106E 01,VCET, 0.310E 00, 0.119E 02 ,
J1,  0.230E-10 )
RBB, B-1 =  0.200E-01
RE,  1-E =  0.500E 06
RC,  1-2 =  0.500E 06
RCC, 2-C =  0.200E-02
JA, E-1 = 0.800E 00 * J2
JB, 2-1 = 0.978E 00 * J1
FUNCTIONS
Q1(A,B,C,D,E,F,G)=((A/ABS(B-C)**D)+E*(F+G))
```

```
MODEL  2N2453 (PERM) (B-C-E)
SUPPLIED BY CONVAIR DIVISION, GENERAL DYNAMICS, SAN DIEGO CALIF.92112
UNITS  OHMS,VOLTS,AMPS,FARADS,HENRIES,SECONDS
ELEMENTS
RB,B-1=100.
RC,C-2=20.
RE,E-3=10.
RCC,1-2=5.E7
REE,3-1=5.E7
CE,1-3=EQUATION1(4.5E-12,.70,VCE,.414,46.1,2.79E-9,JE,2.85E-15)
CC,1-2=EQUATION1(5.6E-12,.70,VCC,.225,36.0,5.31E-8,JC,1.232E-14)
JE,1-3=DIODE EQUATION(2.85E-15,46.1)
JC,1-2=DIODE EQUATION(1.232E-14,36.0)
J1,3-1=.50*JC
JN,2-1=.995*JE
JP1,3-1=0.
JP2,2-1=0.
FUNCTIONS
EQUATION1(A,B,C,D,E,F,G,H)=(A/(B-C)**D+E*F*(G+H))
```

```
MODEL  2N2484     (PERM) (E-B-C)
SUPPLIED BY  DAVID N. HARSTAD, SANDIA CORPORATION
SANDIA BASE,ALBUQUERQUE,NEW MEXICO 87115
EBERS MOLL TRANSISTOR EQUIVALENT CIRCUIT  NPN
UNITS-MILLIAMPS-VOLTS-KOHMS-PF-MICROHENRY-NANOSEC
ELEMENTS
CET,1-E = G1   (   0.600E 01,  0.500E 00,VCET, 0.330E 00, 0.683E 04 ,
J1,  0.141E-05 )
CCT,1-2 = G1   (   0.105E 02,  0.500E 00,VCCT, 0.330E 00, 0.306E 05 ,
J2,  0.104E-04 )
J1, 1-E = DIODE EQUATION (  0.141E-05 ,  0.215E 02 )
J2, 1-2 = DIODE EQUATION (  0.104E-04 ,  0.192E 02 )
RBB, B-1 =  0.200E-02
RE,  1-E =  0.600E 04
RC,  1-2 =  0.400E 04
RCC, 2-C =  0.186E-03
JA, E-1 = 0.627E 00 * J2
JB, 2-1 = 0.996E 00 * J1
FUNCTIONS
G1(A,B,C,D,E,F,G)=((A/ABS(B-C)**D)+E*(F+G))
```

```
MODEL     GE2646   (PERM) (E1-E-B2)
SUPPLIED BY  DAVID N. HARSTAD, SANDIA CORPORATION
SANDIA BASE,ALBUQUERQUE,NEW MEXICO 87115
UNIJUNCTION TRANSISTOR MODEL-M E DANIEL
UNITS  MILLIAMPS-VOLTS-KOHMS-PF-MICROHENRIES-NANOSECONDS
ELEMENTS
R1,I-E1=EQUATION3(.60,0.0018,VJ1,10,8.26,.054,1.,PIE,PIE1,.6044)
C1,E-I=1000
JB,B2-I=EQUATION 2(.1391,PIE)
RC,E-I=6.E5
JE,E-I=DIODE EQUATION(1.16E-7,24.76)
R2,B2-I=EQUATION 1(1.,.60,.0018,VJ1,10,8.26,.054)
J1,B2-E1=0
FUNCTIONS
EQUATION 1(A,B,C,D,E,F,G)=((A-B+C*(D-E))*(F+G*(D-E)))
EQUATION 2(A,B)=(A*B)
EQUATION 3(A,B,C,D,E,F,G,C,F,C,)=(((A-B*(C-D))*(E+F*(C-D)))/((G+(C/F))
**G))
```

```
MODEL  MU2647 (PERM) (E1-E-B2)
SUPPLIED BY  DAVID N. HARSTAD, SANDIA CORPORATION
SANDIA BASE,ALBUQUERQUE,NEW MEXICO 87115
UNIJUNCTION TRANSISTOR MODEL-M E DANIEL
UNITS  MILLIAMPS-VOLTS-KOHMS-PF-MICROHENRIES-NANOSECONDS
ELEMENTS
R1,I-E1=G3(.85,.0025,VJ1,10.,6.3,102.,1.,PIE,PIE1,.5633)
JB,B2-I=G2(.1759,PIE)
C1,E-I=1000
RC,E-I=6.E5
JE,E-I= DIODE EQUATION(3.433E-9,35.5)
R2,B2-I=G1(1.,85.,.0025,VJ1,10.,6.3,102.)
FUNCTIONS
EQUATION 1(A,B,C,D,E,F,G)=((A-B+C*(D-E))*(F+G*(D-E)))
EQUATION 2(A,B)=(A*B)
EQUATION 3(A,B,C,D,E,F,G,C,F,C,)=(((A-B*(C-D))*(E+F*(C-D)))/((G+(C/F))
**G))
```

```
MODEL 2N2658   (PERM) (E-B-C)
TAKEN FROM AFWL-TR-69-44
UNITS-VOLTS-KOHMS-PFARADS-MICROHENRIES-MILLIAMPS-NANOSEC
NPN   EBERS-MOLL EQUIVALENT CIRCUIT
ELEMENTS
RC,C-3=.019
RB,B-1=.035
CC,3-1=EQUATION 1(5.8,.8,VCC,.31,12540.,JC,3.E-11)
CE,E-1=EQUATION 1(4.3,.9,VCE,.37,15.83,JE,4.85E-12)
JE,1-E=DIODE EQUATION(4.85E-12,40.)
JC,1-3=DIODE EQUATION(3.E-11,38.)
JI,E-1=.50*JC
JN,3-1=.9896*JE
JP1,3-1=0.
JP2,E-1=0.
FUNCTIONS
EQUATION 1 (A,B,C,D.E,F,G)=(A/(B-C)**D+E*(F+G))
```

```
MODEL 2N2708 (PERM) (B-C-E)
SUPPLIED BY CONVAIR DIVISION, GENERAL DYNAMICS, SAN DIEGO CALIF.92112
UNITS  OHMS,VOLTS,AMPS,FARADS,HENRIES,SECONDS
ELEMENTS
RB,B-1=80.
RC,C-2=15.
RE,E-3=1.0
RCC,1-2=5.E7
REE,3-1=5.E7
CE,1-3=EQUATION1(1.2E-12,.90,VCE,.14,39.,1.6E-10,JE,1.61E-16)
CC,1-2=EQUATION1(1.5E-12,.90,VCC,.15,34.,5.8E-8,JC,7.2E-15)
JE,1-3=DIODE EQUATION(1.61E-16,39.)
JC,1-2=DIODE EQUATION(7.2E-15,34.0)
JI,3-1=.834*JC
JN,2-1=.991*JE
JP1,3-1=0.
JP2,2-1=0.
FUNCTIONS
EQUATION1(A,B,C,D,E,F,G,H)=(A/(B-C)**D+E*F*(G+H))
```

```
MODEL 2N2808    (PERM) (E-B-C)
TAKEN FROM AFWL-TR-69-44
UNITS-VOLTS-KOHMS-PFARADS-MICROHENRIES-MILLIAMPS-NANOSEC
NPN   EEERS-MOLL EQUIVALENT CIRCUIT
ELEMENTS
RC,C-3=.036
RE,E-1=.08
CC,3-1=EQUATION 1(1.3,1.,VCC,.23,876.1,JC,9.02E-12)
CE,E-1=EQUATION 1(0.9,1.,VCE,.08,5.3,JE,4.17E-14)
JE,1-E=DIODE EQUATION(4.17E-14,41.)
JC,1-3=DIODE EQUATION(9.02E-12,35.)
JI,E-1=.5650*JC
JN,3-1=.9580*JE
JP1,3-1=0.
JP2,E-1=0.
FUNCTIONS
EQUATION 1 (A,B,C,D,E,F,G)=(A/(B-C)**D+E*(F+G))
```

```
MODEL 2N2887    (PERM) (E-B-C)
TAKEN FROM AFWL-TR-69-44
UNITS-VOLTS-KOHMS-PFARADS-MICROHENRIES-MILLIAMPS-NANOSEC
NPN   EEERS-MOLL EQUIVALENT CIRCUIT
ELEMENTS
RC,C-3=3.E-4
RE,E-1=.013
CC,3-1=EQUATION 1(103.,.8,VCC,.49,18050.,JC,1.155E-9)
CE,E-1=EQUATION 1(440.,.9,VCE,.39,10.6,JE,2.851E-10)
JE,1-E=DIODE EQUATION(2.851E-10,37.3)
JC,1-3=DIODE EQUATION(1.155E-9,37.3)
JI,E-1=.2424*JC
JN,3-1=.9797*JE
JP1,3-1=0.
JP2,E-1=0.
FUNCTIONS
EQUATION 1 (A,B,C,D,E,F,G)=(A/(B-C)**D+E*(F+G))
```

```
MODEL   2N2905     (PERM) (E-B-C)
SUPPLIED BY  DAVID N. HARSTAD, SANDIA CORPORATION
SANDIA BASE,ALBUQUERQUE,NEW MEXICO 87115
A MODIFIED EBERS-MOLL EQUIVALENT CIRCUIT        PNP
UNITS-MILLIAMPS-VOLTS-KOHMS-PF-MICROHENRY-NANOSEC
ELEMENTS
CET,E-1 = Q1   (    0.180E 02,   0.100E 01,VCET, 0.420E 00, 0.196E 04 ,
J1,   0.232E-10 )
CCT,2-1 = Q1   (    0.130E 02,   0.800E 00,VCCT, 0.380E 00, 0.308E 04 ,
J2,   0.613E-10 )
J1, E-1 = DIODE EQUATION (   0.232E-10 ,   0.406E 02 )
J2, 2-1 = DIODE EQUATION (   0.613E-10 ,   0.387E 02 )
RBB, 1-B =  0.220E-01
RE,  E-1 =  0.100E 07
RC,  2-1 =  0.100E 07
RCC, C-2 =  0.200E-02
JA, 1-E = 0.583E 00 * J2
JB, 1-2 = 0.993E 00 * J1
FUNCTIONS
Q1(A,B,C,D,E,F,G)=((A/ABS(B-C)**D)+E*(F+G))
```

```
MODEL   2N3019     (PERM) (E-B-C)
SUPPLIED BY  DAVID N. HARSTAD, SANDIA CORPORATION
SANDIA BASE,ALBUQUERQUE,NEW MEXICO 87115
EBERS MOLL TRANSISTOR EQUIVALENT CIRCUIT  NPN
UNITS-MILLIAMPS-VOLTS-KOHMS-PF-MICROHENRY-NANOSEC
ELEMENTS
CET,1-E = Q1   (    0.331E 01,   0.105E 01,VCET, 0.310E 00, 0.119E 02 ,
J1,   0.273E-09 )
CCT,1-2 = Q1   (    0.293E 01,   0.110E 01,VCCT, 0.794E-01, 0.747E 02 ,
J2,   0.123E-05 )
J1, 1-E = DIODE EQUATION (   0.273E-09 ,   0.373E 02 )
J2, 1-2 = DIODE EQUATION (   0.123E-05 ,   0.235E 02 )
RBB, B-1 =  0.200E-01
RE,  1-E =  0.500E 06
RC,  1-2 =  0.500E 06
RCC, 2-C =  0.200E-02
JA, E-1 = 0.833E 00 * J2
JB, 2-1 = 0.980E 00 * J1
FUNCTIONS
Q1(A,B,C,D,E,F,G)=((A/ABS(B-C)**D)+E*(F+G))
```

```
MODEL 2N3026 (PERM) (B-C-E)
SUPPLIED BY CONVAIR DIVISION, GENERAL DYNAMICS, SAN DIEGO CALIF.92112
UNITS  OHMS,VOLTS,AMPS,FARADS,HENRIES,SECONDS
ELEMENTS
RB,1-B=7.8
RC,2-C=.40
RE,3-E=.20
RCC,2-1=5.E7
REE,1-3=5.E7
CE,3-1=EQUATION1(715.E-12,1.22,VCE,.475,46.,1.E-10,JE,9.52E-11)
CC,2-1=EQUATION1(332.E-12,.614,VCC,.434,46.,1.66E-7,JC,2.14E-10)
JE,3-1=DIODE EQUATION(9.52E-11,46.)
JC,2-1=DIODE EQUATION(2.14E-10,46.)
JI,1-3=.666*JC
JN,1-2=.992*JE
JP1,1-3=0.
JP2,1-2=0.
FUNCTIONS
EQUATION1(A,B,C,D,E,F,G,H)=(A/(B-C)**D+E*F*(G+H))
```

```
MODEL   2N3108     (PERM) (E-B-C)
SUPPLIED BY  DAVID N. HARSTAD, SANDIA CORPORATION
SANDIA BASE,ALBUQUERQUE,NEW MEXICO 87115
EBERS MOLL TRANSISTOR EQUIVALENT CIRCUIT  NPN
UNITS-MILLIAMPS-VOLTS-KOHMS-PF-MICROHENRY-NANOSEC
ELEMENTS
CET,1-E = Q1   (   0.800E 02,  0.500E 00,VCET, 0.330E 00, 0.501E 02 ,
J1,  0.177E-07 )
CCT,1-2 = Q1   (   0.435E 02,  0.500E 00,VCCT, 0.330E 00, 0.502E 03 ,
J2,  0.163E-07 )
J1, 1-E = DIODE EQUATION (  0.177E-07 ,  0.302E 02 )
J2, 1-2 = DIODE EQUATION (  0.163E-07 ,  0.303E 02 )
RBB, B-1 =  0.144E-08
RE,  1-E =  0.128E 08
RC,  1-2 =  0.238E 08
RCC, 2-C =  0.100E-03
JA, E-1 = 0.800E 00 * J2
JB, 2-1 = 0.976E 00 * J1
FUNCTIONS
Q1(A,B,C,D,E,F,G)=((A/ABS(B-C)**D)+E*(F+G))
```

```
MODEL   2N3119     (PERM) (E-B-C)
SUPPLIED BY  DAVID N. HARSTAD, SANDIA CORPORATION
SANDIA BASE,ALBUQUERQUE,NEW MEXICO 87115
EBERS MOLL TRANSISTOR EQUIVALENT CIRCUIT  NPN
UNITS-MILLIAMPS-VOLTS-KOHMS-PF-MICROHENRY-NANOSEC
ELEMENTS
CBT,1-B = G1   (   0.130E 03,  0.800E 00,VCBT, 0.290E 00, 0.172E 05 ,
J1,  0.156E-08 )
CCT,1-2 = G1   (   0.655E 02,  0.230E 00,VCCT, 0.110E 00, 0.918E 05 ,
J2,  0.976E-08 )
J1, 1-B = DIODE EQUATION (  0.156E-08 ,  0.314E 02 )
J2, 1-2 = DIODE EQUATION (  0.976E-08 ,  0.288E 02 )
RBB, B-1 =  0.600E-03
RE,  1-E =  0.400E 07
RC,  1-2 =  0.400E 08
RCC, 2-C =  0.100E-05
JA, E-1 = 0.500E 00 * J2
JB, 2-1 = 0.976E 00 * J1
FUNCTIONS
G1(A,B,C,D,E,F,G)=((A/ABS(B-C)**D)+E*(F+G))
```

```
MODEL 2N3244   (PERM) (E-B-C)
TAKEN FROM AFWL-TR-69-44
UNITS-VOLTS-KOHMS-PFARADS-MICROHENRIES-MILLIAMPS-NANOSEC
PNP   EBERS-MOLL EQUIVALENT CIRCUIT
ELEMENTS
RC,C-2=1.4E-3
RB,B-1=.02
CC,2-1=EQUATION 1 (43.,.6,VCC,.41,1953.8,JC,1.02E-10)
CE,E-1=EQUATION 1 (55.,1.,VCE,.45,16.64,JE,1.71E-10)
JE,E-1=DIODE EQUATION(1.71E-10,38.)
JC,2-1=DIODE EQUATION(1.02E-10,39.)
JI,1-E=.6776*JC
JN,1-2=.9914*JE
JP1,1-2=0.
JP2,1-E=0.
FUNCTIONS
EQUATION 1 (A,B,C,D,E,F,G)=(A/(B-C)**D+E*(F+G))
```

```
MODEL 2N3252 (PERM) (B-C-E)
SUPPLIED BY CONVAIR DIVISION, GENERAL DYNAMICS, SAN DIEGO CALIF.92112
UNITS  OHMS,VOLTS,AMPS,FARADS,HENRIES,SECONDS
ELEMENTS
RB,B-1=18.0
RC,C-2=1.8
RE,E-3=.10
RCC,1-2=5.E7
REE,3-1=5.E7
CE,1-3=EQUATION1(55.E-12,.90,VCE,.40,40.,5.E-10,JE,5.22E-14)
CC,1-2=EQUATION1(14.E-12,.80,VCC,.23,37.,5.E-8,JC,4.2E-13)
JE,1-3=DIODE EQUATION(5.22E-14,40.)
JC,1-2=DIODE EQUATION(4.2E-13,37.)
JI,3-1=.50*JC
JN,2-1=.976*JE
JP1,3-1=0.
JP2,2-1=0.
FUNCTIONS
EQUATION1(A,B,C,D,E,F,G,H)=(A/(B-C)**D+E*F*(G+H))
```

```
MODEL 2N3309 (PERM) (B-C-E)
SUPPLIED BY CONVAIR DIVISION, GENERAL DYNAMICS, SAN DIEGO CALIF.92112
UNITS  OHMS,VOLTS,AMPS,FARADS,HENRIES,SECONDS
ELEMENTS
RB,B-1=3.0
RC,C-2=1.6
RE,E-3=.10
RCC,1-2=5.E7
REE,3-1=5.E7
CE,1-3=EQUATION1(24.E-12,1.0,VCE,.33,40.,4.1E-10,JE,8.40E-15)
CC,1-2=EQUATION1(16.E-12,1.0,VCC,.25,37.,1.3E-8,JC,8.4E-14)
JE,1-3=DIODE EQUATION(8.4E-15,40.)
JC,1-2=DIODE EQUATION(8.4E-14,37.)
JI,3-1=.50*JC
JN,2-1=.976*JE
JP1,3-1=0.
JP2,2-1=0.
FUNCTIONS
EQUATION1(A,B,C,D,E,F,G,H)=(A/(B-C)**D+E*F*(G+H))
```

```
MODEL 2N3498 (PERM) (E-C-E)
SUPPLIED BY CONVAIR DIVISION, GENERAL DYNAMICS, SAN DIEGO CALIF.92112
UNITS  OHMS,VOLTS,AMPS,FARADS,HENRIES,SECONDS
ELEMENTS
RB,B-1=12.7
RC,C-2=1.66
RE,E-3=1.3
RCC,1-2=5.E7
REE,3-1=5.E7
CE,1-3=EQUATION1(74.4E-12,1.26,VCE,.518,43.7,2.E-9,JE,6.98E-15)
CC,1-2=EQUATION1(19.E-12,.90,VCC,.378,43.0,6.4E-7,JC,1.49E-14)
JE,1-3=DIODE EQUATION(6.98E-15,43.7)
JC,1-2=DIODE EQUATION(1.49E-14,43.0)
JI,3-1=.666*JC
JN,2-1=.988*JE
JP1,3-1=0.
JP2,2-1=0.
FUNCTIONS
EQUATION1(A,B,C,D,E,F,G,H)=(A/(B-C)**D+E*F*(G+H))
```

```
MODEL   2N3501    (PERM) (E-B-C)
SUPPLIED BY  DAVID N. HARSTAD, SANDIA CORPORATION
SANDIA BASE,ALBUQUERQUE,NEW MEXICO 87115
EBERS MOLL TRANSISTOR EQUIVALENT CIRCUIT  NPN
UNITS-MILLIAMPS-VOLTS-KOHMS-PF-MICROHENRY-NANOSEC
ELEMENTS
CET,1-E = G1   (   0.800E 02,  0.500E 00,VCET, 0.330E 00, 0.688E 02 ,
J1,  0.399E-07 )
CCT,1-2 = G1   (   0.543E 02,  0.500E 00,VCCT, 0.330E 00, 0.241E 03 ,
J2,  0.366E-07 )
J1, 1-E = DIODE EQUATION (  0.399E-07 ,  0.302E 02 )
J2, 1-2 = DIODE EQUATION (  0.366E-07 ,  0.303E 02 )
REE, E-1 =  0.144E-02
RE,  1-E =  0.128E 09
RC,  1-2 =  0.238E 08
RCC, 2-C =  0.100E-03
JA, E-1 = 0.909E 00 * J2
JB, 2-1 = 0.993E 00 * J1
FUNCTIONS
G1(A,B,C,D,E,F,G)=((A/ABS(B-C)**D)+E*(F+G))
```

```
MODEL   2N3503    (PERM) (E-B-C)
SUPPLIED BY  DAVID N. HARSTAD, SANDIA CORPORATION
SANDIA BASE,ALBUQUERQUE,NEW MEXICO 87115
A MODIFIED EBERS-MOLL EQUIVALENT CIRCUIT      PNP
UNITS-MILLIAMPS-VOLTS-KOHMS-PF-MICROHENRY-NANOSEC
ELEMENTS
CET,B-1 = Q1   (   0.218E 02,  0.163E 01,VCET, 0.496E 00, 0.217E 03 ,
J1,  0.232E-07 )
CCT,2-1 = Q1   (   0.107E 02,  0.936E 00,VCCT, 0.361E 00, 0.885E 03 ,
J2,  0.234E-07 )
J1, E-1 = DIODE EQUATION (  0.232E-07 ,  0.272E 02 )
J2, 2-1 = DIODE EQUATION (  0.234E-07 ,  0.278E 02 )
RBB, 1-B =  0.650E-02
RC,  2-1 =  0.300E 07
RCC, C-2 =  0.760E-02
JA, 1-E = 0.833E 00 * J2
JB, 1-2 = 0.995E 00 * J1
FUNCTIONS
Q1(A,B,C,D,E,F,G)=((A/ABS(B-C)**D)+E*(F+G))
```

```
MODEL 2N3600 (PERM) (B-C-E)
SUPPLIED BY CONVAIR DIVISION, GENERAL DYNAMICS, SAN DIEGO CALIF.92112
UNITS  OHMS,VOLTS,AMPS,FARADS,HENRIES,SECONDS
ELEMENTS
RB,B-1=12.0
RC,C-2=11.5
RE,E-3=8.5
RCC,1-2=5.E7
REE,3-1=5.E7
CE,1-3=EQUATION1(2.E-12,.50,VCE,.15,30.4,2.18E-10,JE,9.18E-10)
CC,1-2=EQUATION1(1.7E-12,.80,VCC,.10,28.4,1.24E-8,JC,2.12E-9)
JE,1-3=DIODE EQUATION(9.18E-10,30.4)
JC,1-2=DIODE EQUATION(2.12E-9,28.4)
JI,3-1=.492*JC
JN,2-1=.977*JE
JP1,3-1=0.
JP2,2-1=0.
FUNCTIONS
EQUATION1(A,B,C D,E,F,G,H)=(A/(B-C)**D+E*F*(G+H))
```

```
MODEL 2N3737 (FERM) (E-C-E)
SUPPLIED BY CONVAIR DIVISION, GENERAL DYNAMICS, SAN DIEGO CALIF.92112
UNITS   OHMS,VOLTS,AMPS,FARADS,HENRIES,SECONDS
ELEMENTS
RB,B-1=20.
RC,C-2=.10
RE,E-3=.42
RCC,1-2=5.E7
REE,3-1=5.E7
CE,1-3=EQUATION1(71.E-12,.75,VCE,.37,38.0,3.14E-11,JE,3.96E-15)
CC,1-2=EQUATION1(9.1E-12,.75,VCC,.27,38.0,7.1E-8,JC,7.89E-15)
JE,1-3=DIODE EQUATION(3.96E-15,38.)
JC,1-2=DIODE EQUATION(7.89E-15,38.)
JI,3-1=.333*JC
JN,2-1=.986*JE
JP1,3-1=0.
JP2,2-1=0.
FUNCTIONS
EQUATION1(A,B,C,D,E,F,G,H)=(A/(B-C)**D+E*F*(G+H))
```

```
MODEL 2N3766 (FERM) (E-C-E)
SUPPLIED BY CONVAIR DIVISION, GENERAL DYNAMICS, SAN DIEGO CALIF.92112
UNITS   OHMS,VOLTS,AMPS,FARADS,HENRIES,SECONDS
ELEMENTS
RB,B-1=23.
RC,C-2=.32
RE,E-3=.41
RCC,1-2=5.E7
REE,3-1=5.E7
CE,1-3=EQUATION1(224.2E-12,.94,VCE,.428,31.3,1.06E-8,JE,7.29E-11)
CC,1-2=EQUATION1(122.5E-12,1.15,VCC,.312,32.0,3.44E-7,JC,8.75E-10)
JE,1-3=DIODE EQUATION(7.29E-11,31.3)
JC,1-2=DIODE EQUATION(8.75E-10,32.0)
JI,3-1=.834*JC
JN,2-1=.990*JE
JP1,3-1=0.
JP2,2-1=0.
FUNCTIONS
EQUATION1(A,B,C,D,E,F,G,H)=(A/(B-C)**D+E*F*(G+H))
```

```
MODEL 2N3828    (PERM) (E-B-C)
TAKEN FROM AFWL-TR-69-44
UNITS-VOLTS-KOHMS-PFARADS-MICROHENRIES-MILLIAMPS-NANOSEC
NPN   EBERS-MOLL EQUIVALENT CIRCUIT
ELEMENTS
RC,C-3=.017
RB,B-1=1.2E-3
CC,3-1=EQUATION 1(6.8,.75,VCC,.23,1541.,JC,1.4E-10)
CE,E-1=EQUATION 1(18.,.8,VCE,.31,1094.,JE,3.1E-11)
JE,1-E=DIODE EQUATION(3.1E-11,38.)
JC,1-3=DIODE EQUATION(1.4E-10,38.)
JI,E-1=.6770*JC
JN,3-1=.9600*JE
JP1,3-1=0.
JP2,E-1=0.
FUNCTIONS
EQUATION 1 (A,B,C,D,E,F,G)=(A/(B-C)**D+E*(F+G))
```

```
MODEL   2N3904     (PERM) (E-B-C)
SUPPLIED BY  DAVID N. HARSTAD, SANDIA CORPORATION
SANDIA BASE,ALBUQUERQUE,NEW MEXICO 87115
EBERS MOLL TRANSISTOR EQUIVALENT CIRCUIT  NPN
UNITS-MILLIAMPS-VOLTS-KOHMS-PF-MICROHENRY-NANOSEC
ELEMENTS
CET,1-E = G1   (    0.804E 01,  0.550E 00,VCET, 0.160E 00, 0.113E 04 ,
J1,  0.297E-10 )
CCT,1-2 = G1   (    0.810E 01,  0.219E 00,VCCT, 0.800E-01, 0.681E 03 ,
J2,   0.396E-08 )
J1, 1-E = DIODE EQUATION (  0.297E-10 ,  0.354E 02 )
J2, 1-2 = DIODE EQUATION (  0.396E-08 ,  0.278E 02 )
REE, E-1 =  0.200E-02
RE,  1-E =  0.900E 07
RC,  1-2 =  0.100E 08
RCC, 2-C =  0.100E-02
JA, E-1 = 0.500E 00 * J2
JB, 2-1 = 0.990E 00 * J1
FUNCTIONS
G1(A,B,C,D,E,F,G)=((A/ABS(B-C)**D)+E*(F+G))
```

```
MODEL 2N3913 (PERM) (B-C-E)
SUPPLIED BY CONVAIR DIVISION, GENERAL DYNAMICS, SAN DIEGO CALIF.92112
UNITS  OHMS,VOLTS,AMPS,FARADS,HENRIES,SECONDS
ELEMENTS
RB,1-B=127.
RC,2-C=34.
RE,E-3=3.8
RCC,2-1=5.E7
REE,3-1=5.E7
CE,3-1=EQUATION1(21.E-12,.80,VCE,.50,43.5,5.05E-9,JE,.824E-15)
CC,2-1=EQUATION1(21.E-12,.80,VCC,.50,43.5,5.70E-8,JC,.896E-15)
JE,3-1=DIODE EQUATION(.824E-15,43.5)
JC,2-1=DIODE EQUATION(.896E-15,43.5)
JI,1-3=.91*JC
JN,1-2=.990*JE
JP1,1-3=0.
JP2,1-2=0.
FUNCTIONS
EQUATION1(A,B,C,D,E,F,G,H)=(A/(B-C)**D+E*F*(G+H))
```

```
MODEL   2N4260    (PERM) (E-B-C)
SUPPLIED BY  DAVID N. HARSTAD, SANDIA CORPORATION
SANDIA BASE,ALBUQUERQUE,NEW MEXICO 87115
EBERS MOLL TRANSISTOR EQUIVALENT CIRCUIT  NPN
UNITS-MILLIAMPS-VOLTS-KOHMS-PF-MICROHENRY-NANOSEC
ELEMENTS
CET,1-E = Q1   (    0.866E 01,  0.122E 01,VCET, 0.100E 00, 0.971E 03 ,
J1,  0.390E-11 )
CCT,1-2 = Q1   (    0.811E 01,  0.300E 00,VCCT, 0.500E-01, 0.149E 04 ,
J2,  0.390E-08 )
J1, 1-E = DIODE EQUATION (  0.390E-11 ,  0.366E 02 )
J2, 1-2 = DIODE EQUATION (  0.390E-08 ,  0.281E 02 )
RBB, B-1 =  0.420E-02
RE,  1-E =  0.500E 09
RC,  1-2 =  0.180E 08
RCC, 2-C =  0.200E-03
JA, E-1 = 0.500E 00 * J2
JB, 2-1 = 0.976E 00 * J1
FUNCTIONS
Q1(A,B,C,D,E,F,G)=((A/ABS(B-C)**D)+E*(F+G))
```

```
MODEL 2N3915 (FERM) (B-C-E)
SUPPLIED BY CONVAIR DIVISION, GENERAL DYNAMICS, SAN DIEGO CALIF.92112
UNITS  OHMS,VOLTS,AMPS,FARADS,HENRIES,SECONDS
ELEMENTS
RB,1-B=131.
RC,2-C=18.
RE,E-3=2.1
RCC,2-1=5.E7
REE,3-1=5.E7
CE,3-1=EQUATION1(21.E-12,.80,VCE,.50,43.5,1.36E-9,JE,2.08E-16)
CC,2-1=EQUATION1(21.E-12,.80,VCC,.50,43.5,2.54E-8,JC,2.14E-16)
JE,3-1=DIODE EQUATION(2.08E-16,43.5)
JC,2-1=DIODE EQUATION(2.14E-16,43.5)
JI,1-3=.969*JC
JN,1-2=.997*JE
JP1,1-3=0.
JP2,1-2=0.
FUNCTIONS
EQUATION1(A,B,C,D,E,F,G,H)=(A/(B-C)**D+E*F*(G+H))
```

```
MODEL 2N3960 (FERM) (B-C-E)
SUPPLIED BY CONVAIR DIVISION, GENERAL DYNAMICS, SAN DIEGO CALIF.92112
UNITS  OHMS,VOLTS,AMPS,FARADS,HENRIES,SECONDS
ELEMENTS
RB,B-1=39.0
RC,C-2=1.5
RE,E-3=2.0
RCC,1-2=5.E7
REE,3-1=5.E7
CE,1-3=EQUATION1(2.8E-12,.76,VCE,.248,38.1,9.8E-11,JE,3.51E-16)
CC,1-2=EQUATION1(3.73E-12,.75,VCC,.261,38.0,8.1E-11,JC,1.22E-15)
JE,1-3=DIODE EQUATION(3.51E-16,38.1)
JC,1-2=DIODE EQUATION(1.22E-15,38.0)
JI,3-1=.834*JC
JN,2-1=.982*JE
JP1,3-1=0.
JP2,2-1=0.
FUNCTIONS
EQUATION1(A,B,C,D,E,F,G,H)=(A/(B-C)**D+E*F*(G+H))
```

```
MODEL   2N4923     (PERM) (E-B-C)
SUPPLIED BY  DAVID N. HARSTAD, SANDIA CORPORATION
SANDIA BASE,ALBUQUERQUE,NEW MEXICO 87115
EBERS MOLL TRANSISTOR EQUIVALENT CIRCUIT  NPN
UNITS-MILLIAMPS-VOLTS-KOHMS-PF-MICROHENRY-NANOSEC
ELEMENTS
J1,   0.293E-07 )
CCT,1-2 = G1   (    0.835E 02,  0.400E 00,VCCT, 0.320E 00, 0.751E 05 ,
J2, 1-2 = DIODE EQUATION (  0.195E-06 ,  0.236E 02 )
RBB, B-1 =  0.400E-03
RC,  1-2 =  0.900E 07
RCC, 2-C =  0.300E-04
CET,1-E = G1   (    0.739E 02,  0.100E 01,VCET, 0.300E 00, 0.325E 04 ,
J2,  0.195E-06 )
J1, 1-E = DIODE EQUATION (  0.293E-07 ,  0.306E 02 )
RE,  1-E =  0.500E 07
JA, E-1 = 0.500E 00 * J2
JB, 2-1 = 0.976E 00 * J1
FUNCTIONS
G1(A,B,C,D,E,F,G)=((A/ABS(B-C)**D)+E*(F+G))
```

APPENDIX J

JFET PARAMETER DERIVATION AND EVALUATION

Two basic programs* are presented here. Figure J.1 illustrates the program used to calculate the "best fit" JFET SCEPTRE parameters. Figure J.2 illustrates the program used to evaluate the triode and pinch-off regions of the ID vs VDS characteristics.

The determinations of the constants G, VP and VPHI is accomplished using the iterative routine outlined in Sec. 3.5.3. Evaluation of five VGS values is accomplished using 0, −.5, −1, −1.5 and −2. The actual values are entered during run time. A range of VPHI is requested; the program makes calculations of .1V intervals between the min. and max. values. A VDS, ID point is picked on the VGS = 0 curve which seems to best represent the "pinch-off" point (end of the triode region, start of the pinch-off region) and is entered. Based on this point four more VDS values are calculated for each additional VGS and a corresponding ID (pinch-off) is entered.

Two print-out formats are available: (1) a summary where only VGS and the percent current error is printed, and (2) all data is printed out. The user may select the desired format for each VDS, ID (pinch-off) point selected. The best set of parameters, G, VP, VDS, ID is selected based on the % error calculated for each set of data. This program is on dial file at DIAL-DATA under file /RFET1/.

The second program, /RFET2/, uses G, VP, VDS (P—0), ID (P—0) determined in /RFET1/ plus a VDS, ID point near VDS (max.) and calculates the JFET ID—VDS characteristic through the triode and pinch-off region.

Sample executions of both programs are illustrated in Fig. J.3 and J.4.

```
1 REM PROGRAM /RFET1/ 20 MAR 70 R REID
5 REM THIS PROGRAM CALCULATES JFET PARAMETERS G .VT
10 DIM B(5,5)
20 PRINT "ENTER 5 VGS VALUES FOR EVALUATION";
30 INPUT A(1),A(2),A(3),A(4),A(5)
40 PRINT "ENTER VPHI(MIN), VPHI(MAX)";
50 INPUT C1,C2
65 IF X>0 THAN 110
80 PRINT "FOR VGS=0, ENTER VDS AND I @ PINCH-OFF";
90 INPUT B(1,1),B(1,2)
95 FOR V1=C1 TO C2 STEP .1
100 REM V3=VP=VPHI+VDS(VGS=0)
110 V3=V1+B(1,1)
120 K=2/(3*SQR(V3))
130 G=B(1,2)/(V3/3-V1+K*V1↑1.5)
135 IF X>0 THEN 170
140 PRINT "ENTER I(PINCH-OFF) FOR EACH VGS, VDS"
150 FOR N=2 TO 5
160 REM VDS=VP-VPHI+VGS
170 B(N,1)=V3-V1+A(N)
175 IF X>0 THEN 210
180 PRINT "FOR VGS=";A(N);"VDS=";B(N,1);"I(P-0)=";
190 INPUT B(N,2)
200 NEXT N
210 FOR N=1 TO 5
220 REM CALCULATE I
```

*From Honeywell memo by J. R. Reid dated 25 March 1970.

```
230 B(N,3)=G*(A(N)+V3/3-V1+K*(V1-A(N))*1.5)
240 REM CALCULATE ERROR
250 B(N,4)=B(N,3)-B(N,2)
260 REM CALCULATE % ERROR
270 B(N,5)=B(N,4)/B(N,2)*100
280 NEXT N
281 IF X > 0 THEN 290
282 PRINT "ENTER 1 FOR SUMMARY, 2 FOR TOTAL PRINT OUT";
284 INPUT Z1
290 PRINT
300 PRINT"G=";G,"VP=";V3,"VPHI=";V1
305 IF Z1=1 THEN 325
310 PRINT " VGS";TAB(10);"VDS";TAB(18);"I(TEST)","I(CAL)",
"ERROR";
320 PRINT TAB(57);"% ERROR"
322 GO TO 330
325 PRINT "   VGS      % ERROR"
330 PRINT
340 FOR N=2 TO 5
345 IF Z1=1 THEN 355
350 PRINT USING 500,A(N),B(N,1),B(N,2),B(N,3),B(N,4),
B(N,5)
352 GO TO 360
355 PRINT USING 510,A(N),B(N,5)
360 NEXT N
370 PRINT
375 X=1
380 NEXT V1
390 PRINT "TYPE 1 FOR NEW VDS,I (PINCH-OFF) FOR VGS=0,
OR 2 TO END";
395 X=0
400 INPUT Z
405 PRINT
406 PRINT
408 PRINT
410 ON Z GO TO 80,999
500:###.### ###.### ###.###      ###.###  ###.###  ####.##
510:###.###      ####.##
999 END
```

Fig. J. 1

```
LLKI[STO

1 REM PROGRAM /RFET2/ 20 MAR 70 R REID
5 REM THIS PROGRAM CHECKS G . VT CALCULATED IN /RFET1/
10 DIM A(17,4),B$(17)
20 PRINT "INPUT VP,G,VPHI";
30 INPUT V1,G,V2
32 PRINT "FOR VGS=0, ENTER VDS AND ID @(P-O), AND VDS AND
ID NEAR VDS(MAX)"
33 INPUT V7,I7,V8,I8
```

```
34 R2=17*(V8-V7)/(I8-I7)
40 K=2/(3*SQR(V1))
45 REM V6=VT=VPHI-VP
50 V6=V2-V1
55 PRINT "TO END SET VGS=99"
60 PRINT "VGS=";
70 INPUT V3
75 IF V3=99 THEN 999
80 READ V4
85 N=N+1
87 A(N,1)=V4
88 REM V5=VGD=VGS-VDS
90 V5=V3-V4
100 IF V5>V2 THEN 300
110 IF V3>= V2 THEN 300
120 IF V5=V3 THEN 300
130 IF V5<V6 THEN 250
140 IF V5>V6 THEN 350
150 IF V3<V6 THEN 500
160 A(N,2)=G*(V3+V1/3-V2+K*(-V3+V2)↑1.5)
165 B$(N)="160"
170 GO TO 600
250 IF V3<V6 THEN 300
260 GO TO 140
270 B(N,5)=B(N,4)/B(N,2)*100
300 A(N,2)=0
305 B$(N)="300"
310 GO TO 600
350 IF V3>V6 THEN 400
360 GO TO 150
400 A(N,2)=G*(V3-V5+K*((-V3+V2)↑1.5-(-V5+V2)↑1.5))
405 B$(N)="400"
410 GO TO 600
500 A(N,2)=-G*(V5+V1/3-V2+K*(-V5+V2)↑1.5)
505 B$(N)="500"
600 R1=R2/A(N,2)
610 A(N,3)=A(N,1)/R1.
620 A(N,4)=A(N,3)+A(N,2)
625 IF V4=16 THEN 640
630 GO TO 80
640 PRINT
650 PRINT "VDS","JD","IRD","ID","STEP"
660 PRINT
670 FOR N=1 TO 17
680 PRINT A(N,1),A(N,2),A(N,3),A(N,4),  B$(N)
690 NEXT N
700 PRINT
710 PRINT
720 N=0
725 RESTORE
730 GO TO 55
740 DATA 0,.5,1,1.5,2,2.5,3,3.5,4,4.5,5,6,8,10,12,14,16
999 END
```

Fig. J. 2

```
>RUN
ENTER 5 VGS VALUES FOR EVALUATION? 0,-.5,-1,-1.5,-2
ENTER VPHI(MIN), VPHI(MAX)? .7,1.1
FOR VGS=0, ENTER VDS AND I @ PINCH-OFF? 2.6,4.37
ENTER I(PINCH-OFF) FOR EACH VGS, VDS
FOR VGS=-.5 VDS= 2.1 I(P-0)=? 2.8
FOR VGS=-1 VDS= 1.6 I(P-0)=? 1.58
FOR VGS=1.5  VDS= 1.1 I(P-0)=.66
FOR VGS=-2  VDS= .6  I(P-0)=? .09
ENTER 1 FOR SUMMARY, 2 FOR TOTAL PRINT OUT? 2

G= 7.1064896 VP= 3.3      VPHI= .7
   VGS   VDS     I(TEST) I(CAL)     ERROR      % ERROR

 -.500  2.100   2.800    2.717     _ -.082      -2.94
-1.000  1.600   1.580    1.516      -.063      -4.00
-1.500  1.100    .660     .693       .033       5.01
-2.000   .600    .090     .200       .110     122.36

G= 7.3812829    VP= 3.4    VPHI= .8
   VGS      VDS       I(TEST)  I(CAL)   ERROR  % ERROR

 -.500  2.100   2.800    2.725    -.074   -2.66
-1.000  1.600   1.580    1.524    -.056   -3.54
-1.500  1.100    .660     .697     .037    5.65
-2.000   .600    .090     .201     .111  123.94

G= 7.6542828   VP= 3.5  VPHI= .9
   VGS      VDS       I(TEST)  I(CAL) ERROR  % ERROR

 -.500  2.100   2.800    2.732    -.067  -2.42
-1.000  1.600   1.580    1.530    -.049  -3.14
-1.500  1.100    .660     .701     .041    6.22
-2.000   .600    .090     .202     .112 125.38

G= 7.9257959    VP= 3.6     VPHI= 1
   VGS      VDS       I(TEST)  I(CAL)    ERROR  % ERROR

 -.500  2.100   2.800    2.738    -.061  -2.20
-1.000  1.600   1.580    1.536    -.044  -2.78
-1.500  1.100    .660     .704     .044    6.74
-2.000   .600    .090     .204     .114 126.69

G= 8.1960578    VP= 3.7     VPHI= 1.1
   VGS      VDS       I(TEST)  I(CAL)    ERROR  % ERROR

 -.500  2.100   2.800    2.743    -.056  -2.00
-1.000  1.600   1.580    1.541    -.038  -2.45
-1.500  1.100    .660     .707     .047    7.22
-2.000   .600    .090     .205     .115 127.89

TYPE 1 FOR NEW VDS,I (PINCH-OFF) FOR VGS=0,OR 2 TO END? 1

FOR VGS=0, ENTER VDS AND I @ PINCH-OFF? 2.7,4.4
ENTER I(PINCH-OFF) FOR EACH VGS, VDS
FOR VGS= -.5 VDS= 2.2 I(P-0)=? 2.85
FOR VGS= -1  VDS= 1.7 I(P-0)=? 1.61
```

```
FOR VGS= -1.5 VDS= 1.2 I(P-0)=? .67
FOR VGS= -2  VDS= .7 I(P-0)=? .1
ENTER 1 FOR SUMMARY, 2 FOR TOTAL PRINT OUT? 1

G= 6.8208606     VP= 3.4      VPHI= .7
   VGS   % ERROR

 -.500   -2.21
-1.000    -.56
-1.500  15.16
-2.000 154.91

G= 7.0779517  VP= 3.5   VPHI= .8
   VGS  % ERROR

 -.500   -1.93
-1.000    -.10
-1.500  15.83
-2.000 156.67

G= 7.3333333  VP= 3.6      VPHI= .9
   VGS   % ERROR

 -.500   -1.69
-1.000     .30
-1.500  16.45
-2.000 158.27

G= 7.5872948     VP= 3.7      VPHI= 1
   VGS  % ERROR

 -.500   -1.48
-1.000     .67
-1.500  17.00
-2.000 159.74

G= 7.8400595     VP= 3.8     VPHI= 1.1
   VGS % ERROR

 -.500   -1.29
-1.000    1.00
-1.500  17.51
-2.000 161.09

TYPE 1 FOR NEW VDS,I (PINCH-OFF) FOR VGS=0,OR 2 TO END? 1

FOR VGS=0, ENTER VDS AND I @ PINCH-OFF? 2.8,4.43
ENTER I(PINCH-OFF) FOR EACH VGS, VDS
FOR VGS=-.5 VDS= 2.3 I(P-0)=? 2.89
FOR VGS= -1 VDS= 1.8 I(P-0)=? 1.64
FOR VGS= -1.5   VDS= 1.3 I(P-0)=? .7
FOR VGS= -2 VDS= .8 I(P-0)=? .09
ENTER 1 FOR SUMMARY, 2 FOR TOTAL PRINT OUT? 1

G= 6.5594029     VP= 3.5     VPHI= .7
```

```
   VGS  % ERROR

 -.500   -1.24
-1.000    2.60
-1.500   21.32
-2.000  247.10

G= 6.8005725      VP= 3.6      VPHI= .8
   VGS  % ERROR

 -.500    -.98
-1.000    3.05
-1.500   22.02
-2.000  249.43

G= 7.0401102      VP= 3.7    VPHI= .9
   VGS  % ERROR

 -.500    -.74
-1.000    3.46
-1.500   22.56
-2.000  251.56

G= 7.27829     VP= 3.8       VPHI= 1
   VGS  % ERROR

 -.500    -.53
-1.000    3.83
-1.500   23.22
-2.000  253.50

G= 7.5153232      VP= 3.9     VPHI= 1.1
   VGS  % ERROR

 -.500    -.34
-1.000    4.17
-1.500   23.74
-2.000  255.29

TYPE 1 FOR NEW VDS,I (PINCH-OFF) FOR VGS=0,OR 2 TO END? 2
```

Fig. J. 3

```
>QUIT

@XBAS

>LOAD
FROM: /RFET2/

>RUN
INPUT VP,G,VPHI? 3.7,7.5873,1
FOR VGS=0, ENTER VDS AND ID @(P-O), AND VDS AND ID NEAR
VDS(MAX)
? 2.7,4.4,15,4.85
```

TO END SET VGS=99
VGS=? 0

VDS	JD	IRD	ID	STEP
0	0	0	0	300
.5	1.5923387	6.6200335E-03	1.5989587	400
1	2.7792077	2.3108711E-02	2.8023164	400
1.5	3.6160459	4.5100351E-02	3.6611463	400
2	4.1402592	6.8851318E-02	4.2091105	400
2.5	4.3793081	9.103329E-02	4.4703414	400
3	4.4000029	.10975617	4.5097591	160
3.5	4.4000029	.12804886	4.5280518	160
4	4.4000029	.14634156	4.5463445	160
4.5	4.4000029	.16463425	4.5646372	160
5	4.4000029	.18292695	4.5829299	160
6	4.4000029	.21951234	4.6195153	160
8	4.4000029	.29268312	4.692686	160
10	4.4000029	.3658539	4.7658568	160
12	4.4000029	.43902468	4.8390276	160
14	4.4000029	.51219546	4.9121984	160
16	4.4000029	.58536624	4.9853692	160

TO END SET VGS=99

VGS=? -.5

VDS	JD	IRD	ID	STEP
0	0	0	0	300
.5	1.186869	4.9343224E-03	1.1918033	400
1	2.0237072	1.6826834E-02	2.040534	400
1.5	2.5479205	3.1778388E-02	2.5796989	400
2	2.7869694	4.6346498E-02	2.8333159	400
2.5	2.8076642	5.8363308E-02	2.8660275	160
3	2.8076642	7.003597E-02	2.8777001	160
3.5	2.8076642	8.1708631E-02	2.8893728	160
4	2.8076642	9.3381293E-02	2.9010455	160
4.5	2.8076642	.10505395	2.9127181	160
5	2.8076642	.11672661	2.9243908	160
6	2.8076642	.14007194	2.9477361	160
8	2.8076642	.18676258	2.9944268	160
10	2.8076642	.23345323	3.0411174	160
12	2.8076642	.28014388	3.0878081	160
14	2.8076642	.32683452	3.1344987	160
16	2.8076642	.37352517	3.1811894	160

TO END SET VGS=99

VGS=? -1

VDS	JD	IRD	ID	STEP
0	0	0	0	300
.5	.83683823	3.4790946E-03	.84031732	400

1	1.3610515	1.1316947E-02	1.3723684	400
1.5	1.6001003	1.9956906E-02	1.6200573	400
2	1.6207952	2.6953357E-02	1.6477485	160
2.5	1.6207952	3.3691696E-02	1.6544869	160
3	1.6207952	4.0430035E-02	1.6612252	160
3.5	1.6207952	4.7168374E-02	1.6679635	160
4	1.6207952	5.3906714E-02	1.6747019	160
4.5	1.6207952	6.0645053E-02	1.6814402	160
5	1.6207952	6.7383392E-02	1.6881785	160
6	1.6207952	8.0860071E-02	1.7016552	160
8	1.6207952	.10781342	1.7286086	160
10	1.6207952	.13476678	1.7555619	160
12	1.6207952	.16172014	1.7825153	160
14	1.6207952	.18867349	1.8094687	160
16	1.6207952	.21562685	1.836422	160

TO END SET VGS=99

VGS=? -1.5

VDS	JD	IRD	ID	STEP
0	0	0	0	300
.5	.52421327	2.1793789E-03	.52639265	400
1	.76326216	6.3464148E-03	.76960857	400
1.5	.78395696	9.7777338E-03	.7937347	160
2	.78395696	1.3036978E-02	.79699394	160
2.5	.78395696	1.6296223E-02	.80025319	160
3	.78395696	1.9555467E-02	.80351243	160
3.5	.78395696	2.2814712E-02	.80677168	160
4	.78395696	2.6073956E-02	.81003092	160
4.5	.78395696	2.9333201E-02	.81329017	160
5	.78395696	3.2592446E-02	.81654941	160
6	.78395696	3.9110935E-02	.8230679	160
8	.78395696	5.2147913E-02	.83610488	160
10	.78395696	6.5184892E-02	.84914186	160
12	.78395696	7.822187E-02	.86217883	160
14	.78395696	9.1258848E-02	.87521581	160
16	.78395696	.10429582	.88825279	160

TO END SET VGS=99
VGS=? -2

VDS	JD	IRD	ID	STEP
0	0	0	0	300
.5	.23904888	9.9382851E-04	.24004271	400
1	.25974369	2.1597313E-03	.26190342	160
1.5	.25974369	3.239597E-03	.26298328	160
2	.25974369	4.3194627E-03	.26406315	160
2.5	.25974369	5.3993284E-03	.26514302	160
3	.25974369	6.4791941E-03	.26622288	160
3.5	.25974369	7.5590597E-03	.26730275	160

```
 4      .25974369 8.6389254E-03    .26838261 160
 4.5    .25974369 9.7187911E-03    .26946248 160
 5      .25974369 1.0798656E-02    .27054234 160
 6      .25974369 1.2958388E-02    .27270208 160
 8      .25974369 1.727785E-02     .27702154 160
10      .25974369 2.1597313E-02    .281341   160
12      .25974369 2.5916776E-02    .28566046 160
14      .25974369 3.0236239E-02    .28997993 160
16      .25974369 3.4555701E-02    .29429939 160

 TO END SET VGS=99
 VGS=? 99

 RUN
 INPUT VP,G,VPHI? 3.8,7.84,1.1
 FOR VGS=0, ENTER VDS AND ID @(P-0), AND VDS AND ID NEAR
 VDS(MAX)
 ? 2.7,4.4,15,4.85
 TO END SET VGS=99
 VGS=? 0

 VDS  JD          IRD             ID          STEP

 0    0           0               0           300
  .5  1.5868852   6.597361E-03    1.5934826   400
 1    2.7738299   2.3063996E-02   2.7968939   400
 1.5  3.6126084   4.5057478E-02   3.6576659   400
 2    4.1388788   6.8828362E-02   4.2077072   400
 2.5  4.3791503   9.103001E-02    4.4701803   400
 3    4.3999666    .10975526      4.5907218   160
 3.5  4.3999666    .1280478       4.5280144   160
 4    4.3999666    .14634035      4.5463069   160
 4.5  4.3999666    .16463289      4.5645995   160
 5    4.3999666    .18292544      4.582892    160
 6    4.3999666    .21951052      4.6194771   160
 8    4.3999666    .2926807       4.6926473   160
10    4.3999666    .36585088      4.7658174   160
12    4.3999666    .43902105      4.8389876   160
14    4.3999666    .51219123      4.9121578   160
16    4.3999666    .58536141      4.985328    160

 TO END SET VGS=99
 VGS=? -.5

 VDS  JD          IRD             ID          STEP

 0    0           0               0           300
  .5  1.1869447   4.9346371E-03   1.1918793   400
 1    2.0257232   1.6843596E-02   2.0425668   400
 1.5  2.5519936   3.1829188E-02   2.5838228   400
 2    2.7922651   4.6434564E-02   2.8386996   400
 2.5  2.8130813   5.8475915E-02   2.8715572   160
 3    2.8130813   7.0171098E-02   2.8832524   160
 3.5  2.8130813   8.1866281E-02   2.8949476   160
```

```
 4    2.8130813  9.3561464E-02  2.9066428  160
 4.5  2.8130813   .10525664     2.9183379  160
 5    2.8130813   .11695183     2.9300331  160
 6    2.8130813   .14034219     2.9534235  160
 8    2.8130813   .18712292     3.0002042  160
10    2.8130813   .23390366     3.046985   160
12    2.8130813   .28068439     3.0937657  160
14    2.8130813   .32746512     3.1405464  160
16    2.8130813   .37424585     3.1873272  160

 TO END SET VGS=99
 VGS=? 99
```

Fig. J. 4

APPENDIX K

BLACK BOX CHARACTERIZATION OF LARGE CIRCUITS

K.1 BLACK BOX CHARACTERIZATION OF LARGE CIRCUITS

Some circuits, particularly integrated circuits, have so many elements that it is not possible to program them on SCEPTRE, or any other program. In other cases, a certain large circuit might "fit", but due to its large size may consume considerable computer time. A solution to this dilemma can many times be satisfactorily accomplished using the black-box characterization approach. In the paragraphs to follow, an analogue and a digital integrated circuit are characterized in this fashion with the results given.

K.2 AN INTEGRATED CIRCUIT OPERATIONAL AMPLIFIER

A typical I.C. OP AMP may have between 15 to 35 internal solid-state devices, along with a number of resistors and capacitors. Obviously, accommodating even one of these on a computer program is difficult. However, the device can be characterized in an end-to-end manner, such that a number of I.C.'s and corresponding circuits can all be analyzed simultaneously and at greatly reduced computer running times. Many individuals* active in the field believe that this black-box approach will become increasingly necessary in the years ahead and the reasons for these conclusions are obvious.

Figure K.1 illustrates a typical schematic for an I.C. OP AMP.

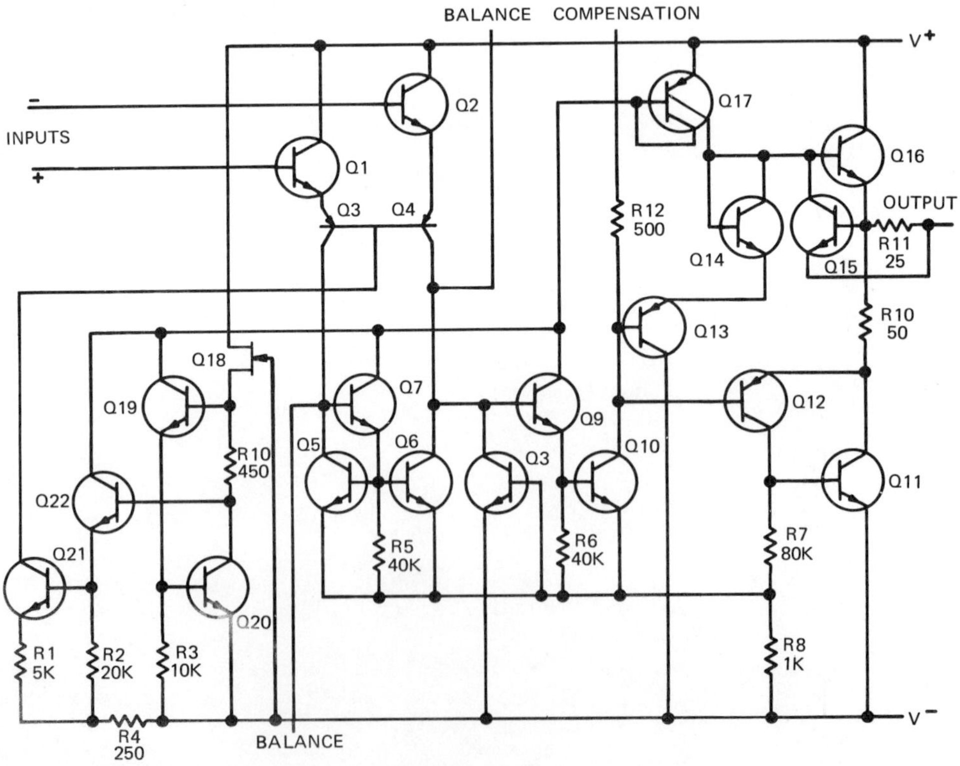

Fig. K.1 LM101A schematic and connection diagram.

The black-box model for the device is illustrated in Fig. K.2, with the effects of radiation included.

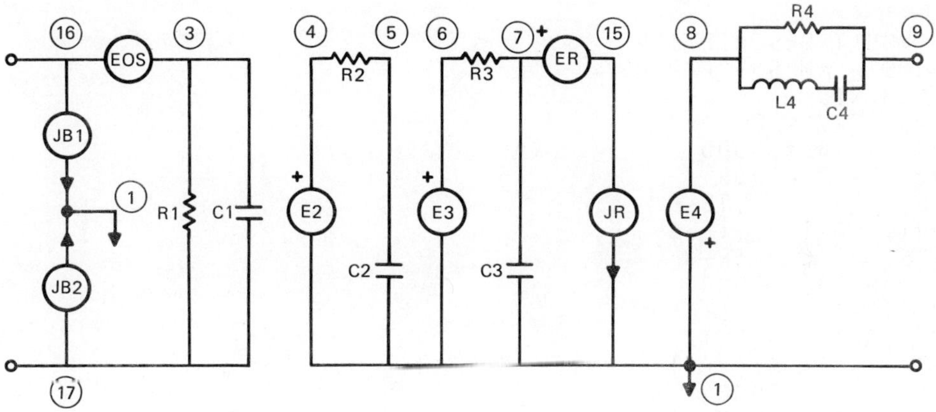

Fig. K.2 OP AMP Simulation diagram.

The symbols are defined as follows:

JB1 and JB2 simulate the input base currents. They are also a function of radiation for correct input characteristics.

EOS simulates the input voltage offset.

R1 and C1 simulate the input Z to the OP AMP.

R2C2 and R3C3 are used to simulate the two significant open-loop break points.

E3 is a non-linear tabulated function of VC2 and simulates the slewing effect.

E4 incorporates the open-loop gain and the power supply output saturation limits.

ER simulates the measured radiation response of the overall closed-loop amplifier.

R4 and C4 simulate the output Z.

A typical SCEPTRE listing is given in Fig. K.3. This particular run has the radiation level set to zero. The amplifier is in a closed loop configuration as shown in Fig. K.4.

Other tests can illustrate the sophistication of the model. For example, when the device is connected in the non-inverting mode with unity feedback, the non-linear slewing effects

* 1st Lt. D. R. Alexander et al, "Systems Approach to Radiation Analysis", IEEE Transactions on Nuclear Science, Dec. 1969.

may be observed for +5 volt inputs. Figure K.5 illustrates the results which are nearly identical to the test data for the actual device.

```
SCEPTRE SYSTEM INPUT-UCC 1/15/69

CIRCUIT DESCRIPTION
OHMS - AMPS - FARADS
ELEMENTS
ER, 15-7 = 0
RX, 1-16 = 1000
RZ, 1-17 = 909
JB2, 17-1 = .1E-6
JB1, 16-1 = .1E-6
EOS, 3-16 = 5E-3
RY, 9-16 = 10000
R1, 3-1 = 4,E6
C1, 3-1 = 3,E-12
E2, 1-4 = Q1(VC1)
R2, 4-5 = 1000
C2, 5-1 = 80,E-12
E3, 1-6 = T1(VC2)
R3, 6-7 = 10,E7
C3, 7-1 = 25,E-12
E4, 8-1 = T2(VJR)
JR, 15-1 = 0
R4, 8-9 = 100
C4, 8-9 = 10,E-12
RL, 9-1 = 10000
DEFINED PARAMETERS
P1 = 1
P2 = 10000
OUTPUTS
   VC1, E2, VC2, E3, VC3, EF, VR1, PLOT
ER, VJR, PLOT
VR4, VRY, PLOT
FUNCTIONS,    T3 = 0, 0, 0, .8, 1, .8
TR = 0, 0, 5E-6, 0, 5E-6, 12E-4, 8E-6, 12E-4, 18E-6, 0, 20E-6
Q1(A) = (A)
T1 = -1, -.15, -.15, -.15, .15, .15, 1, .15, 3, .15, 6, 1
T2 = -1, -10, -1, E-3, -10, 1, 3-3, 10, 1, 10
RUN CONTROLS
STOP TIME = 50E-6
MINIMUM STEP SIZE = 1.E-25
MAXIMUM PRINT POINTS = 100
END
```

Fig. K.3 SCEPTRE OP AMP Listing.

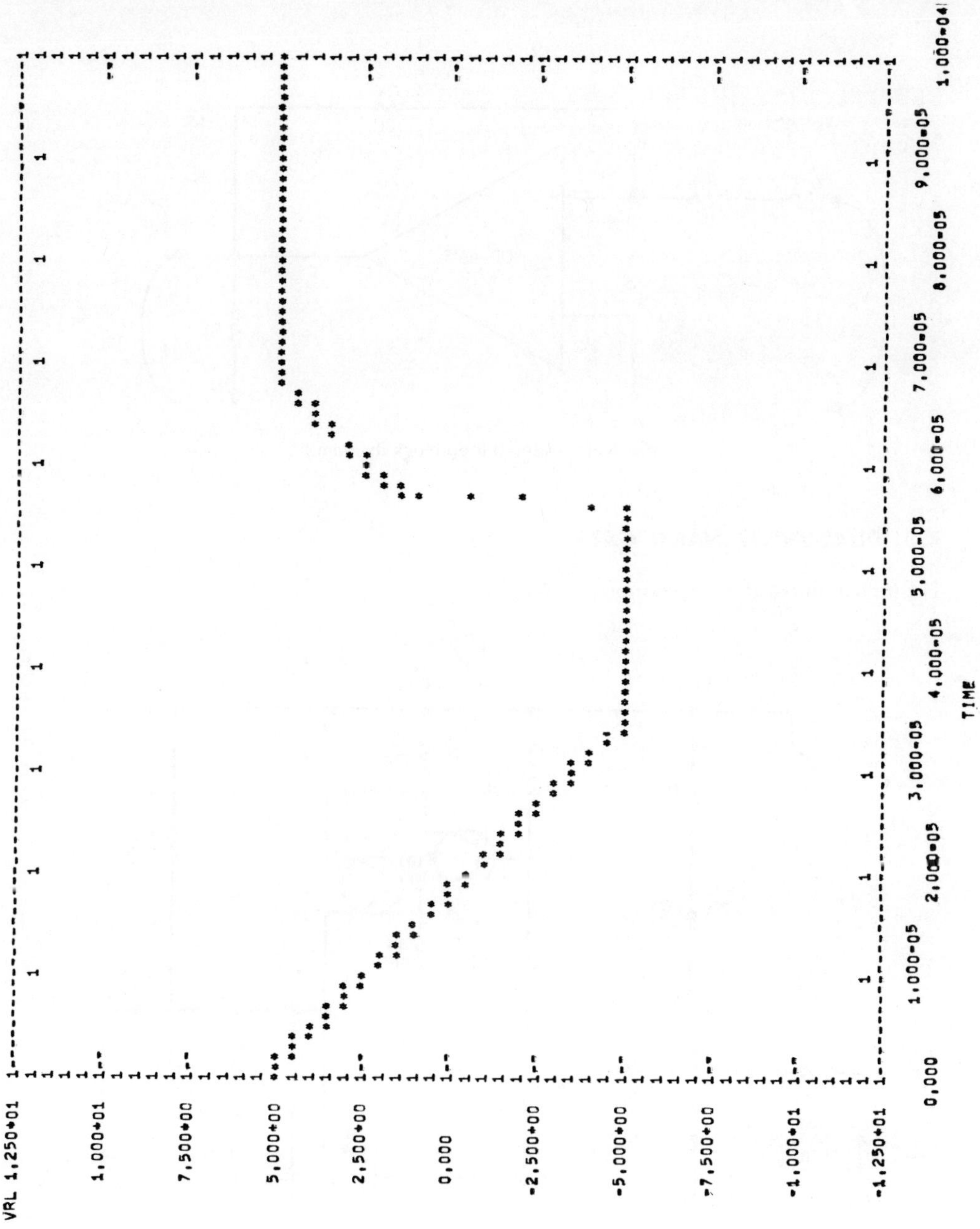
VRL 1,250+01
1,000+01
7,500+00
5,000+00
2,500+00
0,000
-2,500+00
-5,000+00
-7,500+00
-1,000+01
-1,250+01
0,000
1,000-05
2,000-05
3,000-05
4,000-05
5,000-05
6,000-05
7,000-05
8,000-05
9,000-05
1,000-04
TIME

Fig. K.5

It has been found that the computer time saved using the black-box approach is several orders of magnitude.

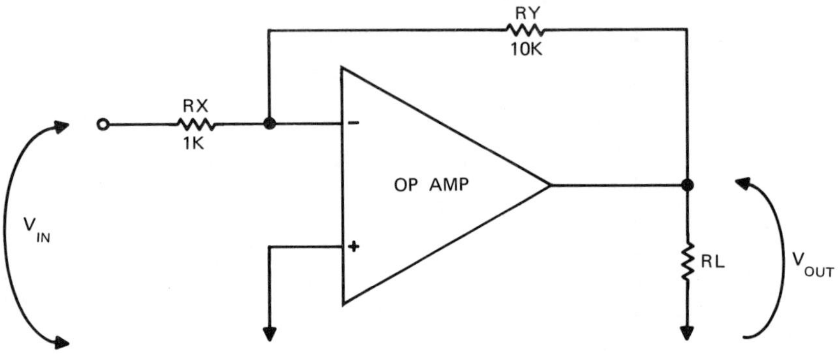

Fig. K.4 Closed loop block diagram.

K.3 DUAL INPUT NAND GATE

Figure K.6 illustrates a dual input NAND gate.

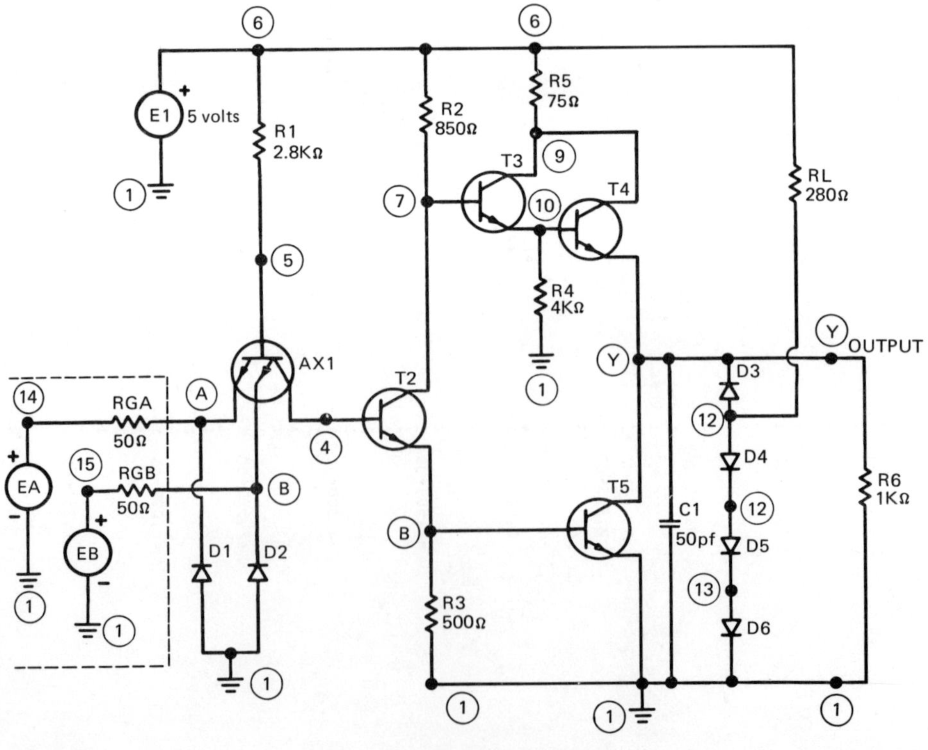

Fig. K.6 Dual input nand gate.

Figure K.7 illustrates the black-box diagram. A complete SCEPTRE listing is given in Fig. K.8 for typical radiation run. The diodes DA and DB simulate the input characteristics as well as the radiation input effects. JA and JB drive controlled currents back through the input generators at the proper levels. Several defined parameter tables, along with E4 simulate the logic. RTCT duplicates the time constant. ERP and ERN simulate the radiation. These generators automatically reproduce the proper effect depending upon the logic state.

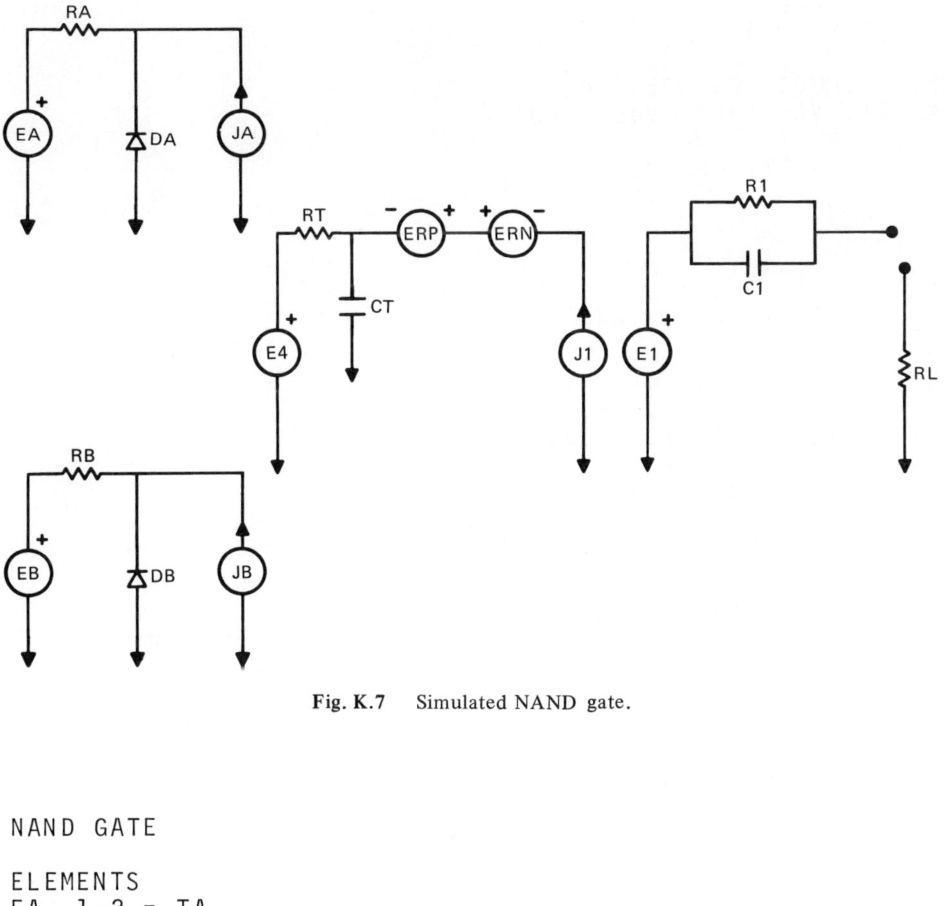

Fig. K.7 Simulated NAND gate.

```
NAND GATE

ELEMENTS
EA, 1-2 = TA
EB, 1-4 = TB
RA, 2-3 = 30
DA, 1-3 = MODEL 1NSPL
JA, 1-3 = TJA (VCODA)
RB, 4-5 = 30
DB, 1-5 = MODEL 1NSPL
JB, 1-5 = T JB(VCODB)
E4, 1-6 = P4
RT, 6-7 = 2000
CT, 7-1 = 1E-12
```

```
ERP, 7-8 = Q1(P3,T1)
ERN, 9-8 = Q1(P4,T2)
J1, 9-1 = 0
E1, 1-10 = T J1(VJ1)
R1, 10-11 = 100
C1, 10-11 =  10, E-12
RL, 11-1  = 10000
DEFINED PARAMETERS
P3 = Q1(P5,P6)
P4 = T5(P3)
P5 = T6(VCODA)
P6 = T7(VCODB)
OUTPUTS
VCODA, VCODB, P5, P6, E4, ERP, PLOT
ERN, E1, VRL, VCT, VJL, PLOT
JA, JB, PLOT
INITIAL CONDITIONS
VCODA = 3
VCODB = 3
FUNCTIONS
Q1(A,B) = (A*B)
TA = 0, 3, 10, 3
TB = 0, 3, 80E-9, 3, 80E-9, 0, 1, 0
TJA = 0, .01, .5, .01, 1, 0, 10, 0
TJB = 0, .01, .5, .01, 1, 0, 10, 0
TJ1 = 0, 0, 1, 4, 1, 25, 5, 10, 5
T5 = 0, 1, .5, 1, 1, o, 5, 0
T6 = 0, 0, 1, 0, 2, 1, 5, 1
T7 = 0, 0, 1, 0, 2, 1, 5, 1
T1 = 0, 0, 30E-9, 0, 35E-9, 1,50E-9, 1.55E-9, 0, 1, 0
T2 = 0, 0, 30E-9, 0, 34E-9, .075, 55E-9, .075, 90E-9, 1, 0
RUN CONTROLS
MAXIMUM PRINT POINTS = 100
MINIMUM STEP SIZE = 1.E-20
STOP TIME = 100.E-9
END
```

Fig. K.8 NAND Gate SCEPTRE Listing.

E1, R1 and C1 simulate the output characteristics. A careful study of Fig. K.7 and K.8 indicates the various linear and non-linear functional relationships. Figure K.9 illustrates an output plot of the run. Since this is a fairly small circuit the computer time is only improved by a four-to-one ratio. In larger digital I.C. models the savings is significant.

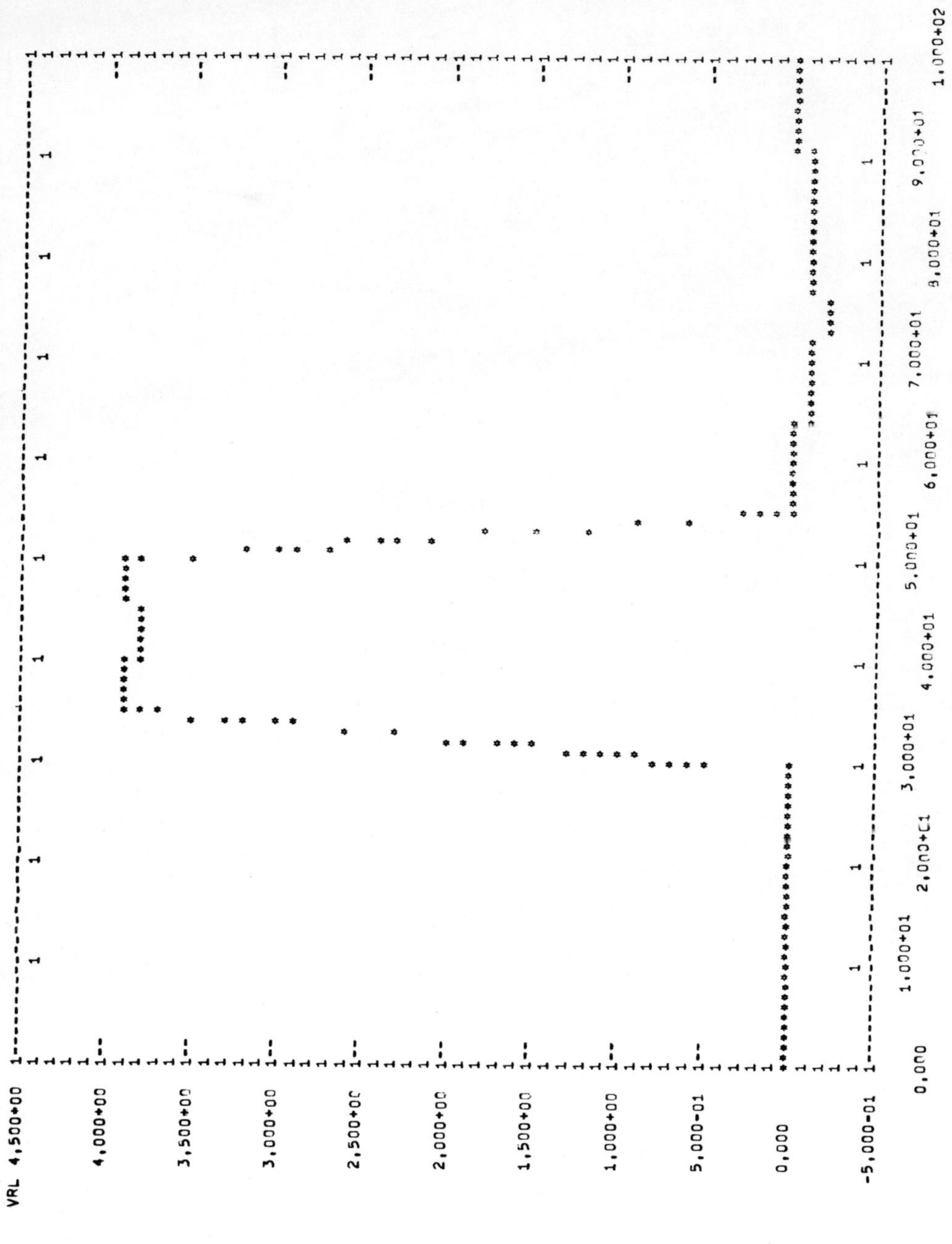
VRL
4,500+00
4,000+00
3,500+00
3,000+00
2,500+00
2,000+00
1,500+00
1,000+00
5,000-01
0,000
-5,000-01
0,000
1,000+01
2,000+01
3,000+01
4,000+01
5,000+01
6,000+01
7,000+01
8,000+01
9,000+01
1,000+02
TIME

Fig. K.9

APPENDIX L

DIGITAL LOGIC

L.1 USING SCEPTRE TO PERFORM DIGITAL LOGIC

SCEPTRE can perform digital logic by using the Defined Parameter mode in conjunction with equations and tables.

Figures L.1, L.2 and L.3 illustrate the principles showing the truth tables, listings and tabular data. The designer determines the "min-max" values acceptable for the logical "1" and "O" quantities. For the examples shown, a logical "1" is equal to .8 to 1 volt.

AND GATE

A	B	Φ
O	1	O
1	O	O
1	1	1
O	O	O

P3 = (E1 + E2)
P4 = T4 (P3)
T4 = O, O, 1.6, O, 2, 1, 3, 1

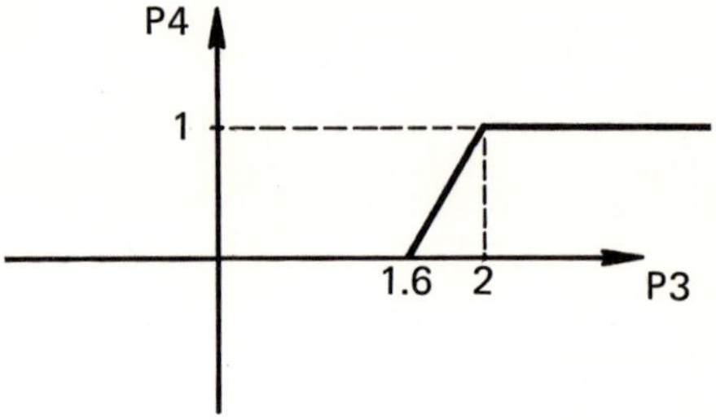

Fig. L.1

OR GATE

A	B	Φ
O	1	1
1	O	1
1	1	1
O	O	O

P1 = (E1 + E2)
P2 = T2 (P1)
T2 = O, O, .8, O, 1, 1, 2, 1

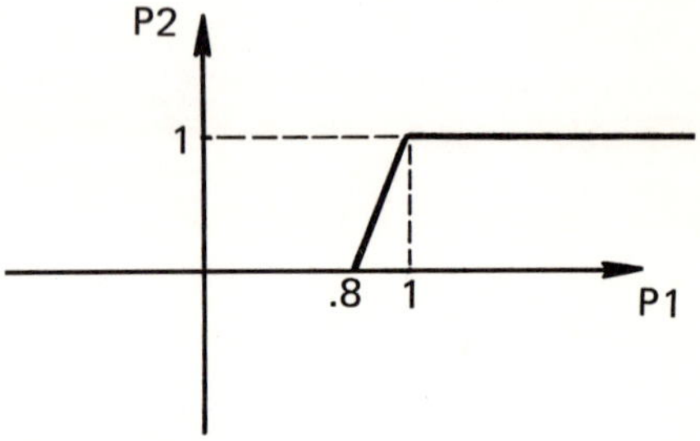

Fig. L.2

NAND GATE

A	B	Φ
O	1	1
1	O	1
1	1	O
O	O	1

P5 = (E1 + E2)
P6 = T6 (P5)
T6 = O, 1, 1.6, 1, 2, O, 3, O

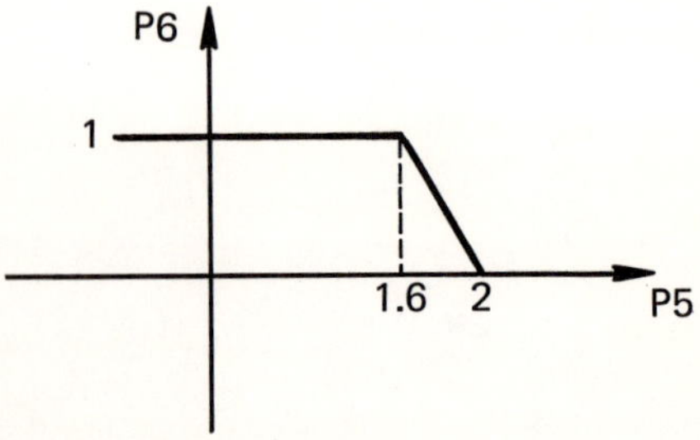

Fig. L.3

To further illustrate the logic flexibility, Fig. L.4 shows a latch circuit* and the corresponding truth table.

Logic Diagram (a)

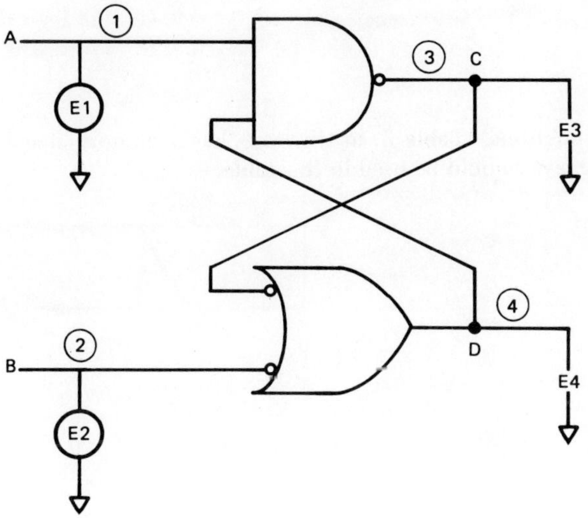

True High Logic (b)

TRUTH TABLE

$\bar{A}$	B	C	$\bar{D}$
A	B	C	$\bar{D}$
A	$\bar{B}$	$\bar{C}$	D
A	B	$\bar{C}$	D
$\bar{A}$	$\bar{B}$	C	D

Fig. L.4

The latch is programmed on SCEPTRE as follows:

E1, 0–1 = Table 1 (Time)

E2, 0–2 = Table 2 (Time)

E3, 0–3 = Table 3 (P1)

E4, 0–4 = Table 3 (P2)

Defined Parameters

P1 = Q1 (E1, E4)

P2 = Q1 (E3, E2)

Functions

Table 1

E1 defined as a function of Time

* Private memo from R. C. Trump, 27 July 1970.

Table 2

E2 defined as a function of Time

Table 3

0, 1, 1.6, 1, 2, 0, 3, 0

Q1 (A, B) = (A + B)

Table 3 and Q1 establish the logic functions. Table 3, in this case, has been normalized to one volt. If desired, actual voltage levels could be used in this table.

ANSWERS TO PROBLEMS

1.1 **(a)** There are six cut sets in this network:
E-R1, R1-C-L, L-R2-J, E-C-R2-J, C-L-E, R1-C-R2-J. These cut sets are independent of the network tree.

(b) There are also six loops:
E-R1-C, E-R1-L-R2, E-R1-L-J, C-L-R2, C-L-J, R2-J. These loops are independent of the network tree.

1.2 There are three fundamental cut sets:
E-R1, R1-C-L, L-R2-J. These fundamental cut sets depend on the network tree.

1.3 There are three fundamental loops:
E-R1-C, C-L-R2, R2-J. These fundamental loops depend on the network tree.

1.4 Neither. A cut set is defined as a group of elements that, when "cut" out of the network, divide the network into two separate groups of nodes. It can be seen that removal of the referenced element set would divide the network into three groups of nodes rather than two.

1.6 A voltage source link could exist only in a loop composed exclusively of voltage sources (otherwise the voltage source in question would be a tree branch, since voltage sources receive first priority in the tree construction). Therefore, the magnitude of this voltage source link could not be independent -- it would have to equal the sum of all other voltage sources in its fundamental loop. The definition of such an element type would do little for the practical value of any automatic analysis program.

1.9 Since the element priority for our specific tree is established, there are only two possible versions that can be constructed. In one, the tree branches are E, C and Cx while all other elements are links. The other possibility utilizes E, C and Cy as the tree branches with all other elements serving as links.

1.11 Each row of Q represents a fundamental cut set and each column represents a network element. By definition, each fundamental cut set must include exactly one tree branch and no tree branch may appear in more than one fundamental cut set. Therefore, the partition of Q that corresponds to the network tree branches must have only a single 1 entry in each row. The signs of these entries are all positive if the reference direction of each fundamental cut set is chosen to agree with that of its associated tree branch. There only remains to appropriately order the columns of Q to insure that the partition forms a unit matrix.

1.14 The maximum number of nodes (8) and tree branches (7) as well as the minimum number of fundamental loops (1) will occur if the eight elements are connected in series. The minimum number of nodes (2) and tree branches (1) as well as the maximum number of fundamental loops (7) arise if the eight elements are connected in parallel.

1.15 After the circuit is drawn, the specific tree priorities require that E1 and C1 be tree branches. In addition, there are eight possible combinations of three resistors each that could be used to complete the tree. Therefore, there must be three resistor tree branches and two resistor links in this network.

1.17 Now M_L^{-1} exists and the solution would be carried out. Unfortunately, this value of M would correspond to a coefficient of coupling $k > 1$, which has the effect of creating a pole in the right half of the complex plane. The solution will be accurate, but the result will be unstable and worthless for a practical network.

2.1

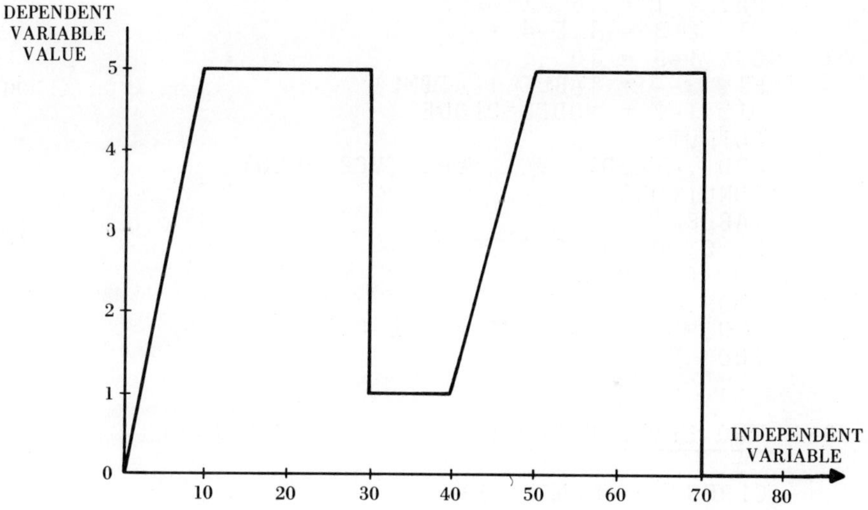

2.2 **(a)** $M_{12} = -$ **(b)** $M_{12} = -$, $M_{13} = -$, $M_{23} = +$ **(c)** $M_{12} < \sqrt{(L1)(L2)}$

2.4 **(b)** Independent variable of tables can not be changed.
(c) Word "CHANGE" in front of CC = 25 was left out.
(d) A table cannot be changed to an equation and vice versa.

2.8 Since the independent variables differ in the two reruns then separate rerun description subheadings must be used.

```
FUNCTIONS
TABLE 1 = 0, 0, 0, 2, 25, 2, 25, 0,
35, 0, 35, 2, 50, 2, 50, 0, 60, 0
RERUN DESCRIPTION (1)
FUNCTIONS
TABLE 1 = 0, 0, 0, 2, 10, 2, 10, 0,
30, 0, 30, 2, 40, 2, 40, 0, 50, 0
```

2.12 The elements subsection is not permitted under CONTINUE because changes to the network are not allowed under this heading.

The initial conditions statement is incorrect because the CONTINUE option is intended only for transient runs and has no control over initial conditions computation.

2.13 (a) DIODE CLAMP

```
MODEL DESCRIPTION
MODEL DIODE (1-3)
ELEMENTS
C, 1-2 = 10
RB, 2-3 = .05
JD, 1-2 = DIODE Q (1.E-7, 30)
CIRCUIT DESCRIPTION
ELEMENTS
R1, 4-2 = .6
R2, 2-1 = .6
C1, 2-3 = 1.E-4
C2, 1-3 = 10
E1, 3-4 = TABLE 1 (TIME)
D1, 1-3 = MODEL DIODE
OUTPUTS
JDD1, IRBD1, VC1, VR2, VC2, PLOT
FUNCTIONS
TABLE 1
0, 0, 0, 5, 200, 5
RUN CONTROLS
STOP TIME = 10
MINIMUM STEP SIZE = 1.E-8
END
```

2.13 (d) TWO LEVEL CLIPPER

```
CIRCUIT DESCRIPTION
ELEMENTS
R1, 3-4 = 90
R2, 6-5 = 180
E1, 1-4 = 25
E2, 1-6 = 100
E1N, 1-2 = TABLE 1 (TIME)
D1, 2-3 = 1N791 (PERM)
D2, 5-3 = 1N791 (PERM)
DEFINED PARAMETERS
P1 = X1(E2 + VR2)
OUTPUTS
P1 (VOUT), PLOT (E1N)
FUNCTIONS
TABLE 1
0, 0, 100, 150, 200, 150
RUN CONTROLS
STOP TIME = 125
MINIMUM STEP SIZE = 1.E-5
END
```

2.13 (f) COMPLEMENTARY EMITTER FOLLOWER

```
CIRCUIT DESCRIPTION
ELEMENTS
```

```
R1, 2-3 = .47
R2, 4-1 = 1.0
C1, 4-1 = 100
EB1, 1-5 = 12
EB2, 6-1 = 12
E1N, 1-2 = TABLE 1 (TIME)
T1, 3-4-5 = MODEL 2N1613 (PERM)
T2, 3-4-6 = MODEL 2N1132 (PERM)
OUTPUTS
E1N, VR2, IR1, PLOT
FUNCTIONS
TABLE 1
0, 0, 0, 12, 250, 12, 250, -12, 500,
-12, 500, 0, 600, 0
RUN CONTROLS
STOP TIME = 600
RUN INITIAL CONDITIONS
MINIMUM STEP SIZE = 1.E-5
END
```

3.1 $\Theta = 35$ Volts^{-1}, IS = 7.5 X 10^{-10}ma, RB = 12Ω

3.2 RS = 3.3 x 10^6 Kilohms

3.3 KD = 5.10 x 10^{12} pf/ma

3.4 ϕ = .9v, CO = 8.5, n = .45

3.7 ϕ = .9v, COC = 4.2, n = .35

3.11 ϕ_B = .9v, V_P = 5v., G_C = 8.7 ma/volt

3.12 RO = 50K, I_T = 8.885 ma, K = .0001 ma

3.13 ϕ_B = .9v, n_G = .45, CG = 10, n_D = .4, CD = 8

3.14 V_T = 3 volts, B = .45 ma/volt^2

3.16 a = .0025 volts^{-1}

3.17

V_{BB}	RBB
5v.	7 Kilohms
10	7.2
15	7.4
20	7.6
25	7.8

$b = .04 \ \frac{\text{Kilohms}}{\text{Volt}}$

3.18 N = .571, IE1 = 7.22 x 10^{-3}ma, α = .148

4.3 ONE SHOT MULTIVIBRATOR

```
CIRCUIT DESCRIPTION
ELEMENTS
EB1, 6-1 = 10
EB2, 1-7 = 10
R1, 3-1 = 1
R2, 4-6, = 50
R3, 7-5 = 1
R4, 9-4 = 10
R5, 7-8 = 20
R6, 7-9 = 1
C1, 2-3 = 225
C2, 5-8 = 680
EIN, 1-2 = TABLE 1 (TIME)
T1, 4-1-5 = MODEL 2N1613 (PERM)
T2, 8-1-9 = MODEL 2N1613 (PERM)
D1, 3-4 = MODEL 1N791 (PERM)
DEFINED PARAMETERS
P1 = X1 (EB2 - VR6)
P2 = X2 (EB2 - VR5)
P3 = X3 (EB2 - VR6 - VR4)
OUTPUTS
EIN, P1(VCT2), P2(VBT1), P3(VBT1), PLOT
FUNCTIONS
TABLE 1
0, 0, 0, 5, 5.E3, 5, 5.E3, 0, 5.5E3, 0
RUN CONTROLS
STOP TIME = 5.E3
RUN INITIAL CONDITIONS
MINIMUM STEP SIZE = 1.E-5
END
```

4.5 ASTABLE MULTIVIBRATOR

```
CIRCUIT DESCRIPTION
ELEMENTS
R1, 6-3 = 1
R2, 6-4 = 20
R3, 6-2 = 20
R4, 6-5 = 1
C1, 3-4 = 650
C2, 5-2 = 650
E1, 1-6 = 10
T1, 2-1-3 = MODEL 2N1613 (PERM)
T2, 4-1-5 = MODEL 2N1613 (PERM)
DEFINED PARAMETERS
P1 = X1 (E1 - VR4)
P2 = X2 (E1 - VR3)
OUTPUTS
P1 (VOUT), P2 (VBT2), PLOT
INITIAL CONDITIONS
VC1 = 9.3
VC2 = -.6
VCET1 = .7
```

```
VCET2 = -9.2
VCCT1 = .6
VCCT2 = -9.3
RUN CONTROLS
MINIMUM STEP SIZE = 1.E-5
STOP TIME = 20E3
END
```

4.9 DIFFERENTIAL AMPLIFIER

```
CIRCUIT DESCRIPTION
ELEMENTS
R1, 9-3 = 9
R2, 9-8 = 9
R3, 4-5 = .1
R4, 6-5 = .1
R5, 1-11 = 9
R6, 10-12 = 1
R7, 11-12 = 2
T1, 2-4-3 = MODEL 2N1613 (PERM)
T2, 1-6-8 = MODEL 2N1613 (PERM)
T3, 11-10-5 = MODEL 2N1613 (PERM)
E2, 1-9 = 15
E3, 12-1 = 15
EIN, 1-2 = TABLE 1 (TIME)
DEFINED PARAMETERS
P1 = X1(E2 - VR2)
OUTPUTS
P1 (VOUT), PLOT
FUNCTIONS
TABLE 1
0, -2, 50, -2, 150, 2, 200, 2
RUN CONTROLS
STOP TIME = 175
RUN INITIAL CONDITIONS
MINIMUM STEP SIZE = 1.E-5
END
```

4.12 BOOTSTRAPPED DARLINGTON

```
CIRCUIT DESCRIPTION
ELEMENTS
R1, 1-2 = 90
R2, 6-2 = 80
R3, 2-3 = 9
R4, 6-7 = 1
R5, 4-3 = .1
C1, 2-4 = 10.E6
C2, 8-1 = 10.E6
EB, 3-6 = 24
EIN, 3-8 = TABLE 1 (TIME)
T1, 1-5-6 = MODEL 2N1613 (PERM)
T2, 5-4-7 = MODEL 2N1613 (PERM)
OUTPUTS
```

```
EIN (VIN), VR5 (VOUT), PLOT
FUNCTIONS
TABLE 1
0, 0, 25, 10, 50, 0, 75, 10, 100, 0, 125, 10, 150, 0
RUN CONTROLS
RUN INITIAL CONDITIONS
STOP TIME = 125
MINIMUM STEP SIZE = 1.E-5
END
```

4.13

```
DEFINED PARAMETERS
DP12 = X1 (VR12 * IR12)
P12 = 0
OUTPUTS
P12, . . .
```

5.1 Figure A-5.1 shows the revised block diagram in SCEPTRE form.

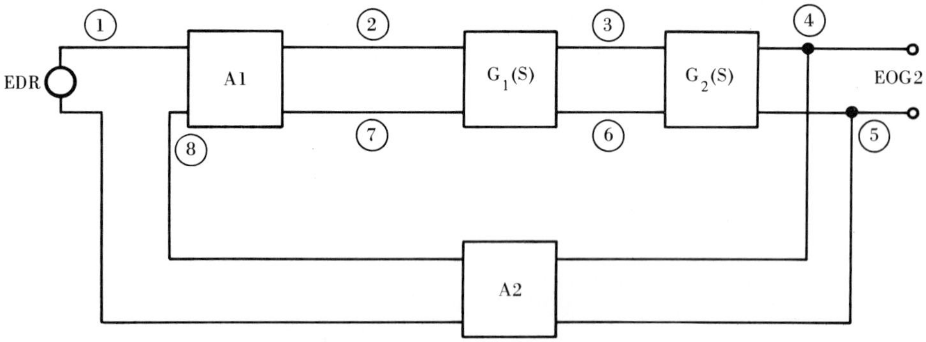

Fig. A-5.1

```
MODEL DESCRIPTION
MODEL GAIN (TEMP) (I-0-GI-GO)
ELEMENTS
JI, I-GI = 0.0
EO, GO-0 = X1 (VJI*PA)
DEFINED PARAMETERS
PA = 100
MODEL 1 ORDER (TEMP) (I-0-GI-GO)
ELEMENTS
JI, I-GI = 0
EO, GO-0 = X1 (PAO*P1)
DEFINED PARAMETERS
DP1 = X2 ((VJI-PBO*P1)/PB1)
PAO = 0, PBO = 0, PB1 = 0, P1 = 0.
MODEL 2 ORDER (TEMP) (I-0-GI-GO)
ELEMENTS
JI, I-GI = 0
EO, GO-0 = X1 (PA1*P2 + PAO*P1)
DEFINED PARAMETERS
DP2 = X2 ((VJI-PB1*P2-PBO*P1)/PB2)
DP1 = X3 (P2)
```

```
PAO = 0, PA1 = 0, PBO = 0, PB1 = 0, PB2 = 0, P1 = 0,
P2 = 0
CIRCUIT DESCRIPTION
EXAMPLE 5.1
DEPTH CONTROL
ELEMENTS
EDR, 9-1 = TABLE 1 (TIME)
A1, 1-2-8-7 = MODEL GAIN (CHANGE PA = 1E-3)
A2, 4-8-5-9 = MODEL GAIN (CHANGE PA = 1.0)
G1, 2-3-7-6 = MODEL 2 ORDER (PA1 = 1.0, PAO = 0.1,
PB2 = 1.0, PBO = 0.02)
G2, 3-4-6-5 = MODEL 1 ORDER (CHANGE PAO = 1.0,
PB1 = 1.0)
OUTPUTS
EOG2, EOA1, EOG1, EOA2, PLOT
FUNCTIONS
TABLE 1
0, 0
0, 20
10, 20
RUN CONTROLS
STOP TIME = 10
END
```

5.2 Figure A-5.2 shows the system block diagram in SCEPTRE form.

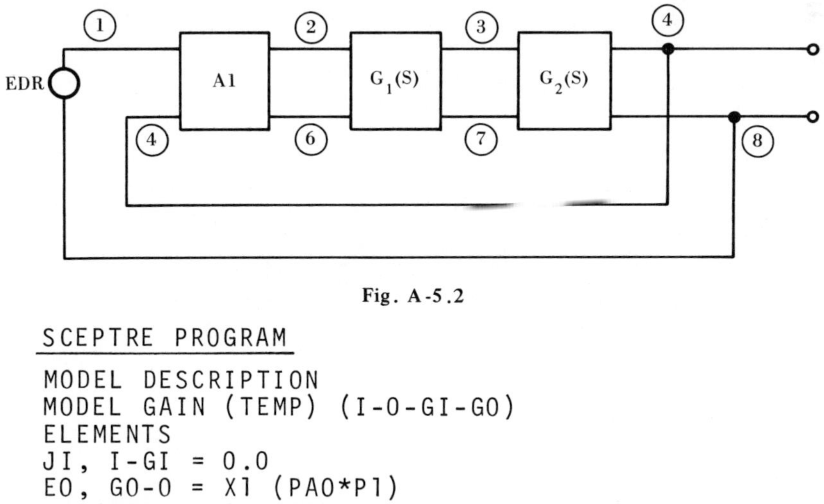

Fig. A-5.2

<u>SCEPTRE PROGRAM</u>

```
MODEL DESCRIPTION
MODEL GAIN (TEMP) (I-O-GI-GO)
ELEMENTS
JI, I-GI = 0.0
EO, GO-O = X1 (PAO*P1)
DEFINED PARAMETERS
PA = 100
MODEL 1 ORDER (TEMP) (I-O-GI-GO)
ELEMENTS
JI, I-GI = 0
EO, GO-O = X1 (PAO*P1)
DEFINED PARAMETERS
DP1 = X2 ((VJI-PBO*P1)/PB1)
PAO = 0, PBO = 0, PB1 = 0, P1 = 0.
```

```
MODEL 2 ORDER (TEMP) (I-0-GI-GO)
ELEMENTS
JI, I-GI = 0
EO, GO-0 = X1 (PA1*P2 + PA0*P1)
DEFINED PARAMETERS
DP2 = X2 ((VJI-PB1*P2-PB0*P1)/PB2)
DP1 = X3 (P2)
PA0 = 0, PA1 = 0, PB0 = 0, PB1 = 0, PB2 = 0, P1 = 0,
P2 = 0
CIRCUIT DESCRIPTION
EXAMPLE 5.2
PHASE DETECTOR
ELEMENTS
EDR, 8-1 = TABLE 1 (TIME)
A1, 1-2-4-6 = MODEL GAIN (CHANGE PA = 10)
G1, 2-3-6-7 = MODEL 2 ORDER (CHANGE PA1 = 1.0 PA0 =
5.0, PB2 = 1.0, PB1 =201.0, PB2 = 200)
G2, 3-4-7-8 MODEL 1 ORDER (CHANGE PA0 = 2.0,
PB1 = 1.0)
OUTPUTS
EOG2, EOA1, EOG1, PLOT
FUNCTIONS
TABLE 1
0,0
1,1
5,5
RUN CONTROLS
STOP TIME = 5
END
```

5.3 (a) Figure A-5.3.1 shows the model for dead zone

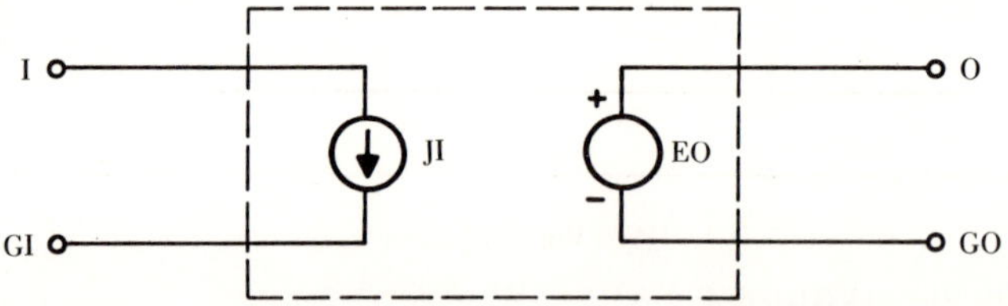

Fig. A-5.3.1

```
MODEL DESCRIPTION
MODEL G DEAD ZONE (TEMP) (I-0-GI-GO)
ELEMENTS
JI, I-GI = 0.0
EO, GO-0 = TABLE 1 (VJI)
FUNCTIONS
TABLE 1
-5, -4, -2, -1, -1, -1, 0, 0, 0, 2, 1, 5, 4
```

(b) Figure A-5.3.2 shows the system block diagram in SCEPTRE form.

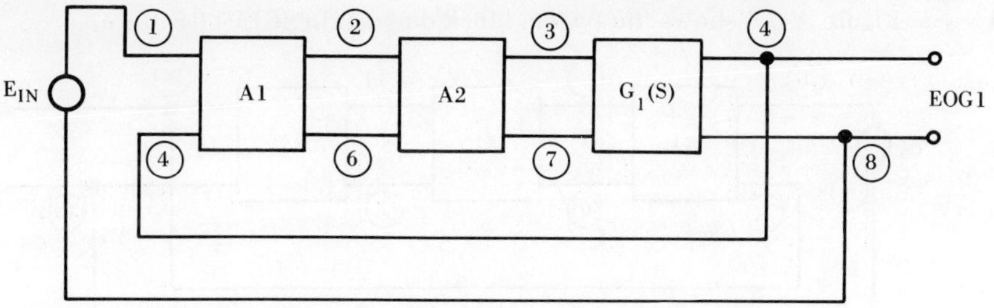

Fig. A-5.3.2

SCEPTRE PROGRAM

```
MODEL DESCRIPTION
MODEL GAIN (TEMP) (I-0-GI-GO)
ELEMENTS
JI, I-GI = 0.0
EO, GO-0 = X1 (VJI*PA)
DEFINED PARAMETERS
PA = 100
MODEL G DEAD ZONE (TEMP) (I-0-GI-GO)
ELEMENTS
JI, I-GI = 0.0
EO, GO-0 = TABLE 1 (VJI)
FUNCTIONS
TABLE 1
-5, -4, -2, -1, -1, 0, 1, 0, 2, 1, 5, 4
MODEL 2 ORDER (TEMP) (I-0-GI-GO)
ELEMENTS
JI, I-GI = 0.0
EO, GO-0 = X1 (4.0*P1)
DEFINED PARAMETERS
DP2 = X2 ((VJI-P2)/2.0)
DP1 = X3 (P2)
P1 = 0, P2 = 0
CIRCUIT DESCRIPTION
EXAMPLE 5.3
INSTRUMENT SERVO
ELEMENTS
EIN, 8-1 = 10
A1, 1-2-4-6 = MODEL G DEAD ZONE
A2, 2-3-6-7 = MODEL GAIN (CHANGE PA = 10)
G1, 3-4-7-8 = MODEL 2 ORDER
DEFINED PARAMETERS
PDEO = X4 (4.0*P2)
OUTPUTS
PDEO (RATE), PLOT (EOG1 (ANGLE))
EOA1, EOA2, PLOT
RUN CONTROLS
STOP TIME = 10
END
```

5.4 Figure A-5.4 shows the system block diagram in SCEPTRE form.

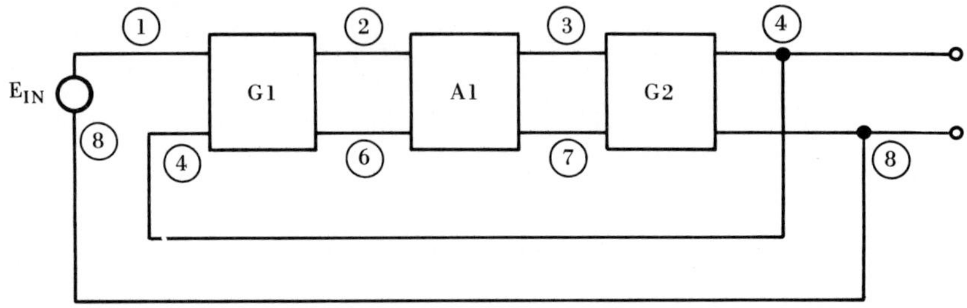

Fig. A-5.4

SCEPTRE PROGRAM

```
MODEL DESCRIPTION
MODEL RELAY (TEMP) (I-0-GI-GO)
ELEMENTS
JI, I-GI = 0.0
EO, GO-0 = TABLE 1 (VJI)
FUNCTIONS
TABLE 1
-50, -1, -1.E-6, -1, 1.E-6, 1, 50, 1
MODEL 3 ORDER (TEMP) (I-0-GI-GO)
ELEMENTS
JI, I-GI = 0.0
EO, GO-0 = X1 (PA2 * P3 + PA1 * P2 + PAO * P1)
DEFINED PARAMETERS
DP3 = X2 ((VJI - PB2 * P3 - PB1 * P2 - PBO * P1)/PB3)
DP2 = X3 (P3)
DP1 = X4 (P2)
PAO = 0, PA1 = 0, PA2 = 0, PBO = 1, PB1 = 0,
PB2 = 0, PB3 = 0, P1 = 0, P2 = 0, P3 = 0
MODEL 1 ORDER (TEMP) (I-0-GI-GO)
ELEMENTS
JI, I-GI = 0
EO, GO-0 = X1 (PAO*P1)
DEFINED PARAMETERS
DP1 = X2 ((VJI-PBO*P1)/PB1)
PAO = 0, PBO = 0, PB1 = 0, P1 = 0
CIRCUIT DESCRIPTION
EXAMPLE 5.4
THIRD ORDER NONLINEAR SYSTEM
ELEMENTS
EIN, 8-1 = 1
G1, 1-2-4-6 = MODEL 3 ORDER (CHANGE PAO = 5.0, PB3 =
1.0, PB2 = 3.0, PB1 = 2.0)
A1, 2-3-6-7 = MODEL RELAY
G2, 3-4-7-8 = MODEL 1 ORDER (CHANGE PAO = 15.0,
PB1 = 1.0, PBO = 3.0)
OUTPUTS
EOG2, VJIA1, EOA1, EOG1, PLOT
RUN CONTROLS
```

```
STOP TIME = 50
END
```

5.5 Figure A-5.5.1 shows the sample hold model and Fig. A-5.5.2 shows R1 timing diagram. System block diagram in SCEPTRE form is shown in Fig. A-5.5.3.

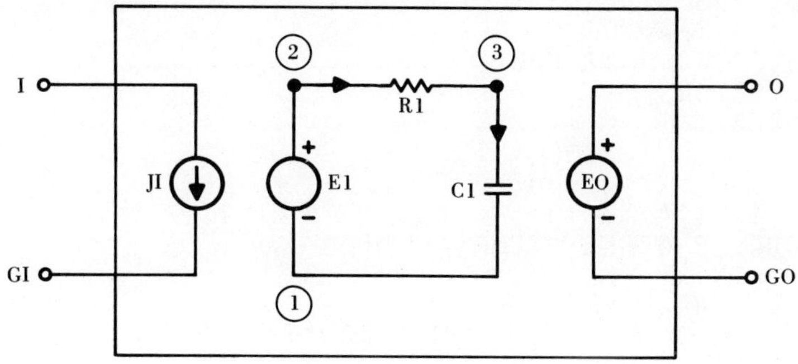

Fig. A-5.5.1

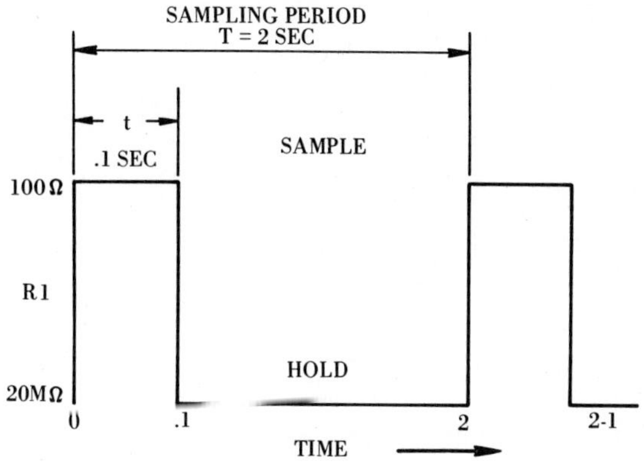

Fig. A-5.5.2

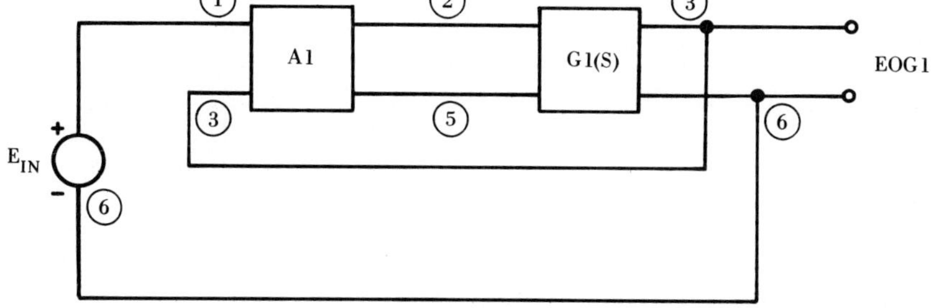

Fig. A-5.5.3

```
SCEPTRE PROGRAM
MODEL DESCRIPTION
MODEL SAMP (TEMP) (I-0-GI-GO)
ELEMENTS
JI, I-GI = 0.0
E1, 1-2 = X1 (VJI)
R1, 2-3 = TABLE 1 (TIME)
C1, 3-1 = 10E-6
EO, GO-0 = X2 (VC1)
FUNCTIONS
TABLE 1
0, 100, .1, 100, .1, 2.E7, 2, 2.E7
2, 100, 2.1, 100, 2.1, 2.E7, 4, 2.E7
4, 100, 4.1, 100, 4.1, 2.E7, 6, 2.E7
MODEL 3 ORDER (TEMP) (I-0-GI-GO)
ELEMENTS
JI, I-GI = 0.0
EO, GO-0 = X1  (PA2*P3 + PA1*P2 + PA0*P1)
DEFINED PARAMETERS
DP3 = X2 ((VJI-PB2*P3-PB1*P2-PB0*P1)/PB3)
DP2 = X3 (P3)
DP1 = X4 (P2)
PA0 = 0, PA1 = 0, PA2 = 0, PB0=0, PB1 = 0, PB2 = 0,
PB3 = 0, P1 = 0, P2 = 0, P3 = 0
CIRCUIT DESCRIPTION
EXAMPLE 5.5
SAMPLED DATA CONTROL SYSTEM
ELEMENTS
EIN, 6-1 = TABLE 1 (TIME)
A1, 1-2-3-5 = MODEL SAMP
G1, 2-3-5-6 = MODEL 3 ORDER (CHANGE PA0 = 2.0, PB3 =
1.0, PB2 = 5.0, PB1 = 6.0)
OUTPUTS
EOG1, EIN, VJIA1, VJIG1, VCIA1, EIA1,
RIA1, PLOT
FUNCTIONS
TABLE 1
0, 0, 0, 2, 5, 2
RUN CONTROLS
STOP TIME = 6
END
```

5.6 Figure A-5.6 shows the system block diagram in SCEPTRE form.

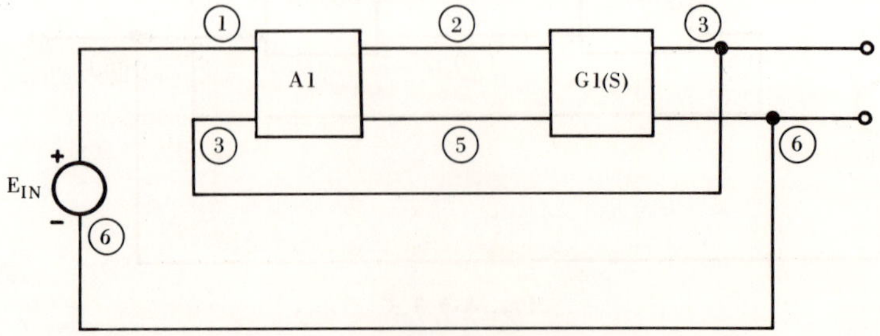

Fig. A-5.6

```
MODEL DESCRIPTION
MODEL SAMP (TEMP) (I-0-GI-GO)
ELEMENTS
JI, I-GI = 0.0
E1, 1-2 = X1 (VJI)
R1, 2-3 = TABLE 1 (TIME)
C1, 3-1 = 100E-6
EO, GO-0 = X2 (VC1)
FUNCTIONS
TABLE 1
0, 50, .2, 50, .2, 1.E6, 1, 1.E6,
1, 50, 1.2, 50, 1.2, 1.E6, 2, 1.E6,
2, 50, 2.2, 50, 2.2, 1.E6, 3, 1.E6,
3, 50, 3.2, 50, 3.2, 1.E6, 4, 1.E6,
4, 50, 4.2, 50, 4.2, 1.E6, 5, 1.E6
MODEL 2 ORDER (TEMP) (I-0-GI-GO)
ELEMENTS
JI, I-GI = 0.0
EO, GO-0 = X1 (1.*P1)
DEFINED PARAMETERS
DP2 = X2 (VJI-2.*P2)
DP1 = X3 (P2)
P1 = 0, P2 = 0
CIRCUIT DESCRIPTION
EXAMPLE 5.6
SAMPLED DATA CONTROL SYSTEM
ELEMENTS
EIN, 6-1 = TABLE 1
A1, 1-2-3-5 = MODEL SAMP
G1, 2-3-5-6 = MODEL 2 ORDER
OUTPUTS
EOG1, EIN, VJIA1, VJIG1, VCIA1, EIA1, RIA1, PLOT
FUNCTIONS
TABLE 1
0, 0, 0, 5, 10, 5
RUN CONTROLS
STOP TIME = 5
END
```

6.1 Figure A-6.1 shows the electrical analog, of the spring-mass-damper system, in SCEPTRE form.

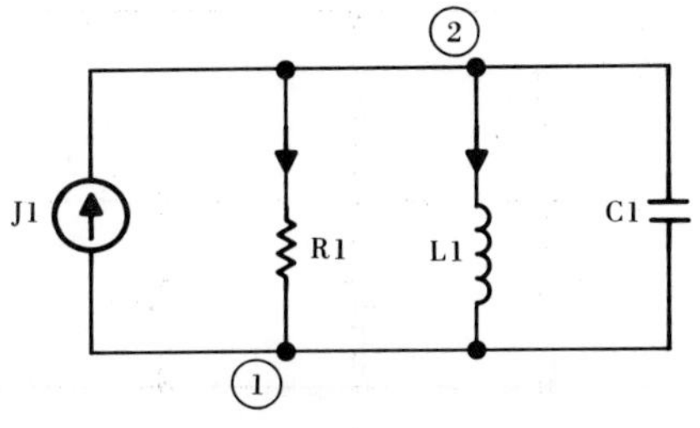

Fig. A-6.1

```
SCEPTRE PROGRAM
CIRCUIT DESCRIPTION
EXAMPLE 6.1
ELEMENTS
J1, 1-2 = 5.0
C1, 2-1 = 10.0
R1, 2-1 = 100.0
L1, 2-1 = 0.02
OUTPUTS
VC1(V1), PLOT
RUN CONTROLS
STOP TIME = 100
END.
```

6.2 Figure A-6.2 shows the electrical analog, of the rotational system, in SCEPTRE form.

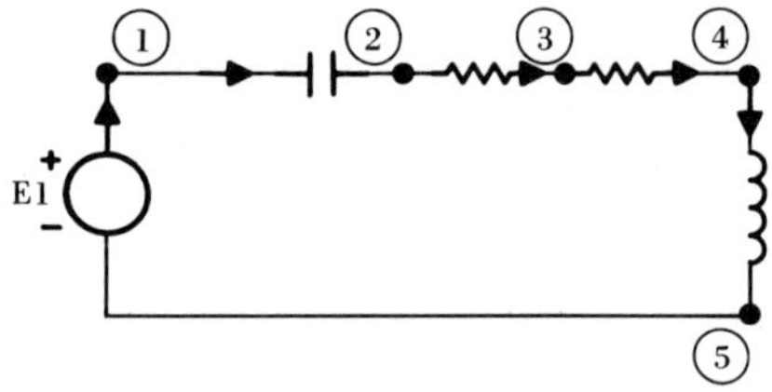


```
SCEPTRE PROGRAM
EXAMPLE 6.2
ELEMENTS
E1, 5-1 = 20.0
C1, 1-2 = 0.1
R1, 2-3 = 0.2
R2, 3-4 = 0.2
L1, 4-5 = 1.E-3
OUTPUTS
IL1 (W1), PLOT
RUN CONTROLS
STOP TIME = 100
END
```

6.3 (a) Figure A-6.3.1 shows the electrical analog of the given mechanical system.

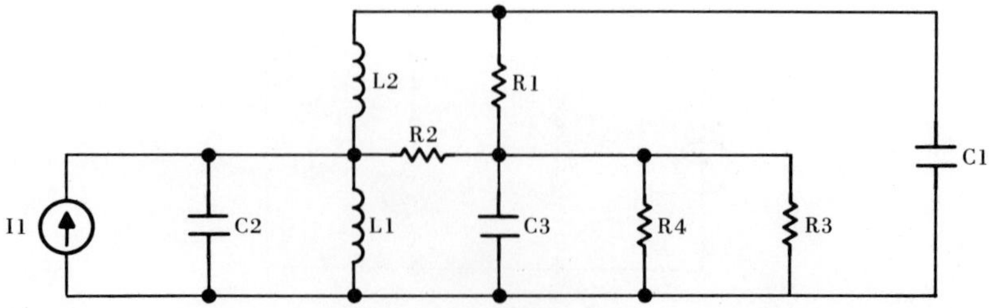

Fig. A-6.3.1

(b) Mechanical-electrical analogies.

MECHANICAL QUANTITY	ELECTRICAL ANALOG	VALUE
Mass M1	C1	0.5 lbs.
Mass M2	C2	2.0 lbs.
Mass M3	C3	1.0 lbs.
Spring Constant K1	1/L1	10 lbs./1n.
Spring Constant K2	1/L2	5 lbs./1n.
Friction Coefficient f1	1/R1	0.001
Friction Coefficient f2	1/R2	0.005
Friction Coefficient f3	1/R3	0.001
Friction Coefficient f4	1/R4	0.002
Force (t)	1(t)	10.0 lbs
Velocity V1(t), V2(t), V3(t)	V1(t), V2(t), V3(t)	To be found

(c) Figure A-6.3.2 shows the circuit in SCEPTRE form.

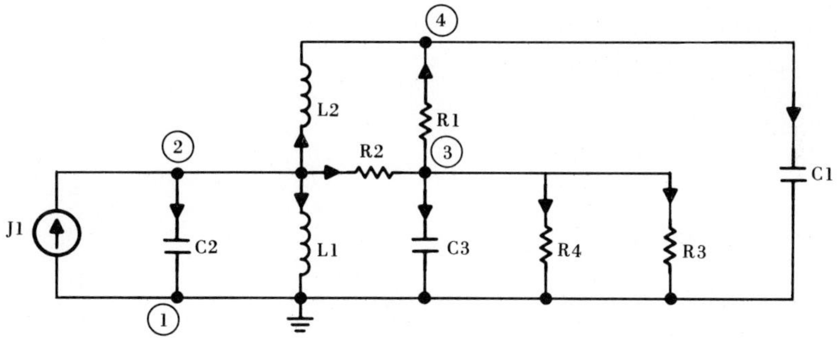

Fig. A-6.3.2

```
SCEPTRE PROGRAM
CIRCUIT DESCRIPTION
EXAMPLE 6.3
ELEMENTS
J1, 1-2 = 10.0
C1, 4-1 = 0.5
C2, 2-1 = 2.0
C3, 3-1 = 1.0
R1, 3-4 = 1000
R2, 2-3 = 200
R3, 3-1 = 1000
R4, 3-1 = 500
L1, 2-1 = 0.1
L2, 3-4 = 0.2
OUTPUTS
VC1(V1), PLOT
```

```
VC2(V2), PLOT
VC3(V3), PLOT
RUN CONTROLS
STOP TIME = 1000
END
```

6.5 Figure A-6.5 shows the electrical analog of the torsionally vibrating mass system in SCEPTRE form.

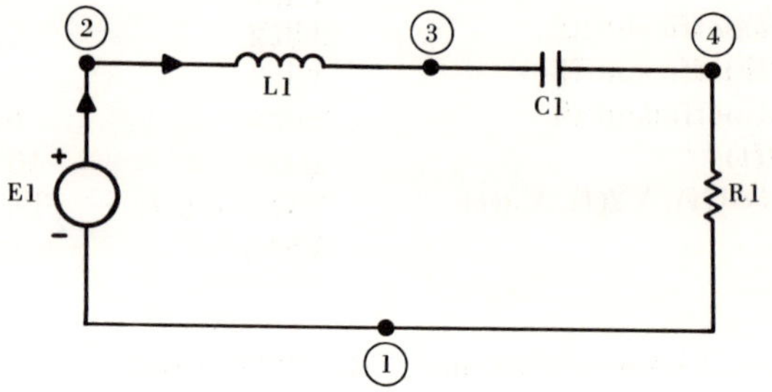


```
SCEPTRE PROGRAM
CIRCUIT DESCRIPTION
EXAMPLE 6.5
ELEMENTS
E1, 1-2 = 10.0
L1, 2-3 = 1E-4
C1, 3-4 = 0.1
R1, 4-5 = 0.05
OUTPUTS
IR1 (W1), PLOT
RUN CONTROLS
STOP TIME = 100
END
```

INDEX

V

W

Z